Raymond Gérard
Hidetoshi Tahara

Singular Nonlinear Partial Differential Equations

Aspects of Mathematics

Edited by Klas Diederich

*A Publication of the Max-Planck-Institut für Mathematik, Bonn

Volumes of the German-language subseries "Aspekte der Mathematik" are listed at the end of the book.

Raymond Gérard
Hidetoshi Tahara

Singular Nonlinear Partial Differential Equations

vieweg

Professor Raymond Gérard
Institut de Recherche Mathématique Alsacien
Université Louis Pasteur
10 rue du Général Zimmer
F-67084 Strasbourg, France

Professor Hidetoshi Tahara
Dept. of Mathematics
Sophia University
Kioicho, Chiyoda – ku
102 Tokyo, Japan

Mathematics Subject Classification:
47 E 05, 47 F 05, 34 A 20, 34 G 20, 35 A 10, 35 F 20

Softcover reprint of the hardcover 1st edition 1996
Vieweg is a subsidiary company of the Bertelsmann Professional Information.

Cover design: Wolfgang Nieger, Wiesbaden

Printed on acid-free paper

ISSN 0179-2156
ISBN-13: 978-3-322-80286-6 e-ISBN-13: 978-3-322-80284-2
DOI: 10.1007/978-3-322-80284-2

Preface

The aim of this book is to put together all the results that are known about the existence of formal, holomorphic and singular solutions of singular non linear partial differential equations.

We study the existence of formal power series solutions, holomorphic solutions, and singular solutions of singular non linear partial differential equations.

In the first chapter, we introduce operators with regular singularities in the one variable case and we give a new simple proof of the classical Maillet's theorem for algebraic differential equations. In chapter 2, we extend this theory to operators in several variables. The chapter 3 is devoted to the study of formal and convergent power series solutions of a class of singular partial differential equations having a linear part, using the method of iteration and also Newton's method. As an application of the former results, we look in chapter 4 at the local theory of differential equations of the form $xy' = f(x,y)$ and, in particular, we show how easy it is to find the classical results on such an equation when $f(0,0) = 0$ and give also the study of such an equation when $f(0,0) \neq 0$ which was never given before and can be extended to equations of the form $\tau y = F(x,y)$ where τ is an arbitrary vector field.

Using the classical method of majorant equations we are able in chapters 5 and 8 to give holomorphic and singular solutions of non linear singular partial differential equations. The chapters 6 and 7 are devoted to Maillet's type theorem for singular non linear partial differential equations, first with a linear part and then without linear part. In chapter 9, we give a very general result for the Cauchy problem for non linear partial differential equations. Finally, in chapter 10, we generalize Maillet's type theorems in chapters 6 and 7 to some non linear integro–differential equations.

During the preparation of this book, the first author (R. G.) stayed twice in Japan and the second author (H. T.) stayed three times in Strasbourg. The first author thanks University of Tokyo and Kyushu University for supporting his stay in Japan. The second author thanks Université Louis Pasteur for supporting his stay in Strasbourg. They also thank their many friends for interesting discussions on the thema of the book.

The original manuscript of this book was written by a local word processor and O. Korten retyped it into the present form. The authors express their hearty gratitude to O. Korten for TEXing this book.

Raymond Gérard
Hidetoshi Tahara

Contents

1 Operators with regular singularities: One variable case

Introduction

In this chapter, we introduce linear and non linear singular operators D acting on formal power series and we study the operators having the property of "regular singularity". This means that we give conditions on D to have the following property:

> "if $\hat{u}$ is a formal power series such that $D\hat{u}$ converges then $\hat{u}$ is a convergent power series".

This study gives us then very interesting applications to differential equations and gives new proofs for classical results.

In a second part, as an application of our operators theory, we give a new proof of the classical Maillet's theorem; that is, if an algebraic differential equation has a formal power series solution then this series is in some formal Gevrey class.

1.1 Notations, definitions, examples

Let us denote by

- $\mathbb{C}[[x]]$ the ring of formal power series with complex coefficients,
- $\mathfrak{M}_f$ its maximal ideal,
- $\mathbb{C}\{x\}$ the ring of holomorphic functions at the origin of $\mathbb{C}$ which is also the ring of convergent power series at the origin of $\mathbb{C}$,
- $\mathfrak{M}_c$ its maximal ideal,
- ***(Op.l.f)*** (resp. ***(Op.l.an)***) (read linear formal operators (resp. linear convergent or analytic operators)) the $\mathbb{C}$–vector space of $\mathbb{C}$–linear operators from $\mathbb{C}[[x]]$ (resp. $\mathbb{C}\{x\}$) into $\mathbb{C}[[x]]$ (resp. $\mathbb{C}\{x\}$).

Let L be a vector subspace of ***(Op.l.f)*** (resp. ***(Op.l.an)***).

Definition 1.1.1 *An operator*

$$D : \mathbb{C}[[x]] \longrightarrow \mathbb{C}[[x]] \quad (\textit{resp. } \mathbb{C}\{x\} \longrightarrow \mathbb{C}\{x\})$$

*is called L–*analytic *if there exist*

1. *an integer* $n \in \mathbb{N}$ *and a function* $F(x, X_0, X_1, \ldots, X_n)$ *holomorphic in a polydisk centered at the origin of* $\mathbb{C} \times \mathbb{C}^{n+1}$ *satisfying*

$$F(0, 0, \ldots, 0) = 0,$$

2. *a finite sequence* $\theta_1, \theta_2, \ldots, \theta_n$ *of elements of* L *such that for all* $u \in \mathfrak{M}_f$ *(resp.* $\mathfrak{M}_c$*)*

$$Du = F(x, u, \theta_1 u, \theta_2 u, \ldots, \theta_n u).$$

Let us denote in the following by $\mathfrak{A}(L)$ the set of L–analytic operators.

Definition 1.1.2 *An operator* $D \in \mathfrak{A}(L)$ *is called* singular *if*

$$D(\mathbb{C}[[x]]) \subset \mathfrak{M}_f.$$

Definition 1.1.3 *A singular operator* $D \in \mathfrak{A}(L)$ *is called* regular singular *if for all* $f \in \mathbb{C}\{x\}$ *every formal power series* $\hat{u}$ *satisfying* $D\hat{u} = f(x)$ *is convergent.*

Definition 1.1.4 *A singular operator* $D \in \mathfrak{A}(L)$ *is called* irregular singular *if there exist* $f \in \mathbb{C}\{x\}$ *and a non convergent formal power series* $\hat{u}$ *such that*

$$D\hat{u} = f(x).$$

Every L–analytic operator can be written in the form

$$Du = \sum_{r+|s|\geq 1} a_{r,s} x^r u^{s_0} (\theta_1 u)^{s_1} \cdots (\theta_n u)^{s_n}$$

where

$$\begin{aligned} r &\in \mathbb{N}, \\ s_i &\in \mathbb{N} \quad \text{for all } i = 0, 1, \ldots, n, \\ |s| &= s_0 + s_1 + \cdots + s_n, \\ a_{r,s} &\in \mathbb{C}. \end{aligned}$$

The *affine part* of D is by definition the operator

$$D_1 u = \sum_{r+|s|=1} a_{r,s} x^r u^{s_0} (\theta_1 u)^{s_1} \cdots (\theta_n u)^{s_n}.$$

An L–analytic operator is called *affine* if it reduces to its affine part. In general, D_1 is not a linear operator but we can consider its *linear part* which is the $\mathbb{C}$–linear operator

$$D_{01} u = \sum_{|s|=1} a_{0,s} u^{s_0} (\theta_1 u)^{s_1} \cdots (\theta_n u)^{s_n}$$

and called the *linearized operator* of D.

Examples 1.1.5 *1. The operator*

$$\theta : \mathbb{C}\{x\} \longrightarrow \mathfrak{M}_c$$
$$u \longmapsto x\frac{du}{dx}$$

is linear, singular, and moreover regular singular.

2. The operator

$$Du = \sum_{0 \le i \le n} a_i(\theta)^{n-i}u$$

where $a_i \in \mathbb{C}$ for all i and $a_0 \neq 0$ is linear and also regular singular.

3. Consider now the operator

$$Du = \sum_{0 \le i \le n} a_i(x)(\theta)^{n-i}u$$

where $a_i(x) \in \mathbb{C}\{x\}$ for all i and $a_0(x) \neq 0$. This operator is ***(Op.l.f)****-analytic and the linear part of it is*

$$D_{01}u = \sum_{0 \le i \le n} a_i(0)(\theta)^{n-i}u.$$

We know that D is regular singular if and only if D_{01} is regular singular that is if $a_0(0) \neq 0$.

4. Let

$$F(x, X_0, X_1, \ldots, X_n) = \sum_{r+|s| \ge 1} a_{r,s}x^r (X_0)^{s_0} (X_1)^{s_1} \cdots (X_n)^{s_n}$$

be a holomorphic function in a polydisk centered at the origin of $\mathbb{C} \times \mathbb{C}^{n+1}$. Consider

$$Du = F(x, u, \theta u, \ldots, \theta^n u) \quad \text{where } \theta = x\frac{d}{dx}.$$

We shall see that if the linear part of D, that is

$$D_{01}u = \sum_{|s|=1} a_{0,s}u^{s_0}(\theta u)^{s_1} \ldots (\theta^n u)^{s_n},$$

is regular singular then the same is true for D.

We denote the affine part of D by

$$D_1 u = \sum_{r+|s|=1} a_{r,s}x^r u^{s_0} (\theta u)^{s_1} \ldots (\theta^n u)^{s_n}.$$

1.2 The good operators

Consider the vector subspace $(\boldsymbol{Op.l.f})_t$ (read linear formal triangular operators) of $(\boldsymbol{Op.l.f})$ of operators

$$\varphi : \mathfrak{M}_f \longrightarrow \mathfrak{M}_f$$

of the form

$$\varphi\left(\sum_{m\geq 1} u_m x^m\right) = \sum_{m\geq 1}\left(\sum_{0\leq k\leq m} \varphi_k(m-k)u_{m-k}\right)x^m$$

where $\{\varphi_k\}_{k\in\mathbb{N}}$ is a sequence of complex valued functions defined on $\mathbb{N}$.

Examples 1.2.1 *1. The identity*

$$\text{id} : \sum_{m\geq 1} u_m x^m \longrightarrow \sum_{m\geq 1} u_m x^m,$$

$(\text{id})_k = 0$ *for* $k \neq 0$ *and* $(\text{id})_0 = 1$ *for all* $m \in \mathbb{N}$.

2. The operator $\theta = x\frac{d}{dx}$ *for which* $\theta_k = 0$ *for all* $k \neq 0$ *and* $\theta_0(m) = m$. *The operator* $\theta^{-1} = \left(x\frac{d}{dx}\right)^{-1}$ *is defined by*

$$\theta^{-1}\left(\sum_{m\geq 1} u_m x^m\right) = \sum_{m\geq 1} \frac{u_m}{m} x^m,$$

$\left(\theta^{-1}\right)_k = 0$ *for all* $k \neq 0$ *and* $\left(\theta^{-1}\right)_0(m) = \frac{1}{m}$ *for all* $m \in \mathbb{N}$.

3. The multiplication operator by an element $a \in \mathbb{C}[[x]]$ *with* $a = \sum_{m\geq 0} a_m x^m$

$$\begin{aligned} \varphi : \mathfrak{M}_f &\longrightarrow \mathfrak{M}_f, \\ \varphi(u) &= a \cdot u = \sum_{m\geq 1}\left(\sum_{0\leq k\leq m} a_k u_{m-k}\right)x^m. \end{aligned}$$

For all $m \in \mathbb{N}$ *and all* $k \in \{0, 1, \ldots, m\}$ *we have* $\varphi_k(m-k) = a_k$. *If* $a \in \mathbb{C}\{x\}$ *then* φ *takes also its values in* $\mathfrak{M}_f$, *but if* $u \in \mathbb{C}\{x\}$ *then* $au \in \mathfrak{M}_c$.

Definition 1.2.2 *An operator* $\varphi \in (\boldsymbol{Op.l.f})_t$ *is called* good operator *if for all* $m \in \mathbb{N}$ *and all* $k \in \{0, 1, \ldots m\}$ *we have*

$$|\varphi_k(m-k)| \leq c_k\, |\varphi_0(m-k)|$$

where the sequence $\{c_k\}_{k\geq 0} \subset \mathbb{R}^+$ *is such that there exists a real number* $\rho > 0$ *such that*

$$c(\rho) = \sum_{k\geq 0} c_k \rho^k < +\infty.$$

Examples 1.2.3 *1. The identity operator is a good operator. To see this it is enough to take $c_k = 0$ for all $k \neq 0$ and $c_0 = 1$ then $c(\rho) = 1$.*

2. The multiplication by a power series $a \in \mathbb{C}\{x\}$ is a good operator

$$\varphi(u) = a \cdot u = \sum_{m \geq 1} \left(\sum_{0 \leq k \leq m} a_k u_{m-k} \right) x^m$$

satisfying $|\varphi_k(m-k)| = |a_k| \leq c_k |a_0|$, and there exists ρ such that

$$c(\rho) = \sum_{k \geq 0} c_k \rho^k < +\infty.$$

Definition 1.2.4 *An operator $\varphi \in (\boldsymbol{Op.l.f})_t$ is called* diagonal *if $\varphi_k = 0$ for all $k \neq 0$.*

It is clear that *every diagonal operator is a good operator.* Among the diagonal operators let us mention the most important and classical operators.

- *Borel operator*

$$B\left(\sum_{m \geq 1} u_m x^m \right) = \sum_{m \geq 1} \frac{u_m}{m!} x^m,$$

- *Gevrey operator*

$$G\left(\sum_{m \geq 1} u_m x^m \right) = \sum_{m \geq 1} \frac{u_m}{(m!)^\lambda} x^m \quad \text{where } \lambda > 0.$$

Definition 1.2.5 *Let $\varphi \in (\boldsymbol{Op.l.f})_t$ and $\psi \in (\boldsymbol{Op.l.f})_t$. We say that φ is* dominating *ψ if for all $m \in \mathbb{N}$ and all $k \in \{0, 1, \ldots, m\}$ there exist $d_k \in \mathbb{R}^+$ such that*

$$|\psi_k(m-k)| \leq d_k |\varphi_0(m-k)|$$

and $\rho > 0$ such that

$$\sum_{k \geq 0} d_k \rho^k < +\infty.$$

A good operator is an operator which is dominating itself.

The *diagonal part* of an operator φ defined by

$$\varphi\left(\sum_{m \geq 1} u_m x^m \right) = \sum_{m \geq 1} \left(\sum_{0 \leq k \leq m} \varphi_k(m-k) u_{m-k} \right) x^m$$

is the operator φ_0 defined by

$$\varphi_0\left(\sum_{m\geq 1} u_m x^m\right) = \sum_{m\geq 1} \varphi_0(m) u_m x^m.$$

Now we see that an operator φ is dominating an operator ψ if ψ is dominated by an operator Φ which is the composition of two operators; the first one is the diagonal part of φ and the second is a multiplication operator by a convergent power series.

Remark 1.2.6 *If φ is a good operator which is dominating the operators $\psi_1, \psi_2, \ldots,$ ψ_n then for all $m \in \mathbb{N}$ and all $k \in \{0, 1, \ldots, m\}$ there exist $c_k \in \mathbb{R}^+$ such that for all $j = 1, 2, \ldots, n$*

$$\begin{aligned} |\psi_{j,k}(m-k)| &\leq c_k\,|\varphi_0(m-k)|\,, \\ |\varphi_k(m-k)| &\leq c_k\,|\varphi_0(m-k)| \end{aligned}$$

and $\rho > 0$ satisfying

$$\sum_{k\geq 0} c_k \rho^k < +\infty.$$

Definition 1.2.7 *We say that an operator $\varphi \in (\boldsymbol{Op.l.f})_t$ is* strictly dominating *an operator $\psi \in (\boldsymbol{Op.l.f})_t$ if*

1. *φ dominates ψ,*
2. $\lim\limits_{m\to\infty} \frac{\psi_0(m)}{\varphi_0(m)} = 0.$

Example 1.2.8
1. *The operator $\theta = x\frac{d}{dx}$ is a good operator and the same is true for θ^p for all $p \in \mathbb{N}$ and θ^p is strictly dominating θ^q for all $q < p$.*
2. *Consider $\varphi = \theta^{-1}$. For all $p \in \mathbb{N}$ φ^p is a good operator which is strictly dominating φ^q for all $q > p$.*
3. *Consider the operator*

$$B^{-1}\left(\sum_{m\geq 1} u_m x^m\right) = \sum_{m\geq 1} u_m m! x^m$$

   called the inverse Borel operator. *It is a good operator. Moreover, $\left(B^{-1}\right)^p$ is strictly dominating $\left(B^{-1}\right)^q$ for all $q < p$.*
4. *The same kinds of remarks are true for the Gevrey operator.*

1.3 A class of operators with a regular singularity

Let $F(x, X_0, X_1, \ldots, X_n)$ be a holomorphic function near the origin of $\mathbb{C} \times \mathbb{C}^{n+1}$. We have

$$F(x, X_0, X_1, \ldots, X_n) = \sum_{r+|s|\geq 1} a_{r,s} x^r (X_0)^{s_0} (X_1)^{s_1} \cdots (X_n)^{s_n}$$

or

$$F(x, X_0, X_1, \ldots, X_n) = \sum_{p\geq 1} F_p(x, X)$$

where for all p

$$F_p(x, X) = \sum_{r+|s|=p} a_{r,s} x^r (X_0)^{s_0} (X_1)^{s_1} \cdots (X_n)^{s_n}.$$

Let $\theta_0 = \mathrm{id}, \theta_1, \theta_2, \ldots, \theta_n$ be a sequence of operators belonging to $(\boldsymbol{Op.l.f})_t$. This means that

$$\theta_j(u) = \sum_{m\geq 1} \left(\sum_{0\leq k\leq m-1} \theta_{j,k}(m-k) u_{m-k} \right) x^m$$

is a map from $\mathfrak{M}_f$ into $\mathfrak{M}_f$.

Consider now the operator

$$Du = F(x, \theta_0 u, \theta_1 u, \ldots, \theta_n u) \quad \text{from } \mathfrak{M}_f \text{ into } \mathfrak{M}_f.$$

The operator D is $(\boldsymbol{Op.l.f})_t$–analytic. Let q be the integer such that $F_p = 0$ for all $p < q$ and $F_q \neq 0$. To the operator D let us associate the $(n+2)$ polynomials

$$C(D)(X) = \sum_{r+|s|=q} a_{r,s} (\theta_{0,0}(1))^{s_0} \cdots (\theta_{n,0}(1))^{s_n} X^{|s|}$$

and for all $j = 0, 1, \ldots, n$

$$C_j(D)(X) = \sum_{r+|s|=q} a_{r,s} s_j (\theta_{0,0}(1))^{s_0} \cdots (\theta_{j,0}(1))^{s_j - 1} \cdots (\theta_{n,0}(1))^{s_n} X^{|s|-1}.$$

Theorem 1.3.1 *If among the operators $\theta_0 = \mathrm{id}, \theta_1, \theta_2, \ldots, \theta_n$ there exists a good operator θ_{j_0} satisfying $C_{j_0}(D)(X) \neq 0$ which is strictly dominating the other operators and if moreover $C(D)(X)$ and $C_{j_0}(D)(X)$ have no common roots then each formal solution of the equation*

$$Du = F(x, \theta_0 u, \theta_1 u, \ldots, \theta_n u) = 0$$

is convergent.

Remark 1.3.2 *The condition $C_{j_0}(D)(X) \neq 0$ implies that θ_{j_0} is explicitly in $F_q(x, \theta_0 u, \theta_1 u, \ldots, \theta_n u)$.*

Proof : In two steps. See also [11].

1. *Existence of formal solutions*

   The equation $Du = 0$ can be written in the form

$$F_q(x, \theta_0 u, \theta_1 u, \ldots, \theta_n u) = R_{q+1}(x, \theta_0 u, \theta_1 u, \ldots, \theta_n u) \tag{1.3.1}$$

where the series $R_{q+1}(x, X)$ contains only monomials of total degree strictly greater than q,

$$R_{q+1}(x, X) = -\sum_{p>q} F_p(x, X).$$

Let us introduce in equation (1.3.1) a formal power series

$$u = \sum_{m \geq 1} u_m x^m.$$

Then

$$\theta_j(u) = \sum_{m \geq 1} \Theta_{j,m} x^m$$

where

$$\Theta_{j,m} = \sum_{0 \leq k \leq m-1} \theta_{j,k}(m-k) u_{m-k}.$$

We have by identification the equation

$$C(D)(u_1) = \sum_{r+|s|=q} a_{r,s} (\theta_{0,0}(1))^{s_0} (\theta_{1,0}(1))^{s_1} \cdots (\theta_{n,0}(1))^{s_n} u_1^{|s|} = 0.$$

This means that u_1 is a root of the polynomial $C(D)(X)$.

To compute u_m for $m > 1$ let us identify the terms in x^{m+q-1}. We obtain

$$\sum_{r+|s|=q} a_{r,s} \sum_{0 \leq j \leq n} s_j \Theta_{j,m} (\Theta_{j,1})^{s_j - 1} \prod_{l \neq j} (\Theta_{l,1})^{s_l}$$
$$= f_m\big(\Theta_{0,1}, \Theta_{0,2}, \ldots, \Theta_{0,m-1}, \Theta_{1,1}, \ldots, \Theta_{1,m-1}, \ldots, \Theta_{n,1}, \Theta_{n,2}, \ldots, \Theta_{n,m-1}, (a)\big)$$

where f_m is a polynomial in all its arguments, (a) stands for the sequence of coefficients of $R_{q+1}(x, X)$, but only a finite number of them are occuring explicitly in f_m.

This equation gives us

$$\left[\sum_{r+|s|=q,|s|\neq 0} a_{r,s} \sum_{0\le j\le n} s_j\theta_{j,0}(m)\,(\theta_{j,0}(1))^{s_j-1}\prod_{l\neq j}(\theta_{l,0}(1))^{s_l}\,u_1^{|s|-1}\right] u_m$$

$$+\sum_{r+|s|=q} a_{r,s}\sum_{0\le j\le n} s_j\left(\sum_{1\le k\le m-1}\theta_{j,k}(m-k)u_{m-k}\right)(\theta_{j,0}(1))^{s_j-1}\times$$

$$\times\prod_{l\neq j}(\theta_{l,0}(1))^{s_l}\,u_1^{|s|-1}$$

$$= f_m\big(u_1,u_2,\dots,u_{m-1},\theta_{1,0}(1)u_1,\theta_{1,0}(2)u_2+\theta_{1,1}(1)u_1,\dots,$$
$$\theta_{1,0}(m-1)u_{m-1}+\theta_{1,1}(m-2)u_{m-2}+\dots+\theta_{1,m-2}(1)u_1,\dots,$$
$$\theta_{n,0}(1)u_1,\theta_{n,0}(2)u_2+\theta_{n,1}(1)u_1,\dots,$$
$$\theta_{n,0}(m-1)u_{m-1}+\dots+\theta_{n,m-2}(1)u_1,(a)\big).$$

Let us denote by $F_{q,m}(u_1)$ the expression

$$\left[\sum_{r+|s|=q,|s|\neq 0} a_{r,s} \sum_{0\le j\le n} s_j\theta_{j,0}(m)\,(\theta_{j,0}(1))^{s_j-1}\prod_{l\neq j}(\theta_{l,0}(1))^{s_l}\,u_1^{|s|-1}\right].$$

We can conclude that for every root u_1 of the polynomial $C(D)(X)$ satisfying $F_{q,m}(u_1)\neq 0$ the equation $Du=0$ has a formal solution with first term u_1x. But $F_{q,m}(u_1)=0$ for some m does not exclude the existence of a formal solution.

2. *Convergence of the formal solutions*

   By our hypothesis among the operators $\theta_0=\mathrm{id},\theta_1,\theta_2,\dots,\theta_n$ there is a good operator which is strictly dominating the others. Without loss of generality we can assume that this operator is θ_n. The two polynomials $C(D)(X)$ and $C_n(D)(X)$ have no common roots and $C_n(D)(X)\neq 0$.

   The operator θ_n dominates the identity. This means that we have

$$1\le c_{0,0}\,|\theta_{n,0}(m)|\quad\text{for all } m\in\mathbb{N} \text{ with } c_{0,0}\neq 0.$$

   By normalizing θ_n, we can assume that $c_{0,0}=1$.

   We have

$$\lim_{m\to\infty}\frac{|F_{q,m}(u_1)|}{\theta_{n,0}(m)} = \sum_{r+|s|=q} a_{r,s}s_n\,(\theta_{n,0}(1))^{s_n-1}\prod_{l\neq n}(\theta_{l,0}(1))^{s_l}\,|u_1|^{|s|-1}$$
$$= C_n(D)(u_1)\neq 0.$$

   There exist $\sigma>0$ and $N\in\mathbb{N}$ such that for all $m\ge N$

$$|F_{q,m}(u_1)|\ge\sigma\,|\theta_{n,0}(m)|$$

   and if $F_{q,m}(u_1)\neq 0$ for all $m<N$ there exists $\sigma>0$ such that

$$|F_{q,m}(u_1)|\ge\sigma\,|\theta_{n,0}(m)|\quad\text{for all } m.$$

Remark 1.3.3 *If there exist integers $m < N$ such that $F_{q,m}(u_1) = 0$ and if the equation $Du = 0$ has a formal solution it is easy to see that the following proof of convergence remains valid. In fact, we have only to change a finite number of terms in a convergent power series.*

The operator θ_n is dominating the operators $\theta_0, \theta_1, \ldots, \theta_n$. This means that we have for all m and $j = 0, 1, \ldots, n$

$$|\theta_{j,k}(m-k)| \leq c_{k,j}\, |\theta_{n,0}(m-k)| \quad \text{for all } k \in \{0, 1, \ldots, m\}.$$

There exists $\rho > 0$ such that

$$c^j(\rho) = \sum_{k \geq 0} c_{k,j}\rho^k < +\infty.$$

Remember that $c^0(\rho) = 1$.

Introduce for all $j = 0, 1, \ldots, n$ the convergent power series for $|x| \leq \rho$

$$c^j(x) = \sum_{k \geq 0} c_{k,j}x^k \quad \text{for } j \neq 0 \text{ and } c^0(x) = 1$$

and consider the equation

$$\sigma_1 x^{q-1} Y = \mu x^q + \sum_{r+|s|=q, |s| \neq 0} |a_{r,s}|\, x^{q-|s|} \prod_{0 \leq j \leq n} \left(c^j(x)Y\right)^{s_j} + |R_{q+1}|\left(x, Y, c^1(x)Y, \ldots, c^n(x)Y, |a|\right)$$

where $|R_{q+1}|$ is a majorant convergent power series of R_{q+1}, σ_1 and μ are real numbers that we have to choose later in a suitable way.

Setting $Y = xZ$ we can see easily using the implicit function theorem that this equation has a unique solution

$$Y = \sum_{m \geq 1} Y_m x^m$$

convergent in a disk centered at the origin of $\mathbb{C}$.

Let us now prove that if σ_1 and μ are chosen in a suitable way then Y is a majorant power series of $\sum_{m \geq 1} u_m x^m$.

The coefficients Y_m are given by

$$\sigma_1 Y_1 = \mu + \sum_{r+|s|=q} |a_{r,s}| \prod_{0 \leq j \leq n} (c_{0,j}Y_1)^{s_j}$$

and for $m > 1$ by

$$\left(\sigma_1 - \sum_{r+|s|=q} |a_{r,s}|\,|s| \prod_{0\le j\le n} (c_{0,j})^{s_j}\, Y_1^{|s|-1}\right) Y_m$$

$$= \sum_{r+|s|=q} |a_{r,s}| \sum_{0\le j\le n} s_j \left(\sum_{1\le k\le m-1} c_{k,j} Y_{m-k}\right) (c_{0,j})^{s_j-1} \prod_{l\ne j} (c_{0,l})^{s_l}\, Y_1^{|s|-1}$$
$$+ f_m^*\big(Y_1, Y_2, \dots, Y_{m-1}, c_{0,1}Y_1, c_{0,1}Y_2 + c_{1,1}Y_1, \dots,$$
$$c_{0,1}Y_{m-1} + c_{1,1}Y_{m-2} + \cdots + c_{m-2,1}Y_1, \dots,$$
$$c_{0,n}Y_1, c_{0,n}Y_2 + c_{1,n}Y_1, \dots, c_{0,n}Y_{m-1} + \cdots + c_{m-2,n}Y_1, (|a|)\big)$$

where f_m^* is obtained from f_m by replacing the coefficients of the family (a) by their majorants denoted by $(|a|)$.

Let us prove by induction that for all $m \geq 1$

$$|u_m| \leq |\theta_{n,0}(m)|\,|u_m| \leq Y_m.$$

Take Y_1 such that

$$|u_1| \leq |\theta_{n,0}(1)|\,|u_1| \leq Y_1$$

and σ_1 such that

$$0 < \sigma_1 - \sum_{r+|s|=q} |a_{r,s}|\,|s| \prod_{0\le j\le n} (c_{0,j})^{s_j}\, Y_1^{|s|-1} < \sigma.$$

Finally

$$\mu = \sigma_1 Y_1 - \sum_{r+|s|=q} |a_{r,s}| \prod_{0\le j\le n} (c_{0,j} Y_1)^{s_j}\,.$$

Assume that for all $p < m$

$$|u_p| \leq |\theta_{n,0}(p)|\,|u_p| \leq Y_p$$

and let us prove that this is also true for $p = m$.

We have for all k with $0 \leq k \leq m-1$ and all $j = 0, 1, \dots, n$

$$\begin{aligned} |\theta_{j,k}(m-k) u_{m-k}| &\leq |\theta_{j,k}(m-k)|\,|u_{m-k}| \\ &\leq c_{k,j}\,|\theta_{n,0}(m-k)|\,|u_{m-k}| \leq c_{k,j} Y_{m-k}. \end{aligned}$$

Coming back to the formula giving u_m we obtain

$$\begin{aligned}
|u_m| &\leq \frac{1}{|F_{q,m}(u_1)|}\left\{\sum_{r+|s|=q} |a_{r,s}| \sum_{0\leq j\leq n} s_j \left(\sum_{1\leq k\leq m-1} c_{k,j} Y_{m-k}\right) \times \right. \\
&\qquad \times (c_{0,j})^{s_j-1} \prod_{l\neq j} (c_{0,l})^{s_l} Y_1^{|s|-1} + \\
&\qquad \left. + f_m^* (Y_1, Y_2, \ldots, Y_{m-1}, c_{0,1} Y_1, \ldots\ldots) \right\} \\
&\leq \frac{1}{|F_{q,m}(u_1)|} Y_m \left(\sigma_1 - \sum_{r+|s|=q} |a_{r,s}| \, |s| \prod_{0\leq j\leq n} (c_{0,j})^{s_j} Y_1^{|s|-1}\right) \\
&\leq \frac{Y_m}{|\theta_{n,0}(m)|}.
\end{aligned}$$

In fact,

$$\sigma_1 - \sum_{r+|s|=q} |a_{r,s}| \, |s| \prod_{0\leq j\leq n} (c_{0,j})^{s_j} Y_1^{|s|-1} < \sigma$$

and

$$|F_{q,m}(u_1)| \geq \sigma \, |\theta_{n,0}(m)| .$$

Finally,

$$|u_m| \leq |\theta_{n,0}(m)| \, |u_m| \leq Y_m$$

which proves the convergence of the power series

$$\sum_{m\geq 1} u_m x^m$$

and the theorem.

∎

It is easy to see that all the power series $\theta_j(u)$ for $j = 0, 1, \ldots, n$ are also convergent.

Let $F(x, X_0, X_1, \ldots, X_n, Y_1, Y_2, \ldots, Y_N)$ be a function holomorphic near the origin of $\mathbb{C} \times \mathbb{C}^{n+1} \times \mathbb{C}^N$ such that $F(0,0,0) = 0$.

$$\begin{aligned}
F(x, X, Y) &= \sum_{\beta+|\alpha|+|\gamma|\geq 1} a_{\beta,\alpha,\gamma} x^\beta X^\alpha Y^\gamma, \\
\alpha &= (\alpha_0, \alpha_1, \ldots, \alpha_n), \\
\gamma &= (\gamma_0, \gamma_1, \ldots, \gamma_n), \\
F(x, X, Y) &= \sum_{p\geq 1} F_p(x, X, Y)
\end{aligned}$$

where

$$F_p(x,X,Y) = \sum_{\beta+|\alpha|+|\gamma|=p} a_{\beta,\alpha,\gamma} x^\beta X^\alpha Y^\gamma.$$

Let $\theta_0 = \mathrm{id}, \theta_1, \theta_2, \ldots, \theta_n, \varphi_1, \varphi_2, \ldots, \varphi_N$ be a sequence of operators belonging to $(\boldsymbol{Op.l.f})_t$. In the following, we consider the $(\boldsymbol{Op.l.f})_t$–analytic operator

$$D : \mathfrak{M}_f \longrightarrow \mathfrak{M}_f,$$

$$Du = F\left(x, \theta_0 u, \theta_1 u, \ldots, \theta_n u, \varphi_1 u, \varphi_2 u, \ldots, \varphi_N u\right).$$

Let q be the integer such that $F_p = 0$ for all $p < q$ and $F_q \neq 0$. Moreover, assume that $F_q(x,X,Y)$ does not contain Y. This means that

$$F_q(x,X,Y) = \sum_{\beta+|\alpha|=q} a_{\beta,\alpha} x^\beta X^\alpha \quad \text{where } a_{\beta,\alpha} = a_{\beta,\alpha,0}.$$

To the operator D we associate the $n+2$ polynomials

$$C_q(D)(X) = \sum_{\beta+|\alpha|=q} a_{\beta,\alpha} \left(\theta_{0,0}(1)\right)^{\alpha_0} \cdots \left(\theta_{n,0}(1)\right)^{\alpha_n} X^{|\alpha|}$$

and for $j = 0, 1, \ldots, n$

$$C_{q,j}(D)(X) = \sum_{\beta+|\alpha|=q} a_{\beta,\alpha} \alpha_j \left(\theta_{0,0}(1)\right)^{\alpha_0} \cdots \left(\theta_{j,0}(1)\right)^{\alpha_j - 1} \cdots \left(\theta_{n,0}(1)\right)^{\alpha_n} X^{|\alpha|}.$$

Make the following hypothesis **(H)**:

1. among the operators $\theta_0 = \mathrm{id}, \theta_1, \theta_2, \ldots, \theta_n$ there is a good operator which is strictly dominating the other operators, for example θ_ν;
2. the operator θ_ν is dominating all the operators $\varphi_1, \varphi_2, \ldots, \varphi_N$;
3. $C_{q,\nu}(D)(X) \neq 0$,
4. the polynomials $C_q(D)(X)$ and $C_{q,\nu}(D)(X)$ have no common roots.

Remark 1.3.4 *The condition $C_{q,\nu}(D)(X) \neq 0$ says that θ_ν is explicitly in $F_q(x, \theta_0 u, \theta_1 u, \ldots, \theta_n u)$.*

Theorem 1.3.5 *Under the hypothesis* **(H)** *every formal solution of*

$$Du = F\left(x, \theta_0 u, \theta_1 u, \ldots, \theta_n u, \varphi_1 u, \varphi_2 u, \ldots, \varphi_N u\right) = 0$$

converges.

Proof : The proof of theorem 1.3.5 goes in the same way as the proof of theorem 1.3.1.

1. *Existence of a formal solution*

The equation $Du = 0$ can be written in the form

(1.3.2)
$$F_q(x, u, \theta_1 u, \dots, \theta_n u) = R_{q+1}(x, u, \theta_1 u, \dots, \theta_n u, \varphi_1 u, \dots, \varphi_N u)$$

where the power series $R_{q+1}(x, X, Y)$ contains only monomials of degree strictly greater than q. Introduce now in the equation (1.3.2) a formal power series

$$u = \sum_{m \geq 1} u_m x^m.$$

To determine u_1 we get the equation

$$C_q(D)(u_1) = \sum_{\beta+|\alpha|=q} a_{\beta,\alpha} (\theta_{0,0}(1))^{\alpha_0} (\theta_{1,0}(1))^{\alpha_1} \cdots (\theta_{n,0}(1))^{\alpha_n} u_1^{|\alpha|} = 0$$

and for $m > 1$ the coefficient u_m is given by

$$\left[\sum_{\beta+|\alpha|=q, |\alpha|\neq 0} a_{\beta,\alpha} \sum_{0\leq j\leq n} \alpha_j \theta_{j,0}(m) (\theta_{j,0}(1))^{\alpha_j - 1} \prod_{l\neq j} (\theta_{l,0}(1))^{\alpha_l} u_1^{|\alpha|-1}\right] u_m$$
$$+ \sum_{\beta+|\alpha|=q} a_{\beta,\alpha} \sum_{0\leq j\leq n} \alpha_j \left(\sum_{1\leq k\leq m-1} \theta_{j,k}(m-k) u_{m-k}\right) (\theta_{j,0}(1))^{\alpha_j-1} \times$$
$$\times \prod_{l\neq j} (\theta_{l,0}(1))^{\alpha_l} u_1^{|\alpha|-1}$$
$$= f_m\big(u_1, u_2, \dots, u_{m-1}, \theta_{1,0}(1)u_1, \theta_{1,0}(2)u_2 + \theta_{1,1}(1)u_1, \dots,$$
$$\theta_{1,0}(m-1)u_{m-1} + \theta_{1,1}(m-2)u_{m-2} + \cdots + \theta_{1,m-2}(1)u_1, \dots,$$
$$\theta_{n,0}(1)u_1, \theta_{n,0}(2)u_2 + \theta_{n,1}(1)u_1, \dots,$$
$$\theta_{n,0}(m-1)u_{m-1} + \cdots + \theta_{n,m-2}(1)u_1,$$
$$\varphi_{1,0}(1)u_1, \dots, \varphi_{1,0}(m-1)u_{m-1} + \cdots + \varphi_{1,m-2}(1)u_1, \dots,$$
$$\varphi_{N,0}(1)u_1, \dots, \varphi_{N,0}(m-1)u_{m-1} + \cdots + \varphi_{N,m-2}(1)u_1, (a)\big).$$

Introduce

$$F_{q,m}(X) = \sum_{\beta+|\alpha|=q, |\alpha|\neq 0} a_{\beta,\alpha} \sum_{0\leq j\leq n} \alpha_j \theta_{j,0}(m) (\theta_{j,0}(1))^{\alpha_j-1} \prod_{l\neq j} (\theta_{l,0}(1))^{\alpha_l} X^{|\alpha|-1}.$$

We can conclude that for every root u_1 of $C_q(D)(X)$ such that $F_{q,m}(u_1) \neq 0$ for all m the equation $Du = 0$ has a formal solution.

The fact $F_{q,m}(u_1) = 0$ for some m does not exclude the existence of a formal solution.

2. *Convergence of the formal solutions*

The two polynomials $C_q(D)(X)$ and $C_{q,\nu}(D)(X)$ have no common roots and, moreover, $C_{q,\nu}(D)(X) \neq 0$. As in the proof of theorem 1.3.1 using the fact that the operator θ_ν is strictly dominating the operators $\theta_0 = \mathrm{id}, \theta_1, \theta_2, \ldots, \theta_n$, there exists $\sigma > 0$ such that

$$|F_{q,m}(u_1)| \geq \sigma\, |\theta_{\nu,0}(m)| \quad \text{for all } m.$$

Remark 1.3.6 *If there exist integers $m < M$ such that $F_{q,m}(u_1) = 0$ and if the equation $Du = 0$ has a formal solution we can see that the following proof remains valid.*

The operator θ_ν dominates the operators $\theta_0 = \mathrm{id}, \theta_1, \theta_2, \ldots, \theta_n, \varphi_1, \varphi_2, \ldots, \varphi_N$. Then for all m and $j = 0, 1, \ldots, n$

$$|\theta_{j,k}(m-k)| \leq c_{k,j}\, |\theta_{\nu,0}(m-k)| \quad \text{for all } k \in \{0, 1, \ldots, m\}$$

and for $r = 1, 2, \ldots, N$

$$|\varphi_{r,k}(m-k)| \leq c'_{k,r}\, |\theta_{\nu,0}(m-k)| \quad \text{for all } k \in \{0, 1, \ldots, m\}$$

and there exists $\rho > 0$ such that

$$\begin{aligned} c^j(\rho) &= \sum_{k\geq 0} c_{k,j}\rho^k < +\infty, \\ c'^r(\rho) &= \sum_{k\geq 0} c'_{k,r}\rho^k < +\infty. \end{aligned}$$

We recall that we can assume that $c^0(\rho) = 1$.

Introduce for all $j = 0, 1, \ldots, n$ and all $r = 1, 2, \ldots, N$ the series which are convergent for $|x| \leq \rho$

$$c^j(x) = \sum_{k\geq 0} c_{k,j}x^k \quad \text{for } j \neq 0 \text{ and } c^0(x) = 1, \qquad c'^r(x) = \sum_{k\geq 0} c'_{k,r}x^k$$

and consider the equation

$$\begin{aligned} \sigma_1 x^{q-1} Y = \mu x^q + \sum_{\beta+|\alpha|=q, |\alpha|\neq 0} |a_{\beta,\alpha}|\, x^{q-|\alpha|} \prod_{0\leq j\leq n} \left(c^j(x)Y\right)^{\alpha_j} + \\ + |R_{q+1}| \left(x, Y, c^1(x)Y, \ldots, c^n(x)Y, c'^1(x)Y, \ldots, c'^N(x)Y, |a|\right) \end{aligned}$$

where σ_1 and μ are real numbers and $|R_{q+1}|$ is a convergent majorant series for R_{q+1}. Setting $Y = xZ$ we see using the implicit function theorem that this equation has a solution

$$Y = \sum_{m\geq 1} Y_m x^m$$

convergent in a disk centered at the origin of $\mathbb{C}$.

Let us show now that if σ_1 and μ are suitably chosen then Y is a convergent majorant series for $\sum_{m\geq 1} u_m x^m$. Let us compute the coefficients Y_m of the power series Y

$$\sigma_1 Y_1 = \mu + \sum_{\beta+|\alpha|=q} |a_{\beta,\alpha}| \prod_{0\leq j\leq n} (c_{0,j}Y_1)^{\alpha_j}$$

and for all $m > 1$

$$\left[\sigma_1 - \sum_{\beta+|\alpha|=q} |a_{\beta,\alpha}|\,|\alpha| \prod_{0\leq j\leq n} (c_{0,j})^{\alpha_j}\, Y_1^{|\alpha|-1}\right] Y_m$$

$$= \sum_{\beta+|\alpha|=q} |a_{\beta,\alpha}| \sum_{0\leq j\leq n} \alpha_j \left(\sum_{1\leq k\leq m-1} c_{k,j} Y_{m-k}\right) (c_{0,j})^{\alpha_j-1} \prod_{l\neq j} (c_{0,l})^{\alpha_l}\, Y_1^{|\alpha|-1}$$

$$\begin{aligned} &+ f_m^*(Y_1, Y_2, \ldots, Y_{m-1}, c_{0,1}Y_1, c_{0,1}Y_2 + c_{1,1}Y_1, \ldots,\\ &\qquad c_{0,1}Y_{m-1} + c_{1,1}Y_{m-2} + \cdots + c_{m-2,1}Y_1, \ldots,\\ &\qquad c_{0,n}Y_1, c_{0,n}Y_2 + c_{1,n}Y_1, \ldots, c_{0,n}Y_{m-1} + \cdots + c_{m-2,n}Y_1,\\ &\qquad c'_{0,1}Y_1, c'_{0,1}Y_2 + c'_{1,1}Y_1, \ldots, c'_{0,N}Y_1, c'_{0,N}Y_2 + c'_{1,N}Y_1, \ldots,\\ &\qquad\qquad c'_{0,N}Y_{m-1} + \cdots + c'_{m-2,N}Y_1, (|a|)) \end{aligned}$$

where f_m^* is obtained from f_m by replacing the coefficients of the family (a) by their majorants $(|a|)$. We are going to prove by induction on m that

$$|u_m| \leq |\theta_{\nu,0}(m)|\,|u_m| \leq Y_m \quad \text{for all } m \geq 1.$$

Take Y_1 such that

$$|u_1| \leq |\theta_{\nu,0}(1)|\,|u_1| \leq Y_1$$

and σ_1 such that

$$0 < \sigma_1 - \sum_{\beta+|\alpha|=q} |a_{\beta,\alpha}|\,|\alpha| \prod_{0\leq j\leq n} (c_{0,j})^{\alpha_j}\, Y_1^{|\alpha|-1} < \sigma$$

and set

$$\mu = \sigma_1 Y_1 - \sum_{\beta+|\alpha|=q} |a_{\beta,\alpha}| \prod_{0\leq j\leq n} (c_{0,j}Y_1)^{\alpha_j}.$$

Assume

$$|u_p| \leq |\theta_{\nu,0}(p)|\,|u_p| \leq Y_p \quad \text{for all } p < m$$

and let us show that this is also true for $p = m$.

We have for all k with $0 \leq k \leq m-1$ and all $j = 0, 1, \ldots, n$

$$\begin{aligned}|\theta_{j,k}(m-k)u_{m-k}| &\leq |\theta_{j,k}(m-k)|\,|u_{m-k}| \\ &\leq c_{k,j}\,|\theta_{\nu,0}(m-k)|\,|u_{m-k}| \leq c_{k,j}Y_{m-k}\end{aligned}$$

and for all $r = 1, 2, \ldots, N$

$$\begin{aligned}|\varphi_{r,k}(m-k)u_{m-k}| &\leq |\varphi_{r,k}(m-k)|\,|u_{m-k}| \\ &\leq c'_{k,r}\,|\theta_{\nu,0}(m-k)|\,|u_{m-k}| \leq c'_{k,r}Y_{m-k}.\end{aligned}$$

Looking at the formula giving u_m, we have

$$\begin{aligned}|u_m| \leq\ & \frac{1}{|F_{q,m}(u_1)|}\Bigg\{\sum_{\beta+|\alpha|=q}|a_{\beta,\alpha}|\sum_{0\leq j\leq n}\alpha_j \times \\ & \times\left(\sum_{1\leq k\leq m-1}c_{k,j}Y_{m-k}\right)(c_{0,j})^{\alpha_j-1}\prod_{l\neq j}(c_{0,l})^{\alpha_l}\,Y_1^{|\alpha|-1} + \\ & + f^*_m(Y_1, Y_2, \ldots, Y_{m-1}, c_{0,1}Y_1, c_{0,1}Y_2 + c_{1,1}Y_1, \ldots, \\ & c_{0,1}Y_{m-1} + c_{1,1}Y_{m-2} + \cdots + c_{m-2,1}Y_1, \ldots, \\ & c_{0,n}Y_1, c_{0,n}Y_2 + c_{1,n}Y_1, \ldots, c_{0,n}Y_{m-1} + \cdots + c_{m-2,n}Y_1, \\ & c'_{0,1}Y_1, c'_{0,1}Y_2 + c'_{1,1}Y_1, \ldots, c'_{0,N}Y_1, c'_{0,N}Y_2 + c'_{1,N}Y_1, \ldots, \\ & c'_{0,N}Y_{m-1} + \cdots + c'_{m-2,N}Y_1, (|\alpha|))\Bigg\} \\ \leq\ & \frac{1}{|F_{q,m}(u_1)|}\left[\sigma_1 - \sum_{\beta+|\alpha|=q}|a_{\beta,\alpha}|\,|\alpha|\prod_{0\leq j\leq n}(c_{0,j})^{\alpha_j}\,Y_1^{|\alpha|-1}\right]Y_m.\end{aligned}$$

Finally,

$$|u_m| \leq |\theta_{\nu,0}(m)|\,|u_m| \leq Y_m$$

and also for $j = 0, 1, \ldots, n$

$$|\theta_{j,0}(m)u_m| \leq |\theta_{j,0}(m)|\,|u_m| \leq c_{0,j}\,|\theta_{j,0}(m)|\,|u_m| \leq c_{0,j}Y_m$$

and for all $r = 1, 2, \ldots, N$

$$|\varphi_{r,0}(m)u_m| \leq c'_{0,r}Y_m.$$

This proves the convergence of $\sum_{m\geq 1} u_m x^m$ and the theorem 1.3.5.

■

It is easy to see that all series $\theta_j(u)$ for $j = 0, 1, \ldots, n$ and $\varphi_r(u)$ for $r = 1, 2, \ldots, N$ are also convergent.

Corollary 1.3.7 *If among the operators $\theta_0 = \mathrm{id}, \theta_1, \theta_2, \dots, \theta_n$ there is a good operator θ_ν strictly dominating the others such that $C_q(D)(X)$ and $C_{q,\nu}(D)(X)$ have no common roots then the operator*

$$D_q(u) = \sum_{\beta+|\alpha|=q, |\alpha|\neq 0} a_{\beta,\alpha} x^\beta u^{\alpha_0} (\theta_1 u)^{\alpha_1} \cdots (\theta_n u)^{\alpha_n}$$

is regular singular.

Remark 1.3.8 *For every root u_1 of $C_q(D)(X) = 0$ satisfying $F_{q,m}(u_1) \neq 0$ for all m the equation $Du = 0$ has a formal solution with first term $u_1 x$. If moreover $C_{q,\nu}(D)(u_1) \neq 0$ this formal solution is convergent.*

Examples 1.3.9 *1. Let $\theta = x\frac{d}{dx}$ and $\theta_j = \theta^j$ for $j = 2, 3, \dots, n$. Consider the operator*

$$Du = \sum_{|\alpha|=1} a_\alpha(x) u^{\alpha_0} (\theta u)^{\alpha_1} \cdots (\theta^n u)^{\alpha_n} \quad \text{with } a_\alpha \in \mathbb{C}\{x\}$$

which can be written in the form

$$Du = \sum_{0\leq i\leq n} b_i(x)\theta^i u \quad \text{where } b_i(x) \in \mathbb{C}\{x\} \text{ for all } i$$

and also

$$Du = \sum_{0\leq i\leq n} b_i(0)\theta^i u + R_2(x, u, \theta u, \dots, \theta^n u).$$

The operator θ^n is good and strictly dominates the operators θ^q for all $q < n$ and also the identity.

$$C(D)(X) = \sum_{0\leq i\leq n} b_i(0)X, \qquad C_n(D)(X) = b_n(0).$$

If $b_n(0) \neq 0$ the operator D is regular singular. This is the classical criterion of Fuchs ([8], [9]).

2. Let $\theta = \left(x\frac{d}{dx}\right)^{-1}$ be the operator defined by

$$\theta\left(\sum_{m\geq 1} u_m x^m\right) = \sum_{m\geq 1} \frac{u_m}{m} x^m.$$

Take $\theta_j = \theta^j$ and consider

$$\begin{aligned} Du &= \sum_{0\leq i\leq n} b_i(x)\theta^i u \quad \text{where } b_i(x) \in \mathbb{C}\{x\} \text{ for all } i \\ &= \sum_{0\leq i\leq n} b_i(0)\theta^i u + R_2(x, u, \theta u, \dots, \theta^n u). \end{aligned}$$

The identity is a good operator and strictly dominating all the others;

$$C(D)(X) = \sum_{0\le i\le n} b_i(0)X, \qquad C_0(D)(X) = b_0(0).$$

Then if $b_0(0) \neq 0$ *the operator* D *is regular singular.*

1.4 Applications to differential equations

1.4.1 Non linear differential equations

Consider the good operator

$$\begin{aligned} \theta : \mathfrak{M}_f &\longrightarrow \mathfrak{M}_f \\ u = \sum_{m\ge 1} u_m x^m &\longmapsto \theta u = x\frac{du}{dx} = \sum_{m\ge 1} m u_m x^m \end{aligned}$$

and the operator

$$Du = F\left(x, u, \theta u, \dots, \theta^n u\right)$$

where

$$\begin{aligned} F\left(x, X_0, X_1, \dots, X_n\right) &= \sum_{p\ge 1} F_p\left(x, X_0, X_1, \dots, X_n\right), \\ F_p\left(x, X_0, X_1, \dots, X_n\right) &= \sum_{\beta+|\alpha|=p} a_{\beta,\alpha} x^\beta X_0^{\alpha_0} X_1^{\alpha_1}\cdots X_n^{\alpha_n}. \end{aligned}$$

Let q be the integer such that

$$F_q = 0 \text{ for } p < q \text{ and } F_q \neq 0.$$

As $\theta^j u = \sum_{m\ge 1} m^j u_m x^m$ for all $j \in \{0, 1, \dots, n\}$ we know that θ^n strictly dominates all the operators θ^j. Introduce

$$C(D)(X) = \sum_{\beta+|\alpha|=q} a_{\beta,\alpha} X^{|\alpha|}, \qquad C_n(D)(X) = \sum_{\beta+|\alpha|=q} a_{\beta,\alpha}\alpha_n X^{|\alpha|-1}$$

and

$$F_{q,m}(X) = \sum_{\beta+|\alpha|=q} a_{\beta,\alpha} \sum_{0\le j\le n} \alpha_j m^j X^{|\alpha|-1}.$$

By theorem 1.3.1 we have

Theorem 1.4.1.1 *For any root u_1 of $C(D)(X)$ satisfying $C_n(D)(u_1) \neq 0$ and for all $m \in \mathbb{N}$ $F_{q,m}(u_1) \neq 0$ there exists a holomorphic solution*

$$u = \sum_{m \geq 1} u_m x^m$$

of $Du = 0$.

And also

Theorem 1.4.1.2 *If $C(D)(X)$ and $C_n(D)(X)$ have no common roots the operators*

$$D_q u = F_q(x, u, \theta u, \ldots, \theta^n u) \qquad \text{and} \qquad Du = F(x, u, \theta u, \ldots, \theta^n u)$$

are regular singular.

1.4.2 Some particular cases

Linear differential equations

Consider an operator of the form

$$Du = \sum_{0 \leq i \leq n} a_i(x, u) \theta^{n-i} u$$

where the functions $a_i(x, u)$ are holomorphic at the origin of $\mathbb{C}^2$.

$$C(D)(X) = \left(\sum_{0 \leq i \leq n} a_i(0, 0) \right) X, \qquad C_n(D)(X) = a_0(0, 0).$$

Then if $a_0(0, 0) \neq 0$ the operator D is regular singular.

Non linear differential equations

Consider an operator of the form

$$Du = \sum_{0 \leq i \leq n} a_i(x, u) (\theta u)^{n-i}$$

where the functions $a_i(x, u)$ are holomorphic at the origin of $\mathbb{C}^2$.

$$a_i(x, u) = \sum_{p \geq 0} a_i^p(x, u) \quad \text{with } a_i^p(x, u) = \sum_{\beta + \alpha_0 = p} a_{i,\beta,\alpha_0} x^\beta u^{\alpha_0}.$$

Assume that for all $i \in \{0, 1, \ldots, n\}$ $a_i^p = 0$ for $p < i$ and $a_i^i \neq 0$. Then

$$D_n u = a_{0,0,0}(\theta u)^n + (a_{1,1,0}x + a_{1,0,1}u)(\theta u)^{n-1} + \cdots +$$
$$+ \left(\sum_{\beta+\alpha_0=j} a_{j,\beta,\alpha_0} x^\beta u^{\alpha_0} \right) (\theta u)^{n-j} + \cdots + \sum_{\beta+\alpha_0=n} a_{n,\beta,\alpha_0} x^\beta u^{\alpha_0}$$

and

$$C(D)(X) = a_{0,0,0}X^n + (a_{1,1,0} + a_{1,0,1}X)X^{n-1} + \cdots +$$
$$+ \left(\sum_{\beta+\alpha_0=j} a_{j,\beta,\alpha_0} X^{\alpha_0} \right) X^{n-j} + \cdots + \sum_{\beta+\alpha_0=n} a_{n,\beta,\alpha_0} X^{\alpha_0}.$$

The operator θ strictly dominates the identity and

$$C_\theta(D)(X) = na_{0,0,0}X^{n-1} + (n-1)(a_{1,1,0} + a_{1,0,1}X)X^{n-2} + \cdots +$$
$$+ \left(\sum_{\beta+\alpha_0=j} (n-j) a_{j,\beta,\alpha_0} X^{\alpha_0} \right) X^{n-j-1} + \cdots + \sum_{\beta+\alpha_0=n-1} a_{n-1,\beta,\alpha_0} X^{\alpha_0}.$$

If $C(D)(X)$ and $C_\theta(D)(X)$ have no common roots then the operator D is regular singular.

Remark 1.4.2.1 *If for example the valuation of a_0 is greater than or equal to one it is enough to consider the operator*

$$D_n u = (a_{1,1,0}x + a_{1,0,1}u)(\theta u)^{n-1} + \cdots +$$
$$+ \left(\sum_{\beta+\alpha_0=j} a_{j,\beta,\alpha_0} x^\beta u^{\alpha_0} \right) (\theta u)^{n-j} + \cdots + \sum_{\beta+\alpha_0=n} a_{n,\beta,\alpha_0} x^\beta u^{\alpha_0}$$

because in this case $a_0(x,u)(\theta u)^n$ is of order strictly greater than n and can be considered as a perturbation in the sense of theorem 1.3.1.

A theorem of P. Bungartz

Let us give some sentences of the original paper of Bungartz in the original language which we think is better than a translation and long explanations.

> In der üblichen Bewertung von $\mathbb{C}[[z, w, w', \ldots, w^{(l)}]]$ hat jede Variable die Ordnung 1. Statt dessen normieren wir die Bewertung ν folgendermaßen:

$$\begin{aligned}
\nu(z) &:= 1,\\
\nu(w) &:= n,\\
\nu(w') &:= n-1,\\
&\vdots\\
\nu\left(w^{(l)}\right) &:= n-l,\\
&\vdots\\
\nu\left(w^{(n-1)}\right) &:= 1,\\
\nu\left(zw^{(n)}\right) &:= 1,\\
\nu(a) &:= 0
\end{aligned}$$

für $a \in \mathbb{C}$, also gleich so, wie wenn für $w, w', \ldots, w^{(n-1)}, zw^{(n)}$ die Potenzreihen

$$w(z) = c_n z^n + c_{n+1} z^{n+1} + \cdots$$

und ihre Ableitungen bis zur n–ten Ordnung in

$$P\left(z, w, w', \ldots, w^{(l)}, \ldots, zw^{(n)}\right)$$

substituiert werden. Jedes Potenzprodukt

$$z^{\alpha_0} w^{\alpha_1} (w')^{\alpha_2} \cdots \left(w^{(l)}\right)^{\alpha_{l+1}} \cdots \left(zw^{(n)}\right)^{\alpha_{n+1}} \qquad (\alpha_i \geq 0 \text{ ganz})$$

hat damit die wohlbestimmte Ordnung

$$\alpha_0 + n\alpha_1 + (n-1)\alpha_2 + \cdots + \alpha_n + \alpha_{n+1}$$

und schreibt man

$$\begin{aligned}
\alpha &= (\alpha_0, \alpha_1, \ldots, \alpha_{n+1}),\\
|\alpha| &= \alpha_0 + \alpha_1 + \cdots + \alpha_{n+1},\\
P\left(z, w, w', \ldots, zw^{(n)}\right) &= \sum_{|\alpha| \geq 0} c_\alpha z^{\alpha_0} w^{\alpha_1} (w')^{\alpha_2} \cdots \left(zw^{(n)}\right)^{\alpha_{n+1}}
\end{aligned}$$

so sei für $P \not\equiv 0$

$$\nu(P) := \min_{|\alpha| \geq 0, c_\alpha \neq 0} (\alpha_0 + n\alpha_1 + (n-1)\alpha_2 + \cdots + \alpha_n + \alpha_{n+1})$$

und $\nu(P) := \infty$ für $P \equiv 0$.

Definition *Wir bezeichnen diese Bewertung* ν *als* n–Ordnung *von* $\mathbb{C}\left[\left[z, w, w', \ldots, w^{(l)}\right]\right]$ *und schreiben* $n\text{-}ord.P = \nu(P)$.

Satz *([7] p. 19) Für die Funktion $w(z)$ der komplexen Veränderlichen z sei folgende Differentialgleichung $Q+R=0$ vom Typ (M,n) vorgelegt:*

(1.4.1)

$$\lambda z^M + \sum_{\beta_0+n\alpha_0+(n-1)\alpha_1+\cdots+\alpha_n=M} \lambda_{\beta_0\alpha} z^{\beta_0} w^{\alpha_0} (w')^{\alpha_1} \cdots \left(zw^{(n)}\right)^{\alpha_n} + R\left(z,w,w',\ldots,w^{(l)},\ldots,zw^{(n)}\right) = 0$$

mit $M \geq 1, n \geq 1, \beta_0 \geq 0, \alpha_i \geq 0$ ganze Zahlen, $\alpha = (\alpha_0,\alpha_1,\ldots,\alpha_n)$. R sei konvergente Potenzreihe in den angeschriebenen Variablen mit n–ord.$R > M$. Für eine Nullstelle c_n von

$$(1.4.2)\quad \lambda + \sum_{\beta_0+n\alpha_0+(n-1)\alpha_1+\cdots+\alpha_n=M} \lambda_{\beta_0\alpha} c_n^{\alpha_0+\alpha_1+\cdots+\alpha_n} n^{\alpha_1+\cdots+\alpha_n} \times \times (n-1)^{\alpha_2+\cdots+\alpha_n} \cdots (n-l+1)^{\alpha_l+\cdots+\alpha_n} \cdots (1)^{\alpha_n} = 0$$

sei für alle $N > n$, $N \in \mathbb{N}$, mit den Abkürzungen

$$c_{n+k}^{(0)} := c_{n+k}, \qquad c_{n+k}^{(j)} := (n+k)(n+k-1)\cdots(n+k-j+1)c_{n+k}^{(0)} \quad (k=0,1,2,\ldots,j=1,2,\ldots,n),$$

$$(1.4.3)\quad \sum_{\beta_0+n\alpha_0+(n-1)\alpha_1+\cdots+\alpha_n=M} \lambda_{\beta_0\alpha} \left\{ \sum_{1\leq\nu\leq n} \alpha_\nu \left(c_n^{(\nu)}\right)^{\alpha_\nu-1} \times \prod_{\mu\neq\nu} \left(c_n^{(\mu)}\right)^{\alpha_\mu} N(N-1)\cdots(N-\nu+1) \right\} \neq 0.$$

Ferner werde gefordert, daß der Koeffizient der höchsten Potenz von N in (1.4.3) ungleich Null sei, also gelte:

(1.4.4)

$$\sum_{\beta_0+n\alpha_0+(n-1)\alpha_1+\cdots+\alpha_n=M} \lambda_{\beta_0\alpha}\alpha_n \left(c_n^{(n)}\right)^{\alpha_n-1} \prod_{\mu\neq n} \left(c_n^{(\mu)}\right)^{\alpha_\mu} \neq 0.$$

Dann gibt es zu diesem c_n genau eine konvergente Lösung

$$w(z) = c_n z^n + c_{n+1} z^{n+1} + \cdots$$

der obigen Differentialgleichung (1.4.1).

This theorem is an easy consequence of theorem 1.3.1 in its differential equation version. In fact, making in equation (1.4.1) the following transformations:

$$w = z^{n-1}u, \qquad \theta = z\frac{d}{dz}$$

we get

$$\begin{aligned} w' &= z^{n-2}[(n-1)u + \theta u], \\ w'' &= z^{n-3}\left[(n-2)(n-1)u + (2n-3)\theta u + \theta^2 u\right] \end{aligned}$$

and, more generally,

$$w^{(l)} = z^{n-1-l} L_{(l)}\left(u, \theta u, \ldots, \theta^l u\right) \quad \text{for } l = 1, 2, \ldots, n-1$$

where $L_{(l)}$ is a linear form in $u, \theta u, \ldots, \theta^l u$. As well $zw^{(n)} = L_{(n)}\left(u, \theta u, \ldots, \theta^n u\right)$. By this change of unknown function the equation (1.4.1) takes the form

$$\lambda z^M + \sum_{\sigma+|\alpha|=M} a_{\sigma\alpha} z^\sigma u^{\alpha_0} (\theta u)^{\alpha_1} \cdots (\theta^n u)^{\alpha_n} + R_M\left(z, u, \theta u, \ldots, \theta^n u\right)$$

where R_M with the usual valuation has a valuation strictly greater than M.

Let us denote by D the operator that we are just considering. If in this operator we are introducing a formal power series $u = \sum_{m\geq 1} u_m x^m$ we can see easily that the equation (1.4.2) is nothing else as $C(D)(u_1) = 0$ and (1.4.3) expresses the fact $F_{M,m}(u_1) \neq 0$ which means the existence of a formal solution and finally (1.4.4) that $C_n(D)(u_1) \neq 0$. This is the condition for the convergence of a formal solution.

Theorems of Briot–Bouquet type

Among this type of theorems there are the classical Briot–Bouquet theorem and a generalization due to Reich.

Theorem 1.4.2.2 (Briot–Bouquet [5]) *Consider the differential equation*

$$x\frac{dy}{dx} = ay + bx + R_2(x, y) \tag{1.4.5}$$

where $R_2(x, y)$ is a power series convergent near the origin of $\mathbb{C}^2$ with valuation greater than or equal to 2. If $a \notin \mathbb{N} \setminus \{0\}$ this equation has a unique holomorphic solution $\varphi(x)$ such that $\varphi(0) = 0$.

This theorem follows easily from theorem 1.4.1.2, but it says a little more: the operator

$$Du = x\frac{du}{dx} - au - bx - R_2(x, u)$$

is regular singular. And so each formal power series solution of the equation (1.4.5) is convergent. In particular, if a is an integer and if the equation (1.4.5) has a formal solution all the solutions are holomorphic near the origin. We are going to make a more precise study of the behavior of the solutions near the origin in chapter 4.

Theorem 1.4.2.3 (Reich [42]) *Seien m, n, r, s ganze Zahlen, $m \geq 1, n \geq 1, s \geq 1, r \geq 0, s \leq n$. Es sei vorgelegt die Differentialgleichung*

(1.4.6)
$$z^{(m-1)n+r+s}\frac{d^s w}{dz^s} = \lambda_0 z^r w^m + \lambda_1 z^{r+n} w^{n-1} + \cdots + \lambda_m z^{mn+r} + P(z,w),$$

wobei gilt:

1. *In der in einer Umgebung von $z = 0, w = 0$ konvergenten Potenzreihe $P(z,w)$ treten nur hinreichend hohe Potenzprodukte $w^u z^v$ auf, d.h. mit*
$$v + un > mn + r.$$

2. *Für eine Nullstelle c_n der Gleichung*

   (1.4.7)
   $$n(n-1)\cdots(n-s+1)c_n - \lambda_0 (c_n)^m - \lambda_1 (c_n)^{m-1} - \cdots - \lambda_m = 0$$

   sei für alle $N \in \mathbb{N}, N > n$,

   $$(1.4.8) \qquad N(N-1)\cdots(N-s+1) - m\lambda_0 (c_n)^{m-1} - \cdots - \lambda_m \neq 0.$$

Dann existiert genau ein holomorphes Integral $w(z) = c_n z^n + c_{n+1} z^{n+1} + \cdots$ der gegebenen Differentialgleichung.

This theorem is an easy consequence of theorem 1.4.1.2. Introduce for this the operator $\theta = z\frac{d}{dz}$. Then after taking away the factor z^{n-1} in the equation of Reich we see that the polynomial giving c_n is nothing else but $C(D)(X)$ and the condition (1.4.8) is expressing that $C(D)(X)$ and $C_\theta(D)(X)$ have no common roots.

The classical criteria of a regular singularity for linear equations

We shall see that the conditions given in the theorem1.4.1.2 are exactly generalizations of the classical criteria for a regular singularity for linear equations. Consider the operator

$$Du = \sum_{0 \leq i \leq n} a_i(x)\frac{d^{n-i}u}{dx^{n-i}} \quad \text{where } a_i(x) \in \mathbb{C}\{x\}, a_0(x) \neq 0.$$

Criterium of Fuchs [8, 9]

The origin is a regular singularity for D if and only if

$$\forall i = 1, 2, \ldots, n : \frac{a_i(x)}{a_0(x)} = x^{-i} q_i(x) \quad \text{with } q_i(x) \in \mathbb{C}\{x\}.$$

Such an operator with a regular singularity can be written in the form

$$Du = \sum_{0 \le i \le n} b_i(x) x^{n-i} \frac{d^{n-i}u}{dx^{n-i}}$$

and the above condition is expressed by $b_0(0) \neq 0$. This means by introducing the operator $\theta = x\frac{d}{dx}$

$$Du = \sum_{0 \le i \le n} c_i(x) \theta^{n-i} u$$

where $c_i(x) \in \mathbb{C}\{x\}$ for all i and $c_0(x) \neq 0$. The Fuchs's condition is expressed by $c_0(0) \neq 0$. The most interesting form of the Fuchs's criteria is those given by Ince [29](p. 160). Let $\varphi(x) \in \mathbb{C}[[x]], \varphi(0) \neq 0$. The coefficient $P(r)$ of lowest degree in $x^{-r} D\left(x^r \varphi(x)\right)$ is a polynomial in r.

First criterium of Ince [29]

The origin is regular singular for the linear operator D if and only if

$$\text{degree}(P(r)) = n = \textit{(order of the differential operator } D).$$

This criterium is extended to the non linear case in the following way. Let $f(x) \in \mathfrak{M}_f$. To determine a formal power series $\varphi(x) \in \mathfrak{M}_f$ satisfying $D\varphi = f(x)$ we get by the explicit computation of the coefficients φ_q for $q > 1$ of the formal power series $\varphi = \sum_{q \ge 1} \varphi_q x^q$ equations of the form

$$Q(q)\varphi_q = F_q\left(\varphi_1, \varphi_2, \ldots, \varphi_{q-1}\right)$$

where $Q(q)$ is a polynomial in q of degree smaller than or equal to n.

Second criterium of Ince

The operator D is regular singular if and only if

$$\text{degree}(Q(q)) = n = \textit{(order of the differential operator } D).$$

Look at the hypothesis of theorem 1.4.1.2 and introduce in the differential operator $Du = \sum_{0 \le i \le n} c_i(x) \theta^{n-i} u$ a formal power series $u = \sum_{q \ge 1} u_q x^q$ and try to solve the equation

$$Du = f(x) \in \mathfrak{M}_f.$$

We obtain

$$\left(\sum_{0 \le i \le n} c_i(0) \right) u_1 = f_1$$

and for $q > 1$

$$Q(q)u_q = \left(c_0(0)q^n + c_1(0)q^{n-1} + \cdots + c_n(0)\right) u_q = F_q\left(u_1, u_2, \ldots, u_{q-1}\right).$$

Or

$$C_1(D)(X) = \left(\sum_{0\le i\le n} c_i(0)\right) X - f_1 \qquad \text{and} \qquad C_{1,n}(D)(X) = c_0(0).$$

We have equivalence between the following conditions:

(I) $c_0(0) \neq 0$;

(II) degree$(Q(q)) = n = \operatorname{ord}.D$;

(III) $C_1(D)(X)$ and $C_{1,n}(D)(X)$ have no common roots;

(IV) D is regular singular.

The second order differential equations with critical fixed points of P. Painlevé

The second order differential equations with critical fixed points of Painlevé are non linear. Let us show that the simplest non degenerated equation of Painlevé which is the equation (VI) is regular singular at all its singular points. The equation (VI) of the table of Painlevé is the following:

$$\begin{aligned} y'' = \frac{1}{2}\left(\frac{1}{y} + \frac{1}{y-1} + \frac{1}{y-x}\right) y'^2 - \left(\frac{1}{x} + \frac{1}{x-1} + \frac{1}{y-x}\right) y' + \\ + \frac{y(y-1)(y-x)}{x^2(x-1)^2}\left[\alpha + \beta\frac{x}{y^2} + \gamma\frac{x-1}{(y-1)^2} + \delta x\frac{x-1}{(y-x)^2}\right]. \end{aligned}$$

Let us associate to it the following operator:

$$\begin{aligned} Dy = y'' - \frac{1}{2y(y-1)(y-x)}\left[(y-1)(y-x) + y(y-x) + y(y-1)\right] y'^2 + \\ + \frac{1}{x(x-1)(y-x)}\left[(x-1)(y-x) + x(y-x) + x(x-1)\right] y' - \\ - \frac{1}{x^2(x-1)^2 y(y-1)(y-x)}\left[\alpha y^2(y-1)^2(y-x)^2 + \beta x(y-1)^2(y-x)^2 + \right. \\ \left. + \gamma(x-1)y^2(y-x)^2 + \delta x(x-1)y^2(y-1)^2\right]. \end{aligned}$$

Our purpose is to study the equation $Dy = f(x)$ near the singular points $0, 1, \infty$ of the equation (VI). We know that equation (VI) is invariant under the transformations

$$\xi = 1 - x, \qquad \eta = 1 - y, \qquad \beta = -\gamma', \qquad \gamma = -\beta',$$

and

$$\xi = \frac{1}{x}, \qquad \eta = \frac{y}{x}, \qquad \gamma = \frac{1}{2} - \delta', \qquad \delta = \frac{1}{2} - \gamma'.$$

This allows us to limit our study of this equation near $x = 0$. In this case, the equation $Dy = f(x)$ for $f(x) \in \mathbb{C}\{x\}$ can be written

$$\begin{aligned}
(\mathrm{VI}^*)\ & 2x^2(x-1)^2y(y-1)(y-x)y'' - \\
& - x^2(x-1)^2y'^2\left[(y-1)(y-x)+y(y-x)+y(y-1)\right] + \\
& + 2xy'(x-1)y(y-1)\left[(x-1)(y-x)+x(y-x)+x(x-1)\right] - \\
& - 2\left[\alpha y^2(y-1)^2(y-x)^2+\beta x(y-1)^2(y-x)^2 + \right.\\
& \left. + \gamma(x-1)y^2(y-x)^2+\delta x(x-1)y^2(y-1)^2\right] \\
& = 2f(x)x^2(x-1)^2y(y-1)(y-x).
\end{aligned}$$

Let us show that $x = 0$ is a regular singularity. For this set

$$y = y_0 + z \quad \text{with } y_0 \in \mathbb{C}, z(0) = 0, \theta z = x\frac{dz}{dx}, \text{ and } x^2z'' = \theta^2 z - \theta z.$$

We obtain

$$\begin{aligned}
& 2(x-1)^2(y_0+z)(y_0+z-1)(y_0+z-x)\left(\theta^2 z-\theta z\right) - \\
& - (\theta z)^2(x-1)^2\left[(y_0+z-1)(y_0+z-x)+(y_0+z)(y_0+z-x)\times\right.\\
& \left.\times(y_0+z)(y_0+z-1)\right] \\
& + 2(\theta z)(x-1)(y_0+z)(y_0+z-1)\left[(x-1)(y_0+z-x)+\right.\\
& \left.+x(y_0+z-x)+x(x-1)\right] - \\
& - 2\left[\alpha(y_0+z)^2(y_0+z-1)^2(y_0+z-x)^2+\beta x(y_0+z-1)^2(y_0+z-x)^2+\right.\\
& \left.+\gamma(x-1)(y_0+z)^2(y_0+z-x)^2+\delta x(x-1)(y_0+z)^2(y_0+z-1)^2\right] \\
& = 2f(x)x^2(x-1)^2(y_0+z)(y_0+z-1)(y_0+z-x).
\end{aligned}$$

The values for y_0 are given by

$$-2(y_0)^4\left[\alpha(y_0-1)^2-\gamma\right] = 0.$$

This means that $y_0 = 0$ or y_0 is given by $\alpha(y_0-1)^2 = \gamma$.

1. y_0 satisfies $\alpha(y_0-1)^2 = \gamma$ and $y_0 \neq 0$

$$\begin{aligned}
& C_1(D)(X) \\
& = 2(y_0)^2(y_0-1)\left\{X\left(1-2\alpha(y_0)^2\right)-(y_0-1)\left[\beta+\alpha(y_0)^2-\delta\right]\right\},
\end{aligned}$$

$$C_{\theta^2}(D)(X) = 2(y_0)^2(y_0-1).$$

It follows that equation (VI*) has in the generic case a unique holomorphic solution $\varphi(x)$ such that

$$\varphi(0) = y_0, \qquad \varphi'(0) = \frac{(y_0-1)\left[\beta+\alpha(y_0)^2-\delta\right]}{1-2\alpha(y_0)^2}.$$

2. $y_0 = 0$

   In this case,

$$\begin{aligned} C_3(D)(X) &= X^2[-1-2(\beta-\delta)] + 4\beta X - 2\beta, \\ C_{\theta^2}(D)(X) &= -2X(X-1). \end{aligned}$$

   If $C_3(D)(1) = -1+2\delta \neq 0$ and $C_3(D)(0) = -2\beta \neq 0$ then to each root X_1, X_2 of $C_3(D)(X) = 0$ there are associated holomorphic solution φ_1 respectively φ_2 such that

$$\varphi_1(0) = 0, \qquad \varphi_1'(0) = X_1, \qquad \varphi_2(0) = 0, \qquad \varphi_2'(0) = X_2.$$

3. There are some other particular cases left to the reader.

Conclusion: *The equation (VI) of Painlevé has regular singularities at the points* $0, 1, \infty$.

We can also see that the equations (III) and (V) of the table of Painlevé have a regular singularity at the origin of $\mathbb{C}$. The theorem 1.4.1.2 can be used to show the nature of the singularity at ∞ for the equations (I), (II), (III), (IV), and (V) of Painlevé, but these equations are obtained by confluence from equation (VI) and then it is easy to conjecture and also easy to verify that this singular point is irregular singular for these equations.

Remark 1.4.2.4 *Equation (VI) is equivalent to the system (see [39])*

$$\begin{aligned} y' &= \frac{1}{x(x-1)}\big[y(y-1)(y-x)z + \theta_0(y-1)(y-x) + \\ &\qquad + \theta_1 y(y-x) + (\theta_x+1)\, y(y-1)\big], \\ z' &= \frac{-1}{x(x-1)}\left[\left(\frac{3}{2}y^2 - (x+1)y + \frac{1}{2}x\right) z^2 + \right. \\ &\qquad + \left(2\left(\theta_0+\theta_1+\theta_x+1\right)y - (x+1)\theta_0 - x\theta_1 - \theta_x - 1\right) z + \\ &\qquad \left. + \frac{1}{2}\left(\left(\theta_0+\theta_1+\theta_x+1\right)^2 - \left(\theta_\infty\right)^2\right)\right] \end{aligned}$$

where the constants $\theta_0, \theta_1, \theta_x, \theta_\infty$ *are given by*

$$\alpha = \frac{1}{2}\left(\theta_\infty\right)^2, \qquad \beta = \frac{-1}{2}\left(\theta_0\right)^2, \qquad \gamma = \frac{1}{2}\left(\theta_1\right)^2, \qquad \delta = \frac{1}{2} - \frac{1}{2}\left(\theta_x\right)^2.$$

In this form, it is easy to see using theorem 2.3, page 252 of [39] that the equation (VI) has a regular singularity at the points $0, 1$ *and* ∞.

Analogous remarks are valid for the equations (III) and (V) near the origin of $\mathbb{C}$.

Nature of the movable singularities of a first order differential equation solved with respect to y'

Consider a differential equation of the form

$$\frac{dy}{dx} = \frac{P(x,y)}{Q(x,y)}.$$

Our purpose is to study the solutions in the neighborhood of a point (x_0, y_0) satisfying $P(x_0, y_0) \neq 0$ and $Q(x_0, y_0) = 0$. Without loosing generality we can assume that $(x_0, y_0) = (0,0)$. Then assume that $P(0,0) \neq 0$ and $Q(0,0) = 0$ and, moreover, that $Q(0,y) \neq 0$. We can assume that

$$Q(x,y) = xH(x,y) + Q_{0,r}y^r + (RQ)_{r+1}(x,y)$$

where $Q_{0,r} \neq 0$ and $(RQ)_{r+1}(x,y)$ are the terms of $Q(x,y)$ having a valuation strictly greater than r. Introduce the operator $\theta = x\frac{d}{dx}$ and write our equation in the form

$$(xH(x,y) + Q_{0,r}y^r)\,\theta y = xP_{0,0} + x(RP)_1(x,y) - (RQ)_{r+1}(x,y)\theta y.$$

This equation has no formal solutions of the form $\sum a_n x^n$.

Set $x = t^{r+1}$. The transformed equation is

$$\left(t^{r+1}H\left(t^{r+1}, y\right) + Q_{0,r}y^r\right)\frac{\theta^* y}{r+1}$$
$$= t^{r+1}P_{0,0} + t^{r+1}(RP)_1\left(t^{r+1}, y\right) - (RQ)_{r+1}\left(t^{r+1}, y\right)\frac{\theta^* y}{r+1}.$$

The two polynomials considered in theorem 1.4.1.2 are

$$\frac{Q_{0,r}X^{r+1}}{r+1} - P_{0,0} \qquad \text{and} \qquad \frac{Q_{0,r}X^r}{r+1}.$$

The fact $P_{0,0} \neq 0$ implies that these two polynomials have no common roots. The differential equation in the variable t has $r+1$ formal solutions. In fact, we obtain $r+1$ values for the coefficient of t in each formal solution. Moreover, by theorem 1.4.1.2 these formal solutions are convergent.

If we are coming back to the original equation we see that the solutions near the origin can be represented as a Puiseux expansion $\sum_{m\geq 1} a_m x^{\frac{m}{r+1}}$. Moreover, by following the analytic continuation along a loop around the origin, all these solutions are permutating themselves after a finite number of loops. We shall say that the solutions have an *algebraic singularity*. This result was known by Fuchs and Painlevé; the proof was at that time more elaborated.

1.5 The Maillet theorem

1.5.1 A generalization of theorem 1.3.5

Let

1. $$F(x, X_0, X_1, \ldots, X_M) = \sum_{1 \le r+|s|=q \le R} a_{r,s} x^r X^s$$

   a polynomial without constant term, where $s = (s_0, s_1, \ldots, s_M), |s| = s_0 + s_1 + \cdots + s_M$,

   $$X^s = (X_0)^{s_0} (X_1)^{s_1} \cdots (X_M)^{s_M};$$

2. a sequence of diagonal operators $\theta_0 = \mathrm{id}, \theta_1, \theta_2, \ldots, \theta_M$ acting on the maximal ideal of the ring of formal power series in the following way:

   $$\theta_j\left(\sum_{m \ge 1} u_m x^m\right) = \sum_{m \ge 1} \theta_{j,0}(m) u_m x^m.$$

Consider the operator equation

$$F(x, y, \theta_1 y, \theta_2 y, \ldots, \theta_M y) = 0 \tag{1.5.1}$$

and assume that among the operators θ_j for $0 \le j \le M$ which are all good in fact diagonal there is one for example θ_n which is strictly dominating the operators θ_j for $0 \le j < n$.

The equation can be written in the form

$$F_p(x, y, \theta_1 y, \theta_2 y, \ldots, \theta_n y) = R_{p+1}(x, y, \theta_1 y, \theta_2 y, \ldots, \theta_N y)$$

where

$$\begin{aligned} F_p(x, X_0, X_1, \ldots, X_n) &= \sum_{r+|s|=p} a_{r,s} x^r (X_0)^{s_0} (X_1)^{s_1} \cdots (X_n)^{s_n}, \\ R_{p+1}(x, X_0, X_1, \ldots, X_N) &= \sum_{p<q\le R} \sum_{r+|s|=q} a_{r,s} x^r (X_0)^{s_0} (X_1)^{s_1} \cdots (X_N)^{s_N}. \end{aligned}$$

The case where θ_n is strictly dominating the operators θ_j for $0 \le j < n$ and dominates also the operators θ_j for $n \le j \le N$ was studied in section 1.3.

Remark 1.5.1.1 *We have assumed that the polynomial*

$$F(x, X_0, X_1, \ldots, X_M) = \sum_{1 \le r+|s|=q \le R} a_{r,s} x^r X^s$$

is without constant term. If this is not the case and if equation (1.5.1) has a formal solution it is always possible to come back to this case by a translation.

Remark 1.5.1.2 *The theorem that we are going to prove is also true if*

$$F\left(x, X_0, X_1, \dots, X_M\right) = \sum_{1 \leq |s| \leq R} a_s(x) X^s$$

where the coefficients $a_s(x)$ are holomorphic in a disk centered at the origin of the complex plane.

We are adding now to the hypothesis on the operators θ_j the following condition:

(H) for all $m \geq 1$, for all q with $p < q \leq R$, all $(r, s_0, s_1, \dots, s_N)$ with $r+|s| = q$ and $a_{r,s} \neq 0$, all sequences $m_0^1, m_0^2, \dots, m_0^{s_0}, m_1^1, m_1^2, \dots, m_1^{s_1}, \dots, m_N^1, m_N^2, \dots, m_N^{s_N}$ such that $r + m_0 + m_1 + \cdots + m_N = m + p - 1$ where

$$\begin{aligned} m_0^1 + m_0^2 + \cdots + m_0^{s_0} &= m_0, \\ m_1^1 + m_1^2 + \cdots + m_1^{s_1} &= m_1, \\ &\vdots \\ m_N^1 + m_N^2 + \cdots + m_N^{s_N} &= m_N, \end{aligned}$$

$$G_n(m) = \frac{1}{|\theta_{n,0}(m)|} \left\{ \prod_{1 \leq k \leq s_0} \left|\theta_{0,0}\left(m_0^k\right)\right| \times \prod_{1 \leq k \leq s_1} \left|\theta_{1,0}\left(m_1^k\right)\right| \times \cdots \times \right.$$
$$\left. \times \prod_{1 \leq k \leq s_N} \left|\theta_{N,0}\left(m_N^k\right)\right| \right\} \leq h \text{ (constant)}.$$

By normalizing θ_n, we can assume that $h = 1$. By convention in this formula and in the following if $m_j^k = 0$ then $\theta_{j,0}\left(m_j^k\right)$ does not occur explicitly.

Theorem 1.5.1.3 *Under the hypothesis mentioned above, if the two polynomials*

$$C(D)(X) = \sum_{r+|s|=p} a_{r,s} \left(\theta_{0,0}(1)\right)^{s_0} \left(\theta_{1,0}(1)\right)^{s_1} \cdots \left(\theta_{n,0}(1)\right)^{s_n} X^{|s|}$$

and

$$C_n(D)(X) = \sum_{r+|s|=p} a_{r,s} s_n \left(\theta_{n,0}(1)\right)^{s_n - 1} \prod_{l \neq n} \left(\theta_{l,0}(1)\right)^{s_l} X^{|s|-1}$$

have no common roots then each formal solution of the equation (1.5.1) is convergent.

Proof : The proof of this theorem is analogous to the proof of the theorems 1.3.1 and 1.3.5 in section 1.3.

In two steps.

1. *Existence of formal solutions*

Introduce in equation (1.5.1) a formal power series of the form

$$y = \sum_{m \geq 1} y_m x^m.$$

To compute the coefficient y_1 we obtain by identification of the terms in x^p the equation

$$C(D)(y_1) = \sum_{r+|s|=p} a_{r,s} (\theta_{0,0}(1))^{s_0} (\theta_{1,0}(1))^{s_1} \cdots (\theta_{n,0}(1))^{s_n} y_1^{|s|} = 0.$$

To obtain the coefficient y_m identify the terms in x^{m+p-1}. This gives us

$$\begin{aligned}
&\left[\sum_{r+|s|=p, |s| \neq 0} a_{r,s} \sum_{0 \leq j \leq n} s_j \theta_{j,0}(m) (\theta_{j,0}(1))^{s_j - 1} \prod_{l \neq j} (\theta_{l,0}(1))^{s_l} y_1^{|s|-1} \right] y_m \\
&= - \sum_{r+|s|=p} \sum_{r+m_0+\cdots+m_n=m+p-1} a_{r,s} \prod_{1 \leq k \leq s_0, m_0^k \neq m} \theta_{0,0}\left(m_0^k\right) y_{m_0^k} \times \\
&\times \prod_{1 \leq k \leq s_1, m_1^k \neq m} \theta_{1,0}\left(m_1^k\right) y_{m_1^k} \times \cdots \times \prod_{1 \leq k \leq s_n, m_n^k \neq m} \theta_{n,0}\left(m_n^k\right) y_{m_n^k} + \\
&+ \sum_{p<q \leq R} \sum_{r+|s|=q} \sum_{r+m_0+\cdots+m_N=m+p-1} a_{r,s} \prod_{1 \leq k \leq s_0} \theta_{0,0}\left(m_0^k\right) y_{m_0^k} \times \\
&\times \prod_{1 \leq k \leq s_1} \theta_{1,0}\left(m_1^k\right) y_{m_1^k} \times \cdots \times \prod_{1 \leq k \leq s_N} \theta_{N,0}\left(m_N^k\right) y_{m_N^k}
\end{aligned}$$

where

$$\begin{aligned}
m_0^1 + m_0^2 + \cdots + m_0^{s_0} &= m_0, \\
m_1^1 + m_1^2 + \cdots + m_1^{s_1} &= m_1, \\
&\vdots \\
m_N^1 + m_N^2 + \cdots + m_N^{s_N} &= m_N
\end{aligned}$$

with the convention that $y_0 = 0$.

Remark moreover that we must have $m_i \geq s_i$ for all $i = 0, 1, 2, \ldots, N$. And if $m_j^k = 0$ then $\theta_{j,0}\left(m_j^k\right)$ does not occur in the products.

Denote

$$F_{p,m}(y_1) = \sum_{r+|s|=p, |s| \neq 0} a_{r,s} \sum_{0 \leq j \leq n} s_j \theta_{j,0}(m) (\theta_{j,0}(1))^{s_j - 1} \prod_{l \neq j} (\theta_{l,0}(1))^{s_l} y_1^{|s|-1}.$$

Remark 1.5.1.4 *For every root y_1 of $C(D)(X) = 0$ satisfying $F_{p,m}(y_1) \neq 0$ the equation (1.5.1) has a formal solution of the form $y = \sum_{m \geq 1} y_m x^m$, but $F_{p,m}(y_1) = 0$ does not exclude the existence of a formal solution.*

2. *Convergence of the formal solutions*

Since the operator θ_n is strictly dominating the operators θ_j for all $0 \leq j < n$ we have

$$\begin{aligned}\lim_{m\to\infty} \frac{F_{p,m}(y_1)}{\theta_{n,0}(m)} &= \sum_{r+|s|=p} a_{r,s}s_n \left(\theta_{n,0}(1)\right)^{s_n-1} \prod_{l\neq n} \left(\theta_{l,0}(1)\right)^{s_l} |y_1|^{|s|-1} \\ &= C_n(D)(y_1) \neq 0.\end{aligned}$$

This means that there exists a constant $\sigma > 0$ such that

$$|F_{p,m}(y_1)| \geq \sigma \left|\theta_{n,0}(m)\right|.$$

Moreover, as θ_n is dominating the identity θ_n is invertible and $\theta_{n,0}(m) \neq 0$ for all $m \geq 1$.

The condition **(H)** implies

$$\begin{aligned}|y_m| \leq \frac{1}{\sigma}\Bigg\{ &\sum_{r+|s|=p} \sum_{r+m_0+\cdots+m_n=m+p-1} |a_{r,s}| \prod_{1\leq k\leq s_0, m_0^k\neq m} \left|y_{m_0^k}\right| \times \\ &\times \prod_{1\leq k\leq s_1, m_1^k\neq m} \left|y_{m_1^k}\right| \times\cdots\times \prod_{1\leq k\leq s_n, m_n^k\neq m} \left|y_{m_n^k}\right| + \\ &+ \sum_{p<q\leq R} \sum_{r+|s|=q} \sum_{r+m_0+\cdots+m_N=m+p-1} |a_{r,s}| \prod_{1\leq k\leq s_0} \left|y_{m_0^k}\right| \times \\ &\times \prod_{1\leq k\leq s_1} \left|y_{m_1^k}\right| \times\cdots\times \prod_{1\leq k\leq s_N} \left|y_{m_N^k}\right| \Bigg\}.\end{aligned}$$

Consider now the algebraic equation

(1.5.2)
$$\sigma_1 x^{p-1} Y = \mu x^p + \sum_{r+|s|=p} |a_{r,s}|\, x^r Y^{|s|} + \sum_{p<q\leq R} \sum_{r+|s|=q} |a_{r,s}|\, x^r Y^{|s|}$$

where σ_1 and μ are strictly positive constants that we have to choose suitably. Compute for this equation a formal solution of the form $Y = \sum_{m\geq 1} Y_m x^m$. To calculate Y_1 we get the equation

$$\sigma_1 Y_1 = \mu + \sum_{r+|s|=p} |a_{r,s}|\, Y_1^{|s|}$$

and for Y_m

$$
\begin{aligned}
Y_m \left[\sigma_1 - \sum_{r+|s|=p} |a_{r,s}|\,|s|\,Y_1^{|s|-1}\right]
&= \sum_{r+|s|=p} \; \sum_{r+m_0+\cdots+m_n=m+p-1} |a_{r,s}| \prod_{1\le k\le s_0, m_0^k\ne m} Y_{m_0^k} \times \\
&\times \prod_{1\le k\le s_1, m_1^k\ne m} Y_{m_1^k} \times\cdots\times \prod_{1\le k\le s_n, m_n^k\ne m} Y_{m_n^k} + \\
&+ \sum_{p<q\le R} \; \sum_{r+|s|=q} \; \sum_{r+m_0+\cdots+m_N=m+p-1} |a_{r,s}| \prod_{1\le k\le s_0} Y_{m_0^k} \times \\
&\times \prod_{1\le k\le s_1} Y_{m_1^k} \times\cdots\times \prod_{1\le k\le s_N} Y_{m_N^k}.
\end{aligned}
$$

Setting $Y = xZ$ we see easily by using the implicit function theorem that equation (1.5.2) has a unique holomorphic solution. This means that the formal power series

$$
Y = \sum_{m\ge 1} Y_m x^m
$$

is convergent in a disk centered at the origin of $\mathbb{C}$ and, moreover, $Y_m \ge 0$ for all $m \ge 1$.

Choose Y_1 such that $|y_1| \le Y_1$ and σ_1 satisfying

$$
0 < \sigma_1 - \sum_{r+|s|=p} |a_{r,s}|\,|s|\,Y_1^{|s|-1} < \sigma,
$$

finally, μ given by

$$
\mu = \sigma_1 Y_1 - \sum_{r+|s|=p} |a_{r,s}|\,Y_1^{|s|}.
$$

Assume that $|y_i| \le Y_i$ for all i with $1 \le i \le m-1$. This is true for $i = 1$ by our assumption.

As

$$
\begin{aligned}
|y_m| \le \frac{1}{\sigma} \Bigg\{ &\sum_{r+|s|=p} \; \sum_{r+m_0+\cdots+m_n=m+p-1} |a_{r,s}| \prod_{1\le k\le s_0, m_0^k\ne m} \left|y_{m_0^k}\right| \times \\
&\times \prod_{1\le k\le s_1, m_1^k\ne m} \left|y_{m_1^k}\right| \times\cdots\times \prod_{1\le k\le s_n, m_n^k\ne m} \left|y_{m_n^k}\right| + \\
&+ \sum_{p<q\le R} \; \sum_{r+|s|=q} \; \sum_{r+m_0+\cdots+m_N=m+p-1} |a_{r,s}| \prod_{1\le k\le s_0} \left|y_{m_0^k}\right| \times \\
&\times \prod_{1\le k\le s_1} \left|y_{m_1^k}\right| \times\cdots\times \prod_{1\le k\le s_N} \left|y_{m_N^k}\right| \Bigg\}
\end{aligned}
$$

we have

$$|y_m| \le \frac{1}{\sigma}\Bigg\{ \sum_{r+|s|=p} \sum_{r+m_0+\cdots+m_n=m+p-1} |a_{r,s}| \prod_{1\le k\le s_0, m_0^k\neq m} Y_{m_0^k} \times$$
$$\times \prod_{1\le k\le s_1, m_1^k\neq m} Y_{m_1^k} \times\cdots\times \prod_{1\le k\le s_n, m_n^k\neq m} Y_{m_n^k} +$$
$$+ \sum_{p<q\le R} \sum_{r+|s|=q} \sum_{r+m_0+\cdots+m_N=m+p-1} |a_{r,s}| \prod_{1\le k\le s_0} \left|Y_{m_0^k}\right| \times$$
$$\times \prod_{1\le k\le s_1} \left|Y_{m_1^k}\right| \times\cdots\times \prod_{1\le k\le s_N} \left|Y_{m_N^k}\right| \Bigg\}$$

which implies

$$|y_m| \le Y_m$$

and the convergence of the formal power series $y = \sum_{m\ge 1} y_m x^m$ associated to y_1 and proves theorem 1.5.1.3.

■

1.5.2 Application: The theorem of E. Maillet

Let $G(x,X) = \sum_{1\le r+|s|=q\le R} b_{r,s}x^r X^s$ be a polynomial of degree R in the variables $x, X_0, X_1, \ldots, X_M$.

Consider the algebraic differential equation

$$G\left(x, y, y', \ldots, y^{(M)}\right) = 0. \tag{1.5.3}$$

By multiplication by a suitable power of x this equation takes the form

(1.5.4)
$$F\left(x, y, \theta y, \theta^2 y, \ldots, \theta^M y\right) = \sum_{1\le r+|s|=q\le R} a_{r,s} x^r y^{s_0} (\theta y)^{s_1} \cdots \left(\theta^M y\right)^{s_M} = 0$$

where $\theta = x\frac{d}{dx}$. And we can write it in the form

$$F_p\left(x, y, \theta y, \theta^2 y, \ldots, \theta^n y\right) = R_{p+1}\left(x, y, \theta y, \theta^2 y, \ldots, \theta^N y\right)$$

where

$$F_p\left(x, X_0, X_1, \ldots, X_n\right) = \sum_{r+|s|=p} a_{r,s} x^r (X_0)^{s_0} (X_1)^{s_1} \cdots (X_n)^{s_n},$$
$$R_{p+1}\left(x, X_0, X_1, \ldots, X_N\right) = \sum_{p<q\le R} \sum_{r+|s|=q} a_{r,s} x^r (X_0)^{s_0} (X_1)^{s_1} \cdots (X_N)^{s_N}.$$

The case $n \geq N$ was studied in section 1.3 as an application of theorems 1.3.1 and 1.3.5. We consider here the case $n < N$. Let us recall that the operator θ^n is strictly dominating all the operators θ^j for $0 \leq j < n$.

Let us assume that *the two polynomials*

$$F_p(1, X, X, \ldots, X) \qquad \textit{and} \qquad \frac{\partial F_p}{\partial X_n}(1, X, X, \ldots, X)$$

have no common roots. This condition is automatically satisfied if $p = 1$ because in this case θ^n occurs explicitly in $F_1\left(x, y, \theta y, \theta^2 y, \ldots, \theta^n y\right)$.

Example 1.5.2.1 *The* Euler equation $x^2 y' - y = -x$ *can be written in the form*

$$x\theta y - y = -x$$

and $F_1(x, X) = -X_0 + x$ *and* $R_2(x, X) = xX_1$.

Theorem 1.5.2.2 (Maillet [36]) *Every formal solution of (1.5.3) is in a formal Gevrey class.*

This result means that if $\sum_{m \geq 1} y_m x^m$ is a formal solution of (1.5.3) there exist two constants c and γ such that $|y_m| \leq c(m!)^\gamma$.

The existence of a formal solution for the equation (1.5.3) implies the existence of a formal solution for (1.5.4).

Proof of the theorem of Maillet: (The first proof of the theorem of Maillet using the theory of operators acting on formal power series was given by Gonzalez [25].) Introduce an indeterminate diagonal operator φ (we will have to choose it suitably) then

$$\varphi\left(\sum_{m \geq 1} u_m x^m\right) = \sum_{m \geq 1} \varphi_0(m) u_m x^m.$$

Now set $y = \varphi(u)$ in the equation (1.5.4).

Denoting $\theta_j = \theta^j \circ \varphi$ we obtain the equation

(1.5.5)
$$F_p\left(x, \theta_0(u), \theta_1(u), \theta_2(u), \ldots, \theta_n(u)\right) = R_{p+1}\left(x, \theta_0(u), \theta_1(u), \theta_2(u), \ldots, \theta_N(u)\right).$$

This equation has a formal solution $u = \sum_{m \geq 1} u_m x^m$.

We have $\theta_{k,0}(p) = p^k \varphi_0(p)$ for all k with $0 \leq k \leq N$ and all integers $p \geq 1$. Let us prove that we can apply theorem 1.5.1.3 if we are taking for φ the diagonal operator defined by $\varphi_0(p) = ((p-1)!)^\alpha$ where α is a suitably chosen positive integer. For this it is enough to verify the hypothesis **(H)**. In fact, the other conditions are satisfied. θ_n strictly dominates the operators θ_j for all j with $0 \leq j < n$.

Let us recall the condition **(H)**:

for all $m \geq 1$, for all q with $p < q \leq R$, all $(r, s_0, s_1, \dots, s_N)$ with $r+|s| = q$ and $a_{r,s} \neq 0$, all sequences of integers $m_0^1, m_0^2, \dots, m_0^{s_0}, m_1^1, m_1^2, \dots, m_1^{s_1}, \dots, m_N^1, m_N^2, \dots, m_N^{s_N}$ such that $r + m_0 + m_1 + \cdots + m_N = m + p - 1$ where

$$\begin{aligned} m_0^1 + m_0^2 + \cdots + m_0^{s_0} &= m_0, \\ m_1^1 + m_1^2 + \cdots + m_1^{s_1} &= m_1, \\ &\vdots \\ m_N^1 + m_N^2 + \cdots + m_N^{s_N} &= m_N, \end{aligned}$$

$$G_n(m) = \frac{1}{|\theta_{n,0}(m)|} \left\{ \prod_{1 \leq k \leq s_0} \left|\theta_{0,0}\left(m_0^k\right)\right| \times \prod_{1 \leq k \leq s_1} \left|\theta_{1,0}\left(m_1^k\right)\right| \times \cdots \times \right.$$
$$\left. \times \prod_{1 \leq k \leq s_N} \left|\theta_{N,0}\left(m_N^k\right)\right| \right\} \leq h \text{ (constant)}.$$

In this formula, we make the convention that if $m_j^k = 0$ then $\theta_{j,0}\left(m_j^k\right)$ does not occur in it.

We have for all sequences of integers $p_1, p_2, \dots, p_l$

$$(p_1!)\,(p_2!)\cdots(p_l!) \leq (p_1 + p_2 + \cdots + p_l)!.$$

As

$$\prod_{1 \leq k \leq s_j} \left|\theta_{j,0}\left(m_j^k\right)\right| = \prod_{1 \leq k \leq s_j} \left(\left(m_j^k - 1\right)!\right)^\alpha \left(m_j^k\right)^j$$

and $\theta_{n,0}(m) = ((m-1)!)^\alpha m^n$ we have

$$\prod_{1 \leq k \leq s_j} \left(\left(m_j^k - 1\right)!\right)^\alpha \left(m_j^k\right)^j \leq ((m_j - s_j)!)^\alpha \prod_{1 \leq k \leq s_j} \left(m_j^k\right)^j$$

and

$$\begin{aligned} G_n(m) &\leq \frac{1}{((m-1)!)^\alpha m^n} \prod_{0 \leq j \leq N} ((m_j - s_j)!)^\alpha \prod_{0 \leq j \leq N} \prod_{1 \leq k \leq s_j} \left(m_j^k\right)^j \\ &\leq \frac{1}{((m-1)!)^\alpha m^n} \left(\left(\sum_{0 \leq j \leq N} (m_j - s_j)\right)!\right)^\alpha \prod_{0 \leq j \leq N} \prod_{1 \leq k \leq s_j} \left(m_j^k\right)^j \end{aligned}$$

with

$$\sum_{0 \leq j \leq N} m_j = m + p - 1 - r, \qquad \sum_{0 \leq j \leq N} s_j = q - r$$

and then

$$
\begin{aligned}
G_n(m) &\leq \frac{1}{((m-1)!)^{\alpha} m^n}((m+p-1-q)!)^{\alpha} \prod_{0\leq j\leq N} \prod_{1\leq k\leq s_j} (m_j^k)^j \\
&\leq \frac{1}{((m-1)!)^{\alpha} m^n}((m+p-1-q)!)^{\alpha} \prod_{0\leq j\leq N} (m_j)^{js_j} \\
&\leq \frac{1}{((m-1)!)^{\alpha} m^n}((m+p-1-q)!)^{\alpha} (m+p-1)^{\sum\limits_{0\leq j\leq N} js_j} \\
&\leq \frac{(m+p-1)^{\sum\limits_{0\leq j\leq N} js_j}}{m^n((m+p-q)(m+p-q+1)\cdots(m-1))^{\alpha}}.
\end{aligned}
$$

As m tends to $+\infty$ the second member of this expression is equivalent to

$$
m^{\left(\sum\limits_{0\leq j\leq N} js_j\right)-n-(q-p)\alpha}.
$$

Take

$$
\alpha \geq \sup_{p<q\leq R} \left\{ \sup_{r+|s|=q, a_{r,s}\neq 0} \left(\frac{\left(\sum\limits_{0\leq j\leq N} js_j\right) - n}{q-p} \right) \right\}.
$$

With this choice of α we have for all $m \geq 1$

$$
G_n(m) \leq \text{constant}.
$$

By theorem 1.5.1.3 the formal power series solution $u = \sum\limits_{m\geq 1} u_m x^m$ of the operator equation (1.5.5) is convergent which means that the formal solution $y = \sum\limits_{m\geq 1} y_m x^m$ of the equations (1.5.3) and (1.5.4) has the following property: the power series $\sum\limits_{m\geq 1} \frac{y_m}{((m-1)!)^{\alpha}} x^m$ is convergent. In other words, the series $y = \sum\limits_{m\geq 1} y_m x^m$ is in a formal Gevrey class and the theorem of Maillet is proved. ∎

Example 1.5.2.3 *Consider the Euler equation $x\theta y - y = -x$ and let us compute the number α which gives the formal Gevrey class of the formal solution. We have $F_1(x, X) = -X_0 + x$ and $R_2(x, X) = xX_1$ and so $p = 1, R = 2, n = 0$, and $N = 1$. It is easy to see*

$$
\sup_{1<q\leq 2} \left\{ \sup_{r+|s|=q=2, a_{r,s}\neq 0} \left(\frac{s_1 - 0}{(q-p)=1} \right) \right\} = 1.
$$

We can take $\alpha = 1$. By a direct computation on the Euler equation it is easy to see that it has a unique formal solution

$$y = \sum_{m \geq 1} (m-1)! x^m.$$

Remark 1.5.2.4 *This theorem remains true if we are only assuming that the differential equation is a polynomial in $y, y', \ldots, y^{(n)}$ with holomorphic coefficients near the origin of the complex plane.*

Remark 1.5.2.5 *In chapter 6, we will show that it is enough to choose α as*

$$\alpha = \sup \left\{ 0, \sup_{(r,s) \in \Delta} \left(\frac{m_s - n}{r + |s| - p} \right) \right\}$$

where

$$\begin{aligned} \Delta &= \{(r, s_0, s_1, \ldots, s_N) : |s| \geq 1 \text{ and } a_{r,s} \neq 0\}, \\ m_s &= \max\{i : s_i \neq 0\}. \end{aligned}$$

1.5.3 Some notes on the theorem of E. Maillet

Note 1 *Let us consider*

$$(1.5.6) \qquad \left(\theta^n + c_1 \theta^{n-1} + \cdots + c_n\right) y = ax + R_2\left(x, y, \theta y, \ldots, \theta^N y\right)$$

where $\theta = x\frac{d}{dx}, c_i \in \mathbb{C}$, and $R_2(x, X_0, X_1, \ldots, X_N)$ is a holomorphic function defined near the origin of $\mathbb{C}^{N+2}$ such that

$$R_2(x, X) = \sum_{r+|s| \geq 2} a_{r,s} x^r X^s.$$

Then in chapter 6 we will prove the following:

Every formal solution of (1.5.6) is in a formal Gevrey class.

Note 2 *Let us consider*

$$(1.5.7) \qquad F\left(x, y, \theta y, \ldots, \theta^M y\right) = 0$$

where

(i) $F(x, X_0, X_1, \ldots, X_M)$ *is a holomorphic function defined near the origin of* $\mathbb{C}^{M+2}$*;*

(ii) $F(0, 0, \ldots, 0) = 0$*;*

(iii) $F(x, X_0, X_1, \ldots, X_M) \not\equiv 0$.

Then by using the result in note 1 we can prove the following theorem:

Every formal solution of (1.5.7) is in a formal Gevrey class.

Remark 1.5.3.1 *In subsection 1.5.2, we have assumed that $F(x, X)$ is algebraic in X and satisfies some conditions on the roots of two polynomials. The above theorem asserts that such conditions are not necessary for Maillet's theorem.*

Proof : Let $\hat{u}(x) \in \mathfrak{M}_f$ be a formal solution of (1.5.7). For simplicity we write

$$\Theta y = \left(y, \theta y, \ldots, \theta^M y\right).$$

Case 1: If $\frac{\partial F}{\partial X_i}(x, \Theta\hat{u}(x)) \not\equiv 0$ for some i, which means that (1.5.7) has a linear part, then we can apply the result for (1.5.6).
Case 2: If $\frac{\partial^\alpha F}{\partial X^\alpha}(x, \Theta\hat{u}(x)) \not\equiv 0$ for some $|\alpha| \geq 1$ then we can choose $|\alpha'| \geq 1$ such that

1. $\frac{\partial^{\alpha'} F}{\partial X^{\alpha'}}(x, \Theta\hat{u}(x)) \not\equiv 0$;
2. $\frac{\partial^\beta F}{\partial X^\beta}(x, \Theta\hat{u}(x)) \equiv 0$ for any $|\beta| < |\alpha'|$.

If $|\alpha'| = 1$ this case is nothing but the case 1.
If $|\alpha'| \geq 2$ then we write

$$\alpha' = \beta + e_i$$

where $e_i = (0, \ldots, 0, 1, 0, \ldots, 0)$ with 1 as the i-th component. In this case, we put

$$G(x, X) = \frac{\partial^\beta F}{\partial X^\beta}(x, X)$$

and we have

$$G(x, \Theta\hat{u}(x)) \equiv 0, \tag{1.5.8}$$

$$\frac{\partial G}{\partial X_i}(x, \Theta\hat{u}(x)) \not\equiv 0. \tag{1.5.9}$$

Therefore, by applying the result of Case 1 to the equation (1.5.8), we get the conclusion.
Case 3: If $\frac{\partial^\alpha F}{\partial X^\alpha}(x, \Theta\hat{u}(x)) \equiv 0$ for any $|\alpha| \geq 1$ then we can derive a contradiction in the following way. Take any $\varphi \in \mathfrak{M}_f$. Then by the Taylor expansion we see

$$\begin{aligned} F(x, \Theta\varphi) &= F\left(x, \Theta\hat{u} + \left(\Theta\left(\varphi - \hat{u}\right)\right)\right) \\ &= F\left(x, \Theta\hat{u}\right) + \sum_{|\alpha|\geq 1} \frac{1}{\alpha!}\frac{\partial^\alpha F}{\partial X^\alpha}\left(x, \Theta\hat{u}(x)\right)\left(\Theta\left(\varphi - \hat{u}\right)\right)^\alpha \\ &\equiv 0. \end{aligned}$$

Since $\varphi \in \mathfrak{M}_f$ is taken arbitrarily this implies that $F(x, X) \equiv 0$ holds which contradicts the assumption (iii). ∎

Remark 1.5.3.2 *The way of dividing the problem into three cases is due to Malgrange [37].*

2 Operators with regular singularities: Several variables case

Introduction

In this chapter, we are giving several extensions to several variables case of the theory that we have developped in chapter 1. First in part A we are studying formal linear equations and then we are looking at the analytic case. This means that we are giving conditions under which the formal power series solutions that we obtained in part A are convergent.

Finally, we are giving some applications to partial differential equations. In particular, we obtain as a corollary a result of Kashiwara, Kawai, and Sjöstrand [32].

A Formal theory

2.1 Notations

We denote by

- $\mathbb{C}[[x]] = \mathbb{C}[[x_1, x_2, \ldots, x_n]]$ the ring of formal power series in the variables $x_1, x_2, \ldots, x_n$,
- $\mathfrak{M}_{\mathfrak{f}}$ its maximal ideal,
- $x^m = (x_1)^{m_1}(x_2)^{m_2} \cdots (x_n)^{m_n}, m = (m_1, m_2, \ldots, m_n) \in \mathbb{N}^n$,
- $|m| = m_1 + m_2 + \cdots + m_n$,
- $H_p(x)$ the vector subspace of $\mathbb{C}[[x]]$ of homogeneous polynomials of degree p in the variables $x_1, x_2, \ldots, x_n$.

The $\mathbb{C}$–vector space $H_p(x)$ is of finite dimension $\delta(p)$. A basis $e_p(x)$ of this vector space is given by the homogeneous monomials of degree p in the variables $x_1, x_2, \ldots, x_n$. It is useful to order the elements of $e_p(x)$. The order that we choose is given inductively on p and n in the following manner.

For $n = 1, \delta(p) = 1$ and $e_p(x_1) = (x_1)^p$. For $p = 0$ and for all $n, e_0(x) = 1$. For arbitrary $n, e_1(x_1, x_2, \ldots, x_n) = (x_1, x_2, \ldots, x_n)$. By induction on p and n we define

$$e_p(x_1, x_2, \ldots, x_n) = \Big\{(x_1)^p, (x_1)^{p-1}(e_1(x_2, \ldots, x_n)), (x_1)^{p-2}(e_2(x_2, \ldots, x_n)), \ldots, (x_1)(e_{p-1}(x_2, \ldots, x_n)), (e_p(x_2, \ldots, x_n))\Big\}.$$

This ordered basis of $H_p(x)$ will be denoted by $(e_p(x))$. What we have just done is to put an order on $\mathbb{N}^n$ in the following manner: let $(i) = (i_1, i_2, \ldots, i_n) \in \mathbb{N}^n$ and $(j) = (j_1, j_2, \ldots, j_n)$. Then if $|i| < |j|$ we have $(i) < (j)$. If $|i| = |j| = p$ the order is given inductively in the following manner. For example in the case $n = 3$

$$\begin{aligned}(p,0,0) &< (p-1,1,0) < (p-1,0,1) < (p-2,2,0) < (p-2,1,1) < (p-2,0,2) \\ &< (p-3,3,0) < (p-3,2,1) < (p-3,1,2) < (p-3,0,3) < \cdots < (0,0,p).\end{aligned}$$

Let us denote by $(\mathbb{N}^n)$ (Ord) the set $\mathbb{N}^n$ equipped with this order.

The use of the word "basis" is a little excessive but we can justify it in the following way. The $\mathfrak{M}_f$–adic filtration on $\mathbb{C}[[x]]$

$$\mathfrak{M}_f^0 = \mathbb{C}[[x]] \supset \mathfrak{M}_f \supset \mathfrak{M}_f^2 \supset \cdots \supset \mathfrak{M}_f^p \supset \cdots$$

defines on the $\mathbb{C}$–vector space of infinite dimension $\mathbb{C}[[x]]$ a topology called the *$\mathfrak{M}_f$-adic topology.* For this topology $\mathbb{C}[[x]]$ is separated and complete. In this way, we can use the word "basis" for $(e_p(x))$. Denote by

$$(e(x)) = ((e_0(x)), (e_1(x)), \ldots, (e_p(x)), \ldots)$$

this ordered basis of $\mathbb{C}[[x]]$. $\hat{u} \in \mathbb{C}[[x]]$ has in this basis the following form:

$$\hat{u} = \sum_{p \geq 0} (e_p(x))\, U_p$$

where $U_p \in \mathbb{C}^{\delta(p)}$. This allows us to write

$$\mathbb{C}[[x]] = \bigoplus_{p \geq 0} H_p(x)$$

and all $\hat{u} \in \mathbb{C}[[x]]$ in the form

$$\hat{u} = \sum_{p \geq 0} u_p(x)$$

where for all $p \geq 0$

$$u_p(x) \in H_p(x) \qquad \text{and} \qquad u_p(x) = (e_p(x))\, U_p.$$

2.2 Linear operators on $\mathbb{C}[[x]]$

A linear operator T on $\mathbb{C}[[x]]$ is by definition a $\mathbb{C}$–linear map from $\mathbb{C}[[x]]$ into $\mathbb{C}[[x]]$. We denote by ***(Op.l.f)*** (read linear formal operators) the algebra of these operators.

In the basis $(e(x)) = ((e_0(x)), (e_1(x)), \ldots, (e_p(x)), \ldots)$, such an operator has an infinite matrix formed by blocks that we can write into the form

$$\begin{pmatrix}
 & T(e_0(x)) & T(e_1(x)) & T(e_2(x)) & \dots & T(e_p(x)) & \dots \\
(e_0(x)) & T_{0,0} & T_{0,1} & T_{0,2} & \dots & T_{0,p} & \dots \\
(e_1(x)) & T_{1,0} & T_{1,1} & T_{1,2} & \dots & T_{1,p} & \dots \\
(e_2(x)) & T_{2,0} & T_{2,1} & T_{2,2} & \dots & T_{2,p} & \dots \\
\dots & \dots & \dots & \dots & \dots & \dots & \dots \\
\dots & \dots & \dots & \dots & \dots & \dots & \dots \\
\dots & \dots & \dots & \dots & \dots & \dots & \dots \\
(e_p(x)) & T_{p,0} & T_{p,1} & T_{p,2} & \dots & T_{p,p} & \dots \\
\dots & \dots & \dots & \dots & \dots & \dots & \dots \\
\dots & \dots & \dots & \dots & \dots & \dots & \dots \\
\dots & \dots & \dots & \dots & \dots & \dots & \dots
\end{pmatrix}.$$

If $\hat{u} = \sum\limits_{p\geq 0} (e_p(x))\, U_p$ then

$$T\hat{u} = \sum_{p\geq 0} T\,(e_p(x))\, U_p = \sum_{p\geq 0}\sum_{k\geq 0} (e_k(x))\, T_{k,p} U_p = \sum_{k\geq 0} (e_k(x)) \sum_{p\geq 0} T_{k,p} U_p.$$

This expression has a meaning only if for all k and all $\hat{u} = \sum\limits_{p\geq 0} (e_p(x))\, U_p$ the numerical series $\sum\limits_{p\geq 0} T_{k,p} U_p$ is convergent.

In the following, we are assuming that the operators which we are considering have the following property:

for all k there exists an integer $q(k)$ such that $T_{k,p} = 0$ for all $p > q(k)$.

With this hypothesis all the series $\sum\limits_{p\geq 0} T_{k,p} U_p$ are convergent. In fact, they are finite sums. With the operations addition and multiplication this set of operators is a $\mathbb{C}$-algebra denoted by $(\boldsymbol{Op.l.f})_{\boldsymbol{f}}$ (read linear formal finitely defined operators). In the theory of differential equations or in the theory of partial differential equations or integral equations, we are meeting in general such types of operators. In the operator algebra $(\boldsymbol{Op.l.f})_{\boldsymbol{f}}$, there are many sub–algebras that are very useful in the applications

- $(\boldsymbol{Op.l.f})_{\boldsymbol{f,r}}$ (linear formal finitely defined regular operators) the algebra of finitely defined operators but regular; this means $q(p) = q(0) + p$;

- $(\boldsymbol{Op.l.f})_{\boldsymbol{b,t}}$ (linear formal block triangular operators) the sub–algebra of operators T the matrix $(T_{i,j})$ of which in the basis $(e(x))$ has a block triangular form; that is $q(p) = p$ and $T_{i,j} = 0$ if $j > i$;

- $(\boldsymbol{Op.l.f})_{\boldsymbol{b,d}}$ (linear formal block diagonal operators) the sub–algebra of operators T the matrix $(T_{i,j})$ of which in the basis $(e(x))$ has a block diagonal form; that is the matrix $(T_{i,j})$ of such an operator T satisfies $T_{i,j} = 0$ if $j \neq i$;

- $(\boldsymbol{Op.l.f})_{\boldsymbol{t}}$ (linear formal triangular operators) the sub–algebra of $(\boldsymbol{Op.l.f})_{\boldsymbol{b,t}}$ of operators T the matrix $(T_{i,j})$ of which in the basis $(e(x))$ satisfies moreover that $T_{i,i}$ is triangular;

- $(\boldsymbol{Op.l.f})_{\boldsymbol{d}}$ (linear formal diagonal operators) the sub–algebra of $(\boldsymbol{Op.l.f})_{\boldsymbol{b,d}}$ of operators T such that the matrix $(T_{i,j})$ in the basis $(e(x))$ is such that $T_{i,i}$ is diagonal.

The definitions that we have just given depend clearly on the choice of the parameters in the ring $\mathbb{C}[[x]]$, this means on the choice of the basis in the $\mathbb{C}$–vector space. The problem of changing the parameters is in some sense the same as changing the basis of a vector space and in the same way we can give some more general statements.

If T is a triangular operator and if we write $\hat{u} \in \mathbb{C}[[x]]$ in the form

$$\hat{u} = \sum_{m \in (\mathbb{N}^n)(\mathrm{Ord})} u_m x^m$$

then

$$T(\hat{u}) = \sum_{m \in (\mathbb{N}^n)(\mathrm{Ord})} \left(\sum_{0 \le k \le m} T_k(m-k) u_{m-k} \right) x^m$$

where for all $k \in \mathbb{N}^n$ T_k is a map from $\mathbb{N}^n$ into $\mathbb{C}$. The operator T is diagonal in the basis $(e(x))$ if $T_k = 0$ for all $k \neq 0$, that is

$$T(\hat{u}) = \sum_{m \in (\mathbb{N}^n)(\mathrm{Ord})} (T_0(m) u_m)\, x^m.$$

Examples 2.2.1 *1.* $n = 1, T = \frac{d}{dx}, \frac{d}{dx}\left(\sum_{m \ge 0} u_m x^m\right) = \sum_{m \ge 0} m u_m x^{m-1}$, *then*

$$\mathrm{Mat}\left(\frac{d}{dx}\right) = \begin{pmatrix} & T(1) & T(x) & T(x^2) & T(x^3) & \dots & T(x^m) & \dots \\ 1 & 0 & 1 & 0 & 1 & 0\dots & 0 & \dots \\ x & 0 & 0 & 2 & 0 & \dots & 0 & \dots \\ x^2 & 0 & 0 & 0 & 3 & \dots & 0 & \dots \\ \dots & \dots & \dots & \dots & \dots & \dots & \dots & \dots \\ x^{m-1} & 0 & 0 & 0 & 0 & \dots 0 & m & 0\dots \\ \dots & \dots & \dots & \dots & \dots & \dots & \dots & \dots \end{pmatrix}.$$

2. *In the case $n = 1$, we have considered in chapter 1 a great number of diagonal and triangular operators.*

3. *Consider a linear singular vector field in Jordan form*

$$\tau = \sum_{1 \le i \le n} \lambda_i x_i \frac{\partial}{\partial x_i} + \sum_{1 \le i \le n-1} a_i x_{i+1} \frac{\partial}{\partial x_i}.$$

It belongs to the class $(\boldsymbol{Op.l.f})_{\boldsymbol{b,d}}$. *If this vector field is semi–simple (i.e. $a_i = 0$ for all i) it is a diagonal operator*

$$\tau_0(m) = \sum_{1 \le i \le n} \lambda_i m_i.$$

4. *Consider a differential operator*

$$D = \sum_{0\leq|\alpha|=|\beta|\leq m} a_{\alpha,\beta} x^\alpha \frac{\partial^{|\beta|}}{\partial x^\beta}$$

where $x = (x_1, x_2, \ldots, x_n)$, $x^\alpha = (x_1)^{\alpha_1}(x_2)^{\alpha_2}\cdots(x_n)^{\alpha_n}$ *and*

$$\frac{\partial^{|\beta|}}{\partial x^\beta} = \frac{\partial^{|\beta|}}{\partial x_1^{\beta_1}\partial x_2^{\beta_2}\cdots\partial x_n^{\beta_n}}.$$

Such an operator where the $a_{\alpha,\beta}$ *are constants is in the class* $(\boldsymbol{Op.l.f})_{\boldsymbol{b},\boldsymbol{d}}$.

5. *The Borel operator*

$$B\left(\sum_{|m|\geq 0} u_m x^m\right) = \sum_{|m|\geq 0} \frac{1}{m!} u_m x^m$$

where $\frac{1}{m!} = \prod_{1\leq i\leq n} \frac{1}{m_i!}$ *and* $x^m = \prod_{1\leq i\leq n} (x_i)^{m_i}$.

6. *The Gevrey operator*

$$G_\lambda\left(\sum_{|m|\geq 0} u_m x^m\right) = \sum_{|m|\geq 0} \left(\frac{1}{m!}\right)^\lambda u_m x^m$$

where $\left(\frac{1}{m!}\right)^\lambda = \prod_{1\leq i\leq n} \left(\frac{1}{m_i!}\right)^{\lambda_i}$.

7. *If we have a vector field* $\tau = \sum_{1\leq i\leq n} \lambda_i x_i \frac{\partial}{\partial x_i}$ *we can consider the operator* $\exp(\tau)$ *which is also a diagonal operator.*

8. *Let* $P(X_1, X_2, \ldots, X_n) \in \mathbb{C}[X_1, X_2, \ldots, X_n]$ *that is a polynomial in the variables* $(X_1, X_2, \ldots, X_n)$. *The operator* $T = P\left(x_1\frac{\partial}{\partial x_1}, x_2\frac{\partial}{\partial x_2}, \ldots, x_n\frac{\partial}{\partial x_n}\right)$ *is diagonal with* $T_0(m) = P(m_1, m_2, \ldots, m_n)$.

2.3 Non linear operators on $\mathfrak{M}_f$

Let $F(x, X) = F(x_1, x_2, \ldots, x_n, X_0, X_1, \ldots, X_p)$ be a formal power series with complex coefficients in the variables $x_1, x_2, \ldots, x_n, X_0, X_1, \ldots, X_p$ satisfying $F(0,0) = 0$. We can write

$$F(x, X) = \sum_{j\geq 1} F_j(x, X)$$

where for all j the term $F_j(x, X)$ is a homogeneous polynomial of degree j in the variables $x_1, x_2, \dots, x_n, X_0, X_1, \dots, X_p$. For a sequence of linear operators $T_0 = \mathrm{id}, T_1, \dots, T_p$ on $\mathfrak{M}_f$ consider the operator

$$Du = F(x, u, T_1 u, \dots, T_p u).$$

This expression is well defined if $T_i \in \boldsymbol{(Op.l.f)_f}$ for all $i = 0, 1, \dots, p$. We are assuming this fact in all this chapter.

Example 2.3.1 *Let* $\tau = \sum_{1 \le i \le n} \lambda_i x_i \frac{\partial}{\partial x_i}$ *and*

$$Du = a(x)(\tau u)^2 + b(x) u \tau u + c(x) u^2 + \sum_{1 \le i \le n} d_i (x_i)^2$$

where $a(x), b(x)$, *and* $c(x)$ *are elements of* $\mathbb{C}[[x]]$ *and the* d_i*'s for* $1 \le i \le n$ *are constants.*

Definition 2.3.2 *The valuation of an operator* $Du = F(x, u, T_1 u, \dots, T_p u)$ *is the integer* q *such that*

$$F_i(x, X) = 0 \quad \text{for all } i \text{ with } 1 \le i < q \qquad \text{and} \qquad F_q(x, X) \ne 0.$$

If $q = 1$ then $F_1(x, X) = \sum_{|\alpha| + |\beta| = 1} a_{\alpha,\beta} x^\alpha X^\beta$. We call the *affine part* of D the operator defined by

$$D_1 u = \sum_{|\alpha| + |\beta| = 1} a_{\alpha,\beta} (x_1)^{\alpha_1} (x_2)^{\alpha_2} \cdots (x_n)^{\alpha_n} (u)^{\beta_0} (T_1 u)^{\beta_1} \cdots (T_p u)^{\beta_p}.$$

The *linear part* of D is the operator

$$D_1^0 u = \sum_{|\beta| = 1} a_{0,\beta} (u)^{\beta_0} (T_1 u)^{\beta_1} \cdots (T_p u)^{\beta_p}.$$

Remark 2.3.3 *In the classical theory, a* $\mathbb{C}$*–linear operator is an operator* T *satisfying*

$$T(\lambda u + \mu v) = \lambda T(u) + \mu T(v) \quad \text{for all } \lambda, \mu \in \mathbb{C} \text{ and } u, v \in \mathbb{C}[[x]].$$

In particular, an operator of the form

$$Tu = \sum_{0 \le i \le n} a_i(x) T_i(u)$$

where $T_i \in \boldsymbol{(Op.l.f)_f}$ *and* $a_i(x) \in \mathbb{C}[[x]]$ *for all* i *is linear in the classical sense but in our theory it is considered as non linear, its linear part being*

$$T_1^0 u = \sum_{0 \le i \le n} a_i(0) T_i(u).$$

This difference will be very important in our theory.

Example 2.3.4 *In the case $n = 1$, the operator $T = x^{p+1}\frac{d}{dx}$ is linear in the classical sense but we are considering it sometimes as an $(\boldsymbol{Op.l.f})_{\boldsymbol{f}}$ non linear operator.*

$$x^{p+1}\frac{d}{dx} = x^p \circ x\frac{d}{dx} \quad \textit{where } x\frac{d}{dx} \in (\boldsymbol{Op.l.f})_{\boldsymbol{d}}$$

is the composition of the operator $x\frac{d}{dx}$ with the multiplication by x^p and then

$$F(x, X) = x^p X_1.$$

2.4 Solutions of linear equations

We are looking for solutions of equations of the form

$$Tu = f(x) \in \mathbb{C}[[x]].$$

It is clear that T being invertible implies $u = T^{-1}(f(x))$. If $T \in (\boldsymbol{Op.l.f})_{\boldsymbol{b,t}}$ and if $T_{p,p}$ is invertible for all p then T is invertible. In particular, if $T \in (\boldsymbol{Op.l.f})_{\boldsymbol{t}}$ and if we are writing

$$T(u) = \sum_{m \in (\mathbb{N}^n)(\mathrm{Ord})} \left(\sum_{0 \le k \le m} T_k(m-k)u_{m-k} \right) x^m$$

then T is invertible if $T_0(m) \neq 0$ for all m.

The fact that T is not invertible does not exclude the existence of solutions for the equation $Tu = f(x)$ but, in general, the solutions are not unique. In fact, this is the case when $\ker T \neq 0$.

Examples 2.4.1 *1. Assume $n = 1, T = x\frac{d}{dx} - 2$. Consider the equation*

$$Tu = x\frac{du}{dx} - 2u = f(x) = \sum_{m \ge 1} f_m x^m$$

and look for a formal solution $u = \sum_{m \ge 1} u_m x^m$. The coefficients u_m are given by the equations

$$(m-2)u_m = f_m \quad \textit{for } m = 1, \ldots.$$

If $f_2 \neq 0$ this equation has no formal solution of the form

$$u = \sum_{m \ge 1} u_m x^m$$

but if $f_2 = 0$ this equation has infinitely many formal solutions given by

$$\sum_{m \neq 2, m \ge 1} \frac{f_m}{m-2} x^m + kx^2$$

where k is an arbitrary constant, $\mathbb{C}x^2 \in \ker T$ and the dimension of the kernel of T is one.

2. *Assume* $n = 2, T = x_1 \frac{\partial}{\partial x_1} - 2x_2 \frac{\partial}{\partial x_2}$ *and look at the equation*

$$Tu = f(x_1, x_2) = \sum_{|m| \geq 1} f_{m_1,m_2} (x_1)^{m_1} (x_2)^{m_2}.$$

To determine the coefficients of a formal solution

$$u = \sum_{|m| \geq 1} u_{m_1,m_2} (x_1)^{m_1} (x_2)^{m_2}$$

we get by identification the equations

$$(m_1 - 2m_2) u_{m_1,m_2} = f_{m_1,m_2}.$$

If $f_{2p,p} \neq 0$ *for an integer* p *our equation has no formal solutions but if* $f_{2p,p} = 0$ *for all integer* p *the equation* $Tu = f(x_1, x_2)$ *has infinitely many formal solutions given by*

$$\sum_{|m| \geq 1, m_1 \neq 2m_2} \frac{1}{m_1 - 2m_2} f_{m_1,m_2} (x_1)^{m_1} (x_2)^{m_2} + \sum_{q \geq 1} a_{2q,q} (x_1)^{2q} (x_2)^q$$

where $a_{2q,q}$ *are arbitrary constants. This means that*

$$\sum_{q \geq 1} a_{2q,q} (x_1)^{2q} (x_2)^q \in \ker T.$$

3. *Consider now the case of an operator* $T \in (\boldsymbol{Op.l.f})_{\mathfrak{b},\boldsymbol{t}}$ *such that* $T_{p,p}$ *is invertible for all* $p \geq N$. *Let us study the formal equation*

$$Tu = f(x) \in \mathfrak{M}_f.$$

For this aim write $u = \sum_{p \geq 1} (e_p(x)) U_p$ *and* $f = \sum_{p \geq 1} (e_p(x)) f_p$. *If* u *is a solution of our equation we must have for all* $m \geq 1$

$$\sum_{1 \leq k \leq m} T_{m,k} U_k = f_m.$$

Consider the finite system

$$\sum_{1 \leq k \leq p} T_{p,k} U_k = f_p \quad \text{for } p = 1, 2, \ldots, N - 1. \tag{2.4.1}$$

Each solution of (2.4.1) (if there exist solutions) gives us a unique solution of $Tu = f$.

4. Consider an equation of the form

$$Tu = f(x) \in \mathfrak{M}_f$$

where $T : \mathfrak{M}_f \longrightarrow \mathfrak{M}_f$ *is a linear finitely defined regular operator. The operator* T *has in the basis* $(e(x))$ *of* $\mathbb{C}[[x]]$ *a matrix of the form*

$$\left(\begin{array}{cccccc}
 & T(e_0(x)) & T(e_1(x)) & \dots & T\left(e_{q(0)}(x)\right) & T\left(e_{q(0)+1}(x)\right) \\
(e_0(x)) & T_{0,0} & T_{0,1} & \dots & T_{0,q(0)} & 0 \\
(e_1(x)) & T_{1,0} & T_{1,1} & \dots & T_{1,q(0)} & T_{1,q(0)+1} \\
(e_2(x)) & T_{2,0} & T_{2,1} & \dots & T_{2,q(0)} & T_{2,q(0)+1} \\
\dots & \dots & \dots & \dots & \dots & \dots \\
(e_p(x)) & T_{p,0} & T_{p,1} & \dots & T_{p,q(0)} & T_{p,q(0)+1} \\
\dots & \dots & \dots & \dots & \dots & \dots
\end{array}\right.$$

$$\left.\begin{array}{cccc}
\dots & T\left(e_{q(0)+p}(x)\right) & \dots \\
\dots & 0 & \dots \\
0\dots & 0 & \dots \\
T_{2,q(0)+2}\ \dots & 0 & \dots \\
\dots & \dots & \dots \\
\dots & T_{p,q(0)+p} & 0\dots \\
\dots & \dots & \dots
\end{array}\right).$$

A formal solution $u = \sum (e_p(x))\, U_p$ *of* $Tu = f = \sum (e_p(x))\, f_p$ *is given by the system*

$$\left\{ \sum_{-q_0 \le k \le j} T_{j,j-k} U_{j-k} = f_j \right\} \quad \textit{for } j = 0, 1, \dots. \tag{2.4.2}$$

The equation (2.4.2) for $j = 0$ *gives us* U_{q_0} *in terms of* $U_0, U_1, \dots, U_{q_0-1}$. *The equation (2.4.2) for* $j = 1$ *gives* U_{q_0+1} *in terms of* $U_0, U_1, \dots, U_{q_0}$. *The equation (2.4.2) for arbitrary* j *gives us* U_{q_0+j} *in terms of* $U_0, U_1, \dots, U_{q_0+j-1}$. *As a conclusion* $U_0, U_1, \dots, U_{q_0-1}$ *are arbitrary and if for all* j *the equation* $T_{j,q_0+j} U_{q_0+j} = g_j \in \mathbb{C}^{\delta(j)}$ *has solutions the equation* $Tu = f$ *has also solutions.*

Examples 2.4.2 *1.* $T = \frac{d^3}{dx^3}$, *the matrix of this operator in the basis* $(e(x))$ *is the following:*

$$\left(\begin{array}{cccccccccc}
 & T(1) & T(x) & T\left(x^2\right) & T\left(x^3\right) & T\left(x^4\right) & \dots & T\left(x^p\right) & \dots \\
1 & 0 & 0 & 0 & 3\cdot 2\cdot 1 & 0 & \dots & 0 & \dots \\
x & 0 & 0 & 0 & 0 & 4\cdot 3\cdot 2 & 0\dots & 0 & \dots \\
\dots & \dots & \dots & \dots & \dots & \dots & \dots & \dots & \dots \\
x^{p-3} & 0 & 0 & 0 & 0 & 0 & \dots 0 & p(p-1)(p-2) & 0\dots \\
\dots & \dots & \dots & \dots & \dots & \dots & \dots & \dots & \dots
\end{array}\right).$$

This is a good example of a finitely defined regular operator. It is interesting to remark that $x^3 \frac{d^3}{dx^3}$ *is a diagonal operator. In fact,*

$$x^3 \frac{d^3}{dx^3} = \theta^3 - 3\theta^2 + 2\theta$$

where $\theta = x\frac{d}{dx}$ *and all the powers of* θ *are diagonal operators. This result is also true for* $x^p \frac{d^p}{dx^p}$ *for all* $p \geq 1$.

2. *In the case* $n = 1$, *the operator* T *defined by*

$$T(x^m) = \begin{cases} m(m-1)x^{m-2} & \text{if } m = 2p, \\ mx^{m-1} & \text{if } m = 2p+1 \end{cases}$$

is a finitely defined regular operator. The same is true for the operator defined by

$$T(x^m) = \begin{cases} x^m & \text{if } m \text{ is odd,} \\ 0 & \text{if } m \text{ is even.} \end{cases}$$

3. *Let* $\tau = \sum\limits_{1 \leq i \leq n} \lambda_i x_i \frac{\partial}{\partial x_i}$ *be a semi-simple vector field. We assume that the numbers* λ_i *are all real. We have*

$$\tau\left(\sum_{|m|\geq 1} u_m x^m\right) = \sum_{|m|\geq 1}\left(\sum_{1\leq i\leq n} \lambda_i m_i\right) u_m x^m.$$

To this operator let us associate the following "sub-operators" defined by

$$\tau^+\left(\sum_{|m|\geq 1} u_m x^m\right) = \sum_{\sum \lambda_i m_i > 0}\left(\sum_{1\leq i\leq n} \lambda_i m_i\right) u_m x^m$$

and

$$\tau^-\left(\sum_{|m|\geq 1} u_m x^m\right) = \sum_{\sum \lambda_i m_i < 0}\left(\sum_{1\leq i\leq n} \lambda_i m_i\right) u_m x^m.$$

The operators τ, τ^+, *and* τ^- *are diagonal and play an important role in many problems as we shall see later.*

2.5 Solutions of non linear equations

2.5.1 Non linear equations with an affine part in $(Op.l.f)_{b,t}$

Let be given a formal power series

$$F(x_1, x_2, \ldots, x_n, X_0, X_1, \ldots, X_N) = \sum_{p\geq 1} F_p(x_1, x_2, \ldots, x_n, X_0, X_1, \ldots, X_N)$$

where

$$\begin{aligned} F_p(x,X) &= \sum_{|\alpha|+|\beta|=p} a_{\alpha,\beta} x^\alpha X^\beta , \\ x^\alpha &= (x_1)^{\alpha_1} \cdots (x_n)^{\alpha_n} , \\ X^\beta &= (X_0)^{\beta_0} \cdots (X_N)^{\beta_N} . \end{aligned}$$

Let $T_0 = \mathrm{id}, T_1, \ldots, T_N$ be a finite sequence of operators belonging to $\boldsymbol{(Op.l.f)_{b,t}}$.

We are trying to solve the formal equation

$$F(x, u, T_1 u, \ldots, T_N u) = f(x) \in \mathfrak{M}_f .$$

We assume that $F_1(x,X) \neq 0$. This means that the valuation of the operator $Du = F(x, u, T_1 u, \ldots, T_N u)$ is one. Without loosing generality we can assume that $f(x) = 0$ (include $f(x)$ in $F(x,X)$).

Our equation has the form

$$F_1(x, T_0 u, T_1 u, \ldots, T_N u) + R_2(x, T_0 u, T_1 u, \ldots, T_N u) = 0$$

where

$$F_1(x, T_0 u, T_1 u, \ldots, T_N u) = \sum_{1 \leq i \leq n} a_i x_i + \Theta$$

and

$$\Theta = \sum_{0 \leq k \leq N} A_k T_k u$$

is an operator belonging to $\boldsymbol{(Op.l.f)_{b,t}}$.

Denote by $(A_{i,j})$ the block triangular matrix of Θ. Introduce in the equation $Du = 0$ a formal power series $u = \sum_{p \geq 1} (e_p(x)) U_p$. To compute the matrices U_p we get the infinite system

$$\begin{aligned} A_{1,1} U_1 &= -{}^t(a_1, a_2, \ldots, a_n) , \\ A_{2,1} U_1 + A_{2,2} U_2 &= H_2 \left(U_1, (a_{\alpha,\beta})_{|\alpha|+|\beta| \leq 2} \right) , \\ &\vdots \\ A_{p,1} U_1 + A_{p,2} U_2 + \cdots + A_{p,p} U_p &= H_p \left(U_1, U_2, \ldots, U_{p-1}, (a_{\alpha,\beta})_{|\alpha|+|\beta| \leq p} \right) , \\ &\vdots \quad . \end{aligned}$$

We see that if $A_{p,p}$ is invertible for all p the same is true for the operator Θ and our equation has a unique formal solution.

If Θ is not invertible there exists p_0 such that A_{p_0,p_0} is not invertible. Assume that $A_{p,p}$ is invertible for all $p \neq p_0$ then if the system

$$A_{p_0,1} U_1 + \cdots + A_{p_0,p_0} U_{p_0} = H_{p_0} \left(U_1, U_2, \ldots, U_{p_0-1}, (a_{\alpha,\beta})_{|\alpha|+|\beta| \leq p_0} \right)$$

has solutions the given system has also solutions.

If now A_{p_0,p_0} and A_{q_0,q_0} for $p_0 > q_0$ are not invertible we get two indeterminated systems

$$\text{(2.5.1)}\quad A_{q_0,1}U_1 + \cdots + A_{q_0,q_0}U_{q_0} = H_{q_0}\left(U_1, U_2, \ldots, U_{q_0-1}, (a_{\alpha,\beta})_{|\alpha|+|\beta|\leq q_0}\right),$$

$$\text{(2.5.2)}\quad A_{p_0,1}U_1 + \cdots + A_{p_0,p_0}U_{p_0} = H_{p_0}\left(U_1, U_2, \ldots, U_{p_0-1}, (a_{\alpha,\beta})_{|\alpha|+|\beta|\leq p_0}\right).$$

Then $U_1, U_2, \ldots, U_{q_0-1}$ are completely given and

$$A_{q_0,q_0}U_{q_0} = G_{q_0}\left(U_1, U_2, \ldots, U_{q_0-1}, (a_{\alpha,\beta})_{|\alpha|+|\beta|\leq q_0}\right).$$

If $G_{q_0}\left(U_1, U_2, \ldots, U_{q_0-1}, (a_{\alpha,\beta})_{|\alpha|+|\beta|\leq q_0}\right) \in \operatorname{Im} A_{q_0,q_0}$ the system (2.5.1) has solutions. If $\operatorname{rank} A_{q_0,q_0} = r$ these solutions depend on $\delta(q_0) - r$ arbitrary constants. In some case, it is possible to choose the arbitrary constants in such a way that

$$G_{p_0}\left(U_1, U_2, \ldots, U_{p_0-1}, (a_{\alpha,\beta})_{|\alpha|+|\beta|\leq p_0}\right) \in \operatorname{Im} A_{p_0,p_0}$$

and in this case the equation $Du = 0$ can have solutions.

Example 2.5.1.1 *Consider the linear operator*

$$T = ay^2\frac{\partial^2}{\partial x^2} + bx^2\frac{\partial^2}{\partial y^2} \quad \textit{for } a \neq 0 \textit{ and } b \neq 0.$$

$$T\left(x^m y^n\right) = am(m-1)x^{m-2}y^{n+2} + bn(n-1)x^{m+2}y^{n-2}$$

is of degree $m+n = \deg\left(x^m y^n\right)$*. The matrix of* T *in the basis* $(e(x))$ *of* $\mathfrak{M}_f$ *is block triangular. This means that* $T \in (\boldsymbol{Op.l.f})_{b,t}$*. It is easy to see that* $A_{1,1}, A_{2,2}$*, and* $A_{4,4}$ *are not invertible but* $A_{3,3}$ *is invertible.*

It is clear that if D *is an operator belonging to* $(\boldsymbol{Op.l.f})_{b,t}$ *such that* $A_{p,p}$ *is invertible for all* $p \geq 1$ *then an equation of the form*

$$Du = R_2(x, y, u, Tu)$$

where

$$R_2\left(x, y, X_0, X_1\right) = \sum_{|\alpha|+|\beta|\geq 2} a_{\alpha,\beta}x^{\alpha_0}y^{\alpha_1}\left(X_0\right)^{\beta_0}\left(X_1\right)^{\beta_1}$$

has a unique formal solution.

2.5.2 Non linear equations without affine part

The one variable case was studied in details in chapter 1. In the several variables case, the non linear equations without affine part have seldom formal solutions as it can be shown by the following example.

Let $T = x\frac{\partial}{\partial x} + y\frac{\partial}{\partial y}$ be a diagonal invertible vector field acting on $\mathfrak{M}_f$. Consider the equation

$$(Tu)^2 = \alpha u^2 + ax^2 + bxy + cy^2 + R_3(x, y, u)$$

where a, b, c are constants. Look for a formal solution of this equation of the form $\sum_{m+n\geq 1} u_{m,n}x^m y^n$. To compute the coefficients $u_{m,n}$ we get the equations

$$(1-\alpha)\,(u_{1,0})^2 = a, \qquad 2(1-\alpha)u_{1,0}u_{0,1} = b, \qquad (1-\alpha)\,(u_{0,1})^2 = c.$$

If $\alpha = 1$ our equation has a formal solution if $a = b = c = 0$. If $\alpha \neq 1$ it is necessary that $b^2 = 4ac$. In each step of the formal resolution, that is in each block associated to $(e_p(x, y))$, we obtain a system where the number of equations is greater than the number of unknowns and this system is solvable only if certain conditions are satisfied and in this way we have usually infinitely many conditions. But it can happen that all these conditions are satisfied, for example

$$(Tu)^2 = u^3 + ax^2 + bxy + cy^2 \quad \text{with } b^2 - 4ac = 0.$$

B Analytic theory

2.6 Notations and definitions

Denote by

- $\mathbb{C}\{x\}$ the ring of germs of holomorphic functions at the origin of $\mathbb{C}^n$ which is also the ring of convergent power series at the origin of $\mathbb{C}^n$;
- $\mathfrak{M}_c$ its maximal ideal;
- $(\boldsymbol{Op.f})$ the $\mathbb{C}$–algebra of operators $D : \mathbb{C}[[x]] \longrightarrow \mathbb{C}[[x]]$ which are not necessarily linear.

Definition 2.6.1 *An operator $D \in (\boldsymbol{Op.f})$ is called* singular *if*

$$D(\mathbb{C}[[x]]) \subset \mathfrak{M}_f.$$

Definition 2.6.2 *An operator $D \in (\boldsymbol{Op.f})$ is called* regular singular *if it is singular and, moreover, for all $\hat{u} \in \mathbb{C}[[x]]$*

$$D(\hat{u}) \in \mathfrak{M}_c \text{ implies } \hat{u} \in \mathbb{C}\{x\}.$$

Definition 2.6.3 *Let L be a vector subspace of $(\boldsymbol{Op.l.f})$. An operator $D \in (\boldsymbol{Op.f})$ is called L–*analytic *if there exist*

1. *a function $F(x_1, x_2, \ldots, x_n, X_0, X_1, \ldots, X_N)$ holomorphic at the origin of $\mathbb{C}^n \times \mathbb{C}^{N+1}$ satisfying $F(0, 0, \ldots, 0) = 0$;*
2. *a sequence of $(N+1)$ operators $T_0 = \mathrm{id}, T_1, \ldots, T_N$ belonging to L such that*

$$Du = F(x_1, x_2, \ldots, x_n, T_0 u, T_1 u, T_2 u, \ldots, T_N u) \quad \textit{for all } u \in \mathbb{C}[[x]]\,.$$

We shall denote by $(\boldsymbol{Op.an})(L)$ the set of L–analytic operators.

It is easy to see that to study the regularity of singular operators it is enough to restrict ourselves to operators D acting on $\mathfrak{M}_f$.

We shall see later that the vector subspaces of $(\boldsymbol{Op.l.f})$ having interesting applications are the following: $(\boldsymbol{Op.l.f})_d$, $(\boldsymbol{Op.l.f})_t$, $(\boldsymbol{Op.l.f})_{b,d}$, $(\boldsymbol{Op.l.f})_{b,t}$ and also $(\boldsymbol{Op.l.f})_{f,r}$. Remark that we have

$$(\boldsymbol{Op.l.f})_d \subset (\boldsymbol{Op.l.f})_t \quad \text{and} \quad (\boldsymbol{Op.l.f})_{b,d} \subset (\boldsymbol{Op.l.f})_{b,t} \subset (\boldsymbol{Op.l.f})_{f,r}.$$

2.7 The good operators and the notion of domination

2.7.1 In $(\boldsymbol{Op.l.f})_t$

The set $\mathbb{N}^n$ is ordered as indicated in section 2.1 of this chapter and denoted by $(\mathbb{N}^n)\,(\mathrm{Ord})$. If $m, k \in \mathbb{N}^n$ we denote

$$(m-k) = (m_1 - k_1, m_2 - k_2, \ldots, m_n - k_n)\,.$$

An operator T of the vector space $(\boldsymbol{Op.l.f})_t$ acting on $\mathfrak{M}_f$ is expressed in the basis $(e(x))$ in the way

$$T\left(\sum_{|m|\geq 1} u_m x^m\right) = \sum_{|m|\geq 1}\left(\sum_{0\leq k\leq m} T_k(m-k)\,u_{m-k}\right) x^m \quad \text{with } u_0 = 0.$$

Let φ and ψ be two operators belonging to $(\boldsymbol{Op.l.f})_t$.

Definition 2.7.1.1 *We shall say that φ is* dominating *ψ if for all m and all k with $0 \leq k \leq m$ we have*

$$|\psi_k(m-k)| \leq |\varphi_k(m-k)|\,.$$

Definition 2.7.1.2 *We shall say that φ is* well dominating *ψ if for all m and all k with $0 \leq k \leq m$ we have*

$$|\psi_k(m-k)| \leq d_k\,|\varphi_0(m-k)|$$

where the sequence of positive numbers d_k satisfies

$$\sum_{|k|\geq 0} d_k \rho^{|k|} < +\infty \quad \textit{for some } \rho > 0.$$

For any operator T of $(\boldsymbol{Op.l.f})_t$ denote by $\operatorname{diag}(T)$ the operator the matrix of which in the basis $(e(x))$ is the diagonal part of the matrix of T.

Definition 2.7.1.2 says that ψ is dominated by the composition of two operators (in the following order) $\operatorname{diag}(\varphi)$ and the multiplication by the power series $\sum_{|k|\geq 0} d_k x^k$.

Definition 2.7.1.3 *We shall say that φ is* strictly dominating *ψ if φ is well dominating ψ and, moreover,*

$$\lim_{|m|\to\infty} \frac{\psi_0(m)}{\varphi_0(m)} = 0.$$

2.7.2 In $(\boldsymbol{Op.l.f})_{b,t}$

An operator T of $(\boldsymbol{Op.l.f})_{b,t}$ acting on $\mathfrak{M}_f$ can be written in the basis $(e(x)) = \bigoplus_{p\geq 1} (e_p(x))$ in the form

$$T\left(\sum_{p\geq 1} (e_p(x))\, U_p\right) = \sum_{p\geq 1} (e_p(x)) \left(\sum_{0\leq k\leq p} T_k(p-k) U_{p-k}\right)$$

where $U_0 = 0$.

Let Φ and Ψ be two operators belonging to the vector space $(\boldsymbol{Op.l.f})_{b,t}$.

Definition 2.7.2.1 *We shall say that Φ is* dominating *Ψ if for all p and all k with $0 \leq k \leq p$ we have*

$$\|\Psi_k(p-k)\| \leq \|\Phi_k(p-k)\|.$$

Definition 2.7.2.2 *We shall say that Φ is* well dominating *Ψ if for all p and all k with $0 \leq k \leq p$ we have*

$$\|\Psi_k(p-k)\| \leq d_k \|\Phi_0(p-k)\|$$

where the sequence of positive real numbers d_k is such that

$$\sum_{k\geq 0} d_k \rho^k < +\infty \quad \textit{for some } \rho > 0.$$

The norm that we are using on matrices is one of the classical equivalent norms on the ring of complex matrices.

In the applications, we are often using the notion of *domination from a given rank M*. This means that in the above definitions the inequalities are only at least true for $p \geq M$.

Definition 2.7.2.3 *We shall say that Φ is* strictly dominating *Ψ if Φ is well dominating Ψ and, moreover,*

$$\lim_{p\to\infty} \frac{\|\Psi_0(p)\|}{\|\Phi_0(p)\|} = 0.$$

2.7.3 In $(\boldsymbol{Op.l.f})_{b,d}$

The definitions given for the class $(\boldsymbol{Op.l.f})_{b,t}$ can be applied to $(\boldsymbol{Op.l.f})_{b,d}$. But let us give a more useful form for the applications.

An operator T of $(\boldsymbol{Op.l.f})_{b,d}$ can be written explicitly in the form

$$T\left(\sum_{p\geq 1} u_p(x)\right) = \sum_{p\geq 1} (e_p(x))\, T_0(p)\, (u_p)$$

where $u_p(x) = (e_p(x))\, u_p$ and $T_0(p)$ is for all p a square matrix of order $\delta(p)$ (dimension of the vector space of homogeneous polynomials of degree p) which is the matrix of the restriction of T to the vector space $H_p(x)$ of homogeneous polynomials of degree p.

Let Φ and Ψ be two operators belonging to $(\boldsymbol{Op.l.f})_{b,d}$.

Definition 2.7.3.1 *We shall say that* Φ *is* well dominating Ψ from the rank M *if for all* $p \geq M$ *and all* $v = (e_p(x))\, v_p$ *of* $H_p(x)$ *we have*

$$\|\Psi_0(p)v_p\| \leq C\, \|\Phi_0(p)v_p\| \quad \textit{where } C \textit{ is a constant.}$$

For all the classes of operators defined above the following definition can be used.

Definition 2.7.3.2 *A* good *operator is an operator which is well dominating itself.*

This means that a good operator is an operator which is dominated by the composition of its diagonal part and a multiplication operator by a convergent power series. In the case of an operator belonging to $(\boldsymbol{Op.l.f})_{b,t}$, we have to understand multiplication in the sense of multiplication by blocks. To an operator Φ of $(\boldsymbol{Op.l.f})_{b,t}$ we associate the operator diag(Φ) the matrix of which in the basis used above is formed by blocks located along the principal diagonal of the matrix of Φ. It is clear that a diagonal operator or a block diagonal operator is a good operator.

2.8 A class of operators having a regular singularity

Let $F(x, X, Y) = F(x_1, x_2, \ldots, x_n, X_0, X_1, \ldots, X_p, Y_1, \ldots, Y_q)$ be a holomorphic function defined near the origin of $\mathbb{C}^n \times \mathbb{C}^{p+1} \times \mathbb{C}^q$ satisfying $F(0,0,0) = 0$. We can write it in the form

$$F(x, X, Y) = \sum_{j\geq 1} F_j(x, X, Y)$$

where for all j the term $F_j(x, X, Y)$ is a homogeneous polynomial of degree j in the variables $x_1, x_2, \ldots, x_n, X_0, X_1, \ldots, X_p, Y_1, \ldots, Y_q$.

2.8.1 In $(\boldsymbol{Op.l.f})_d$

Let $\theta_0 = \mathrm{id}, \theta_1, \ldots, \theta_p$ and $\varphi_1, \ldots, \varphi_q$ be sequences of operators belonging to the class $(\boldsymbol{Op.l.f})_d$. Consider the operator

$$Du = F\left(x, \theta_0 u, \theta_1 u, \ldots, \theta_p u, \varphi_1 u, \ldots, \varphi_q u\right).$$

We are assuming that

$$\begin{aligned} D_1 u = F_1 &\left(x, \theta_0 u, \theta_1 u, \ldots, \theta_p u, \varphi_1 u, \ldots, \varphi_q u\right) \\ &= \sum_{|\alpha|+|\beta|=1} a_{\beta,\alpha} x^\beta u^{\alpha_0} \left(\theta_1 u\right)^{\alpha_1} \cdots \left(\theta_p u\right)^{\alpha_p} \not\equiv 0. \end{aligned}$$

This assumption means that *our equation has a non zero linear part.* To simplify the notations we write

$$D_1 u = \sum_{1 \leq i \leq n} b_i x_i + D_1^0 u$$

with

$$D_1^0 u = \sum_{0 \leq j \leq p} a_j \theta_j u \quad \text{where } \theta_0 = \mathrm{id} \text{ and } a_j \in \mathbb{C}.$$

We have

$$Du = D_1 u - R_2\left(x, \theta_0 u, \theta_1 u, \ldots, \theta_p u, \varphi_1 u, \ldots, \varphi_q u\right)$$

where $R_2(x, X, Y)$ is a function holomorphic at the origin of $\mathbb{C}^n \times \mathbb{C}^{p+1} \times \mathbb{C}^q$ with a valuation greater than or equal to two with respect to the variables $x_1, x_2, \ldots, x_n, X_0, X_1, \ldots, X_p, Y_1, \ldots, Y_q$.

The assumption **(H)**

(H) Among the operators $\{\theta_0 = \mathrm{id}, \theta_1, \ldots, \theta_p\}$ there is one which occurs explicitly in D_1^0 and is strictly dominating all the others and dominating the operators of the family $\{\varphi_1, \ldots, \varphi_q\}$.

Theorem 2.8.1.1 *Under the assumption* **(H)** *every formal solution of*

$$Du = 0$$

is convergent.

Proof : Without loss of generality we can assume for example that θ_p is strictly dominating the operators $\theta_0, \theta_1, \ldots, \theta_{p-1}$ and dominates the operators $\varphi_1, \ldots, \varphi_q$ and moreover that $a_p \neq 0$. This means that θ_p occurs explicitly in D_1^0.

We have

$$\lim_{|m| \to \infty} \frac{\sum_{0 \leq i \leq p} a_i \theta_{i,0}(m)}{\theta_{p,0}(m)} = a_p \neq 0,$$

that is for $|m| \geq m_0$

$$\left| \sum_{0 \leq i \leq p} a_i \theta_{i,0}(m) \right| \geq c_1 \left| \theta_{p,0}(m) \right|$$

where $c_1 \in \mathbb{R}^+$. If for all m such that $|m| < m_0$ we have

$$\sum_{0 \leq i \leq p} a_i \theta_{i,0}(m) \neq 0$$

there exists a constant $c \in \mathbb{R}^+$ such that for all m

$$\left| \sum_{0 \leq i \leq p} a_i \theta_{i,0}(m) \right| \geq c \left| \theta_{p,0}(m) \right| .$$

1. *Existence of a formal power series solution*
   Introduce in the equation

$$D_1 u = R_2 \left(x, \theta_0 u, \theta_1 u, \ldots, \theta_p u, \varphi_1 u, \ldots, \varphi_q u \right)$$

   a formal power series

$$u = \sum_{|m| \geq 1} u_m x^m .$$

   By formal identification we get

   a) for $|m| = 1$

$$u_{(1_k)} \left(\sum_{0 \leq i \leq p} a_i \theta_{i,0} \left((1_k) \right) \right) + b_k = 0$$

   where $(1_k) = (0, \ldots, 0, 1, 0, \ldots, 0)$ the 1 being located at the k–th place;

   b) for $|m| > 1$

$$\begin{aligned} u_m & \left(\sum_{0 \leq i \leq p} a_i \theta_{i,0}(m) \right) \\ & = f_m \Big(\left\{ u_1 \right\}, \ldots, \left\{ u_{m-1} \right\}, \left\{ \theta_{1,0}(1) u_1 \right\}, \ldots, \left\{ \theta_{1,0}(m-1) u_{m-1} \right\}, \\ & \qquad \ldots, \left\{ \theta_{p,0}(1) u_1 \right\}, \ldots, \left\{ \theta_{p,0}(m-1) u_{m-1} \right\}, \\ & \qquad \left\{ \varphi_{1,0}(1) u_1 \right\}, \ldots, \left\{ \varphi_{1,0}(m-1) u_{m-1} \right\}, \\ & \qquad \qquad \ldots, \left\{ \varphi_{q,0}(1) u_1 \right\}, \ldots, \left\{ \varphi_{q,0}(m-1) u_{m-1} \right\}, (R_2) \Big) \end{aligned}$$

   where

- $\{\theta_{j,0}(l)u_l\}$ (resp. $\{\varphi_{j,0}(l)u_l\}$) stands for the sequence of elements
$$\theta_{j,0}(l_1,\dots,l_n)\,u_{l_1,\dots,l_n}\ (\text{resp. } \varphi_{j,0}(l_1,\dots,l_n)\,u_{l_1,\dots,l_n})$$
with $l_1+\cdots+l_n=l$;
- (R_2) stands for the set of coefficients of R_2;
- f_m is a polynomial in all its arguments.

The assumption
$$\sum_{0\le i\le p} a_i\theta_{i,0}(m)\neq 0 \quad \text{for all } m\in\mathbb{N}^n$$
implies the existence of a formal solution $\hat{u}$ of the equation $Du=0$.

2. *Convergence of the formal power series solution*

Consider the analytic equation
$$(2.8.1)\quad \sigma Y=\sum_{1\le i\le n}\alpha_i x_i+\sum_{0\le i\le p}|a_i|\,Y+|R_2|\,(x,Y,Y,\dots,Y,Y,\dots,Y)$$
where

- $|R_2|\,(x,X_0,X_1,\dots,X_p,X'_1,\dots,X'_q)$ is a majorant convergent power series of $R_2\,(x,X_0,X_1,\dots,X_p,X'_1,\dots,X'_q)$;
- σ and $\alpha_1,\dots,\alpha_n$ are real non negative constants.

By the implicit function theorem the equation (2.8.1) has a holomorphic solution near the origin of $\mathbb{C}^n$
$$Y=\sum_{|m|\ge 1}Y_m x^m$$
where $Y_m\ge 0$ for all m. We are now going to show that for a good choice of σ and of the sequence $\{\alpha_i\}_{1\le i\le n}$ the equation (2.8.1) has a solution which is a majorant power series of the formal solution $\hat{u}$ of $Du=0$.

Take

(a) σ such that
$$0<\sigma-\sum_{0\le i\le p}|a_i|<c;$$

(b) $Y^*_{(1_i)}>0$ such that
$$|\theta_{p,0}\left((1_i)\right)|\,\left|u_{(1_i)}\right|\le Y^*_{(1_i)};$$

(c) for all $i=1,\dots,n$
$$\alpha_i=\sigma Y^*_{(1_i)}.$$

With this choice of σ and $\alpha_1, \ldots, \alpha_n$ we see that the equation (2.8.1) has a holomorphic solution

$$Y = \sum_{|m|\geq 1} Y_m x^m \tag{2.8.2}$$

such that for all $i = 1, \ldots, n$

$$Y_{(1_i)} = Y^*_{(1_i)}$$

and for $|m| > 1$

$$\left(\sigma - \sum_{0\leq i\leq p} |a_i|\right) Y_m = f^*_m\Big(\{Y_1\}, \ldots, \{Y_{m-1}\}, \ldots, \{Y_1\}, \ldots, \{Y_{m-1}\},$$
$$\ldots, \{Y_1\}, \ldots, \{Y_{m-1}\}, (|R_2|)\Big)$$

where

- $\{Y_k\}$ stands for the sequence of coefficients $Y_{k_1,\ldots,k_n}$ with $\sum_{1\leq i\leq n} k_i = k$;
- f^*_m is the function obtained from f_m by replacing the coefficients of R_2 by those of the majorant series of R_2.

We are proving that the power series (2.8.2) is a majorant series of $\hat{u}$. This means that

$$|u_m| \leq Y_m \quad \text{for all } m \text{ with } |m| \geq 1.$$

We are doing this by induction on $|m|$.

Assume that for all l such that $|l| < m$ we have

$$\begin{aligned} |\theta_{k,0}(l)|\,|u_l| &\leq Y_l \quad \text{for } k = 0, 1, \ldots, p, \\ |\varphi_{k,0}(l)|\,|u_l| &\leq Y_l \quad \text{for } k = 1, \ldots, q. \end{aligned} \tag{2.8.3}$$

Remark 2.8.1.2 *It is enough to assume that $|\theta_{p,0}(l)|\,|u_l| \leq Y_l$ because the operator θ_p is dominating the operators $\{\theta_i\}_{i=0,\ldots,p-1}$ and $\{\varphi_i\}_{i=1,\ldots,q}$.*

Let us prove that (2.8.3) remains true for all n-ples l such that $|l| = m$. We have for all l such that $|l| = m$

$$u_l = \frac{1}{\sum_{0\leq i\leq p} a_i\theta_{i,0}(l)} f_l\Big(\{u_1\}, \ldots, \{u_{m-1}\}, \{\theta_{1,0}(1)u_1\}, \ldots,$$
$$\ldots, \{\theta_{1,0}(m-1)u_{m-1}\}, \ldots, \{\theta_{p,0}(1)u_1\}, \ldots,$$
$$\ldots, \{\theta_{p,0}(m-1)u_{m-1}\}, \{\varphi_{1,0}(1)u_1\}, \ldots,$$
$$\ldots, \{\varphi_{1,0}(m-1)u_{m-1}\}, \ldots, \{\varphi_{q,0}(1)u_1\}, \ldots,$$
$$\ldots, \{\varphi_{q,0}(m-1)u_{m-1}\}, (R_2)\Big)$$

and

$$\begin{aligned} |u_l| &\leq \frac{1}{\left|\sum\limits_{0\leq i\leq p} a_i\theta_{i,0}(l)\right|} f_l^*\Big(\{Y_1\},\ldots,\{Y_{m-1}\},\ldots,\{Y_1\},\ldots,\{Y_{m-1}\}, \\ &\qquad\qquad \ldots,\{Y_1\},\ldots,\{Y_{m-1}\},(|R_2|)\Big) \\ &\leq \frac{c^{-1}}{|\theta_{p,0}(l)|} f_l^*(\ldots) \\ &\leq \frac{c^{-1}}{|\theta_{p,0}(l)|}\left(\sigma - \sum_{0\leq i\leq p} |a_i|\right) Y_l \end{aligned}$$

and so

$$|\theta_{p,0}(l)|\,|u_l| \leq Y_l \quad \text{for all } l \text{ such that } |l| = m.$$

From remark 2.8.1.2 the inequalities (2.8.3) remain true for all $l \in \mathbb{N}^n$ such that $|l| = m$. The choice of the $Y_{(1_i)}$ shows that (2.8.3) are true for all l such that $|l| = 1$. This proves the convergence of the series $\sum\limits_{|m|\geq 1} u_m x^m$ and also the convergence of the power series $\theta_k(\hat{u})$ for $k = 0,1,\ldots,p$ and $\varphi_k(\hat{u})$ for $k = 1,\ldots,q$. The theorem 2.8.1.1 is proved.

■

Remark 2.8.1.3 *The assumption*

$$\sum_{0\leq i\leq p} a_i\theta_{i,0}(m) \neq 0 \quad \textit{for } |m| < m_0$$

is not essential. It implies the existence of a formal solution but we know that the equation $Du = 0$ has a formal solution and so it is enough to modify the coefficients of the majorant series having an index m with $|m| < m_0$ such that $\sum\limits_{0\leq i\leq p} a_i\theta_{i,0}(m) = 0$. This corresponds to the modification of a finite number of terms in a convergent power series which does not change the nature of the series.

Corollary 2.8.1.4 *Under the assumption* **(H)** *the operator*

$$Du = \sum_{|\beta|+|\alpha|+|\alpha'|\geq 1} a_{\beta,\alpha,\alpha'} x^\beta u^{\alpha_0} (\theta_1 u)^{\alpha_1} \cdots (\theta_p u)^{\alpha_p} (\varphi_1 u)^{\alpha'_1} \cdots (\varphi_q u)^{\alpha'_q}$$

is regular singular.

Corollary 2.8.1.5 *If among the operators $\theta_0 = \mathrm{id}, \theta_1, \ldots, \theta_p$ there is one which occurs explicitly in D_1^0 and is strictly dominating the others then the operator*

$$D_1^0 u = \sum_{|\alpha|=1} a_{0,\alpha} u^{\alpha_0} (\theta_1 u)^{\alpha_1} \cdots (\theta_p u)^{\alpha_p}$$

is regular singular.

Let

$$\begin{aligned} F(x_1, \ldots, x_n, X_0, \ldots, X_p) \\ = \sum_{|\beta|+|\alpha|\geq 1} a_{\beta,\alpha} (x_1)^{\beta_1} \cdots (x_n)^{\beta_n} (X_0)^{\alpha_0} (X_1)^{\alpha_1} \cdots (X_p)^{\alpha_p} \end{aligned}$$

be a convergent power series at the origin of $\mathbb{C}^n \times \mathbb{C}^{p+1}$ and consider the operator

$$\begin{aligned} Du &= F(x_1, \ldots, x_n, u, \theta_1 u, \ldots, \theta_p u) \\ &= D_1^0 u + \sum_{|\beta|=1} a_{\beta,0} x^\beta + \sum_{|\beta|+|\alpha|\geq 2} a_{\beta,\alpha} (x_1)^{\beta_1} \cdots (x_n)^{\beta_n} u^{\alpha_0} (\theta_1 u)^{\alpha_1} \cdots (\theta_p u)^{\alpha_p} \end{aligned}$$

where the operators $\theta_0 = \mathrm{id}, \theta_1, \ldots, \theta_p$ belong to $(\boldsymbol{Op.l.f})_{\boldsymbol{d}}$.

Corollary 2.8.1.6 *If among the operators $\theta_0 = \mathrm{id}, \theta_1, \ldots, \theta_p$ there is one which occurs explicitly in D_1^0 and strictly dominates the others then the operator D is regular singular.*

Moreover, if the operator

$$D_1^0 u = \sum_{0\leq j\leq p} a_j \theta_j u \quad \text{with } a_j \in \mathbb{C}$$

satisfies

$$\sum_{0\leq j\leq p} a_j \theta_{j,0}(m) \neq 0 \quad \text{for all } m$$

the equation $Du = 0$ has a unique holomorphic solution near the origin.

2.8.2 In $(\boldsymbol{Op.l.f})_{\boldsymbol{b},\boldsymbol{d}}$

Let $\theta_0 = \mathrm{id}, \theta_1, \ldots, \theta_p$ and $\varphi_1, \ldots, \varphi_q$ be sequences of operators belonging to the class $(\boldsymbol{Op.l.f})_{\boldsymbol{b},\boldsymbol{d}}$. Consider the operator

$$Du = F(x, \theta_0 u, \theta_1 u, \ldots, \theta_p u, \varphi_1 u, \ldots, \varphi_q u).$$

We are assuming that

$$\begin{aligned} D_1 u &= F_1(x, \theta_0 u, \theta_1 u, \ldots, \theta_p u, \varphi_1 u, \ldots, \varphi_q u) \\ &= \sum_{|\alpha|+|\beta|=1} a_{\beta,\alpha} x^\beta u^{\alpha_0} (\theta_1 u)^{\alpha_1} \cdots (\theta_p u)^{\alpha_p} \not\equiv 0. \end{aligned}$$

This assumption means that our equation has a non zero linear part. To simplify the notations we write

$$D_1 u = (e_1(x))\, b_1 + D_1^0 u$$

with

$$D_1^0 u = \sum_{0 \le j \le p} a_j \theta_j u \quad \text{where } \theta_0 = \mathrm{id} \text{ and } a_j \in \mathbb{C}.$$

We have

$$Du = D_1 u - R_2\,(x, \theta_0 u, \theta_1 u, \dots, \theta_p u, \varphi_1 u, \dots, \varphi_q u)$$

where $R_2(x, X, Y)$ is a function holomorphic at the origin of $\mathbb{C}^n \times \mathbb{C}^{p+1} \times \mathbb{C}^q$ with a valuation greater than or equal to two with respect to the variables $x_1, x_2, \dots, x_n$, $X_0, X_1, \dots, X_p, Y_1, \dots, Y_q$.

The assumption **(H)**

(H) Among the operators $\{\theta_0 = \mathrm{id}, \theta_1, \dots, \theta_p\}$ there is one which occurs explicitly in D_1^0 and is strictly dominating all the others and dominating the operators of the family $\{\varphi_1, \dots, \varphi_q\}$.

Theorem 2.8.2.1 *Under the assumption* **(H)** *every formal solution of*

$$Du = 0$$

is convergent.

Proof : Without loss of generality we can assume for example that θ_p is strictly dominating the operators $\theta_0, \theta_1, \dots, \theta_{p-1}$ and dominates the operators $\varphi_1, \dots, \varphi_q$ and moreover that $a_p \neq 0$. This means that θ_p occurs explicitly in D_1^0.

We have

$$\lim_{m \to \infty} \frac{\left\| \sum_{0 \le i \le p} a_i \theta_{i,0}(m) \right\|}{\|\theta_{p,0}(m)\|} = |a_p| \neq 0;$$

that is for $m \ge m_0$

$$\left\| \sum_{0 \le i \le p} a_i \theta_{i,0}(m) \right\| \ge c_1 \, \|\theta_{p,0}(m)\|$$

where $c_1 \in \mathbb{R}^+$. If $\sum_{0 \le i \le p} a_i \theta_{i,0}(m)$ is invertible for all m such that $m < m_0$ there exists a constant $c \in \mathbb{R}^+$ such that for all m

$$\left\| \sum_{0 \le i \le p} a_i \theta_{i,0}(m) \right\| \ge c \, \|\theta_{p,0}(m)\| .$$

1. *Existence of a formal power series solution*

Introduce in the equation

$$D_1 u = R_2\left(x, \theta_0 u, \theta_1 u, \ldots, \theta_p u, \varphi_1 u, \ldots, \varphi_q u\right)$$

a formal power series

$$u = \sum_{m \geq 1} \left(e_m(x)\right) U_m.$$

By formal identification we get

a) for $m = 1$

$$\left(\sum_{0 \leq i \leq p} a_i \theta_{i,0}(1)\right) U_1 + b_1 = 0;$$

b) for $m > 1$

$$\begin{aligned}\left(\sum_{0 \leq i \leq p} a_i \theta_{i,0}(m)\right) & U_m \\ &= f_m\Big(\{U_1\}, \ldots, \{U_{m-1}\}, \{\theta_{1,0}(1)U_1\}, \ldots, \\ &\quad \ldots, \{\theta_{1,0}(m-1)U_{m-1}\}, \ldots, \{\theta_{p,0}(1)U_1\}, \ldots, \\ &\quad \ldots, \{\theta_{p,0}(m-1)U_{m-1}\}, \{\varphi_{1,0}(1)U_1\}, \ldots, \\ &\quad \ldots, \{\varphi_{1,0}(m-1)U_{m-1}\}, \ldots, \{\varphi_{q,0}(1)U_1\}, \ldots, \\ &\quad \ldots, \{\varphi_{q,0}(m-1)U_{m-1}\}, (R_2)\Big)\end{aligned}$$

where

- (R_2) stands for the set of coefficients of R_2;
- f_m is a polynomial in all its arguments.

The assumption that $\sum\limits_{0 \leq i \leq p} a_i \theta_{i,0}(m)$ is invertible for all $m \geq 1$, which is a consequence of the hypothesis, implies the existence of a formal solution $\hat{u}$ of the equation $Du = 0$.

2. *Convergence of the formal power series solution*

Consider the analytic equation

$$\text{(2.8.4)} \quad \sigma Y = (e_1(x))\alpha + \sum_{0 \leq i \leq p} |a_i| Y + |R_2|\left(x, Y, Y, \ldots, Y, Y, \ldots, Y\right)$$

where

- $|R_2|(x, X_0, X_1, \dots, X_p, X'_1, \dots, X'_q)$ is a majorant convergent power series of $R_2(x, X_0, X_1, \dots, X_p, X'_1, \dots, X'_q)$;
- σ and $\alpha = {}^t(\alpha_1, \dots, \alpha_n)$ are real non negative constants.

By the implicit function theorem the equation (2.8.4) has a holomorphic solution near the origin of $\mathbb{C}^n$

$$Y = \sum_{m\geq 1} (e_m(x))Y_m$$

where $Y_m \geq 0$ for all m. We are now going to show that for a good choice of σ and of the sequence $\{\alpha_i\}_{1\leq i\leq n}$ the equation (2.8.4) has a solution which is a majorant power series of the formal solution $\hat{u}$ of $Du = 0$.

Take

(a) σ such that

$$0 < \sigma - \sum_{0\leq i\leq p} |a_i| < c;$$

(b) $Y_1^* > 0$ such that

$$\|\theta_{p,0}((1))\| \, \|U_1\| \leq Y_1^*;$$

(c)

$$\alpha = \sigma Y_1^*.$$

With this choice of σ and $\alpha_1, \dots, \alpha_n$ we see that the equation (2.8.4) has a holomorphic solution

$$Y = \sum_{m\geq 1} (e_m(x))Y_m \tag{2.8.5}$$

such that

$$Y_1 = Y_1^*$$

and for $m > 1$

$$\left(\sigma - \sum_{0\leq i\leq p} |a_i|\right) Y_m = f_m^*\Big(\{Y_1\}, \dots, \{Y_{m-1}\}, \dots, \{Y_1\}, \dots, \{Y_{m-1}\},$$
$$\dots, \{Y_1\}, \dots, \{Y_{m-1}\}, (|R_2|)\Big)$$

where f_m^* is the function obtained from f_m by replacing the coefficients of R_2 by those of the majorant series of R_2.

We are proving that the power series (2.8.5) is a majorant series of $\hat{u}$. This means that

$$\|U_m\| \leq Y_m \quad \text{for all } m \text{ with } m \geq 1.$$

We are doing this by induction on m.

Assume that for all l such that $l < m$ we have

$$\begin{array}{rcll} \|\theta_{k,0}(l)\| \, \|U_l\| & \leq & Y_l & \text{for } k = 0, 1, \ldots, p, \\ \|\varphi_{k,0}(l)\| \, \|U_l\| & \leq & Y_l & \text{for } k = 1, \ldots, q. \end{array} \tag{2.8.6}$$

Remark 2.8.2.2 *It is enough to assume that*

$$\|\theta_{p,0}(l)\| \, \|U_l\| \leq Y_l$$

because the operator θ_p *is dominating the operators* $\{\theta_i\}_{i=0,\ldots,p-1}$ *and the operators* $\{\varphi_i\}_{i=1,\ldots,q}$.

Let us prove that (2.8.6) remains true for $l = m$. We have

$$\begin{aligned} U_m = \left(\sum_{0 \leq i \leq p} a_i \theta_{i,0}(m) \right)^{-1} & f_m\Big(\{U_1\}, \ldots, \{U_{m-1}\}, \{\theta_{1,0}(1)U_1\}, \\ \ldots, \{\theta_{1,0}(m-1)U_{m-1}\}, & \ldots, \{\theta_{p,0}(1)U_1\}, \ldots, \{\theta_{p,0}(m-1)U_{m-1}\}, \\ & \{\varphi_{1,0}(1)U_1\}, \ldots, \{\varphi_{1,0}(m-1)U_{m-1}\}, \\ & \ldots, \{\varphi_{q,0}(1)U_1\}, \ldots, \{\varphi_{q,0}(m-1)U_{m-1}\}, (R_2)\Big) \end{aligned}$$

and

$$\begin{aligned} \|U_m\| & \leq \left\| \left(\sum_{0 \leq i \leq p} a_i \theta_{i,0}(m) \right)^{-1} \right\| f_m^*\Big(\{Y_1\}, \ldots, \{Y_{m-1}\}, \ldots \\ & \qquad \ldots, \{Y_1\}, \ldots, \{Y_{m-1}\}, \ldots, \{Y_1\}, \ldots, \{Y_{m-1}\}, (|R_2|)\Big) \\ & \leq \frac{c^{-1}}{\|\theta_{p,0}(m)\|} f_m^*(\ldots) \\ & \leq \frac{c^{-1}}{\|\theta_{p,0}(m)\|} \left(\sigma - \sum_{0 \leq i \leq p} |a_i| \right) Y_m \end{aligned}$$

and so

$$\|\theta_{p,0}(m)\| \, \|U_m\| \leq Y_m.$$

From remark 2.8.2.2 the inequalities (2.8.6) remain true for $l = m$. The choice of the Y_1 shows that (2.8.6) are true for $l = 1$. This proves the convergence of the series $\sum_{m \geq 1} (e_m(x))U_m$ and also the convergence of the power series $\theta_k(\hat{u})$ for $k = 0, 1, \ldots, p$ and $\varphi_k(\hat{u})$ for $k = 1, \ldots, q$. The theorem 2.8.2.1 is proved.

■

Remark 2.8.2.3 *The assumption that* $\sum_{0\le i\le p} a_i\theta_{i,0}(m)$ *is invertible for* $m < m_0$ *is not essential. It implies the existence of a formal solution but we know that the equation* $Du = 0$ *has a formal solution and so it is enough to modify the coefficients of the majorant series having an index* $m < m_0$ *such that* $\sum_{0\le i\le p} a_i\theta_{i,0}(m)$ *is not invertible. This corresponds to the modification of a finite number of terms in a convergent power series which does not change the nature of the series.*

Corollary 2.8.2.4 *Under the assumption* **(H)** *the operator*

$$Du = \sum_{|\beta|+|\alpha|+|\alpha'|\ge 1} a_{\beta,\alpha,\alpha'} x^\beta u^{\alpha_0} (\theta_1 u)^{\alpha_1} \cdots (\theta_p u)^{\alpha_p} (\varphi_1 u)^{\alpha'_1} \cdots (\varphi_q u)^{\alpha'_q}$$

is regular singular.

Corollary 2.8.2.5 *If among the operators* $\theta_0 = \mathrm{id}, \theta_1, \ldots, \theta_p$ *there is one which occurs explicitly in* D_1^0 *and is strictly dominating the others then the operator*

$$D_1^0 u = \sum_{|\alpha|=1} a_{0,\alpha} u^{\alpha_0} (\theta_1 u)^{\alpha_1} \cdots (\theta_p u)^{\alpha_p}$$

is regular singular.

Let

$$\begin{aligned} &F(x_1, \ldots, x_n, X_0, \ldots, X_p) \\ &\quad = \sum_{|\beta|+|\alpha|\ge 1} a_{\beta,\alpha} (x_1)^{\beta_1} \cdots (x_n)^{\beta_n} (X_0)^{\alpha_0} (X_1)^{\alpha_1} \cdots (X_p)^{\alpha_p} \end{aligned}$$

be a convergent power series at the origin of $\mathbb{C}^n \times \mathbb{C}^{p+1}$ and consider the operator

$$\begin{aligned} Du &= F(x_1, \ldots, x_n, u, \theta_1 u, \ldots, \theta_p u) \\ &= D_1^0 u + \sum_{|\beta|=1} a_{\beta,0} x^\beta + \sum_{|\beta|+|\alpha|\ge 2} a_{\beta,\alpha} (x_1)^{\beta_1} \cdots (x_n)^{\beta_n} u^{\alpha_0} (\theta_1 u)^{\alpha_1} \cdots (\theta_p u)^{\alpha_p} \end{aligned}$$

where the operators $\theta_0 = \mathrm{id}, \theta_1, \ldots, \theta_p$ belong to $\boldsymbol{(Op.l.f)_{b,d}}$.

Corollary 2.8.2.6 *If among the operators* $\theta_0 = \mathrm{id}, \theta_1, \ldots, \theta_p$ *there is one which occurs explicitly in* D_1^0 *and strictly dominates the others then the operator* D *is regular singular.*

Moreover, if the operator

$$D_1^0 u = \sum_{0\le j\le p} a_j \theta_j u \quad \text{with } a_j \in \mathbb{C}$$

satisfies the condition that $\sum_{0\le j\le p} a_j\theta_{j,0}(m)$ is invertible for all m then the equation $Du = 0$ has a unique holomorphic solution near the origin.

2.8.3 In $(\boldsymbol{Op.l.f})_{b,t}$

Let $\theta_0 = \mathrm{id}, \theta_1, \ldots, \theta_p$ and $\varphi_1, \ldots, \varphi_q$ be sequences of operators belonging to the class $(\boldsymbol{Op.l.f})_{b,t}$. Consider the operator

$$Du = F\left(x, \theta_0 u, \theta_1 u, \ldots, \theta_p u, \varphi_1 u, \ldots, \varphi_q u\right).$$

We are assuming that

$$D_1 u = F_1\left(x, \theta_0 u, \theta_1 u, \ldots, \theta_p u, \varphi_1 u, \ldots, \varphi_q u\right) = \sum_{|\alpha|+|\beta|=1} a_{\beta,\alpha} x^\beta u^{\alpha_0} \left(\theta_1 u\right)^{\alpha_1} \cdots \left(\theta_p u\right)^{\alpha_p} \not\equiv 0.$$

This assumption means that our equation has a non zero linear part. To simplify the notations we write

$$D_1 u = \left(e_1(x)\right) b_1 + D_1^0 u$$

with

$$D_1^0 u = \sum_{0 \le j \le p} a_j \theta_j u \quad \text{where } \theta_0 = \mathrm{id} \text{ and } a_j \in \mathbb{C}.$$

We have

$$Du = D_1 u - R_2\left(x, \theta_0 u, \theta_1 u, \ldots, \theta_p u, \varphi_1 u, \ldots, \varphi_q u\right)$$

where $R_2(x, X, Y)$ is a function holomorphic at the origin of $\mathbb{C}^n \times \mathbb{C}^{p+1} \times \mathbb{C}^q$ with a valuation greater than or equal to two with respect to the variables $x_1, x_2, \ldots, x_n$, $X_0, X_1, \ldots, X_p, Y_1, \ldots, Y_q$.

The assumption **(H)**

(H) Among the operators $\{\theta_0 = \mathrm{id}, \theta_1, \ldots, \theta_p\}$ there is one which occurs explicitly in D_1^0 and is strictly dominating all the others and dominating the operators of the family $\{\varphi_1, \ldots, \varphi_q\}$.

Theorem 2.8.3.1 *Under the assumption* **(H)** *every formal solution of*

$$Du = 0$$

is convergent.

Proof : It is easy to see that the proof given for the one variable case and the proof given above for $(\boldsymbol{Op.l.f})_{b,d}$ can be easily adapted to this case. The explicit proof is left to the reader. ■

2.9 Applications to partial differential equations

2.9.1 With a Poincaré vector field

Let

$$\theta = \sum_{1\leq i\leq n} \lambda_i x_i \frac{\partial}{\partial x_i}$$

be a vector field having a singularity at the origin which is semi–simple in Jordan normal form and consider the operator

$$Du = \sum_{0\leq i\leq p} a_i(x,u)\theta^{p-i}u.$$

Denote by

$$D_1^0 u = \sum_{0\leq i\leq p} a_i(0,0)\theta^{p-i}u.$$

Assume that the operator D_1^0 is not identically zero and that $a_0(0,0) \neq 0$. If θ satisfies a Poincaré condition

$$\left|\sum_{1\leq i\leq n} \lambda_i m_i\right| \geq c_0\,|m| \quad \text{for all } m = (m_1,\ldots,m_n) \in \mathbb{N}^n$$

the operator θ^p strictly dominates the operators θ^q for all $q < p$. This implies that D_1^0 is regular singular and by theorem 2.8.1.1 D is also regular singular.

Moreover, every operator of the form

$$Du = \sum_{0\leq i\leq p} a_i\theta^{p-i}u + R_2\left(x,u,\theta u,\ldots,\theta^p u\right)$$

with $a_i \in \mathbb{C}$ for all i, $a_0 \neq 0$ and $R_2\left(x,X_0,X_1,\ldots,X_p\right)$ being a holomorphic function in $x, X_0, X_1, \ldots, X_p$ near the origin of $\mathbb{C}^n \times \mathbb{C}^{p+1}$ with a valuation greater than or equal to two in the variables $x, X_0, X_1, \ldots, X_p$ is regular singular.

2.9.2 The partial differential equations of the hypergeometric functions of P. Appell

If $\theta_1 = x_1\frac{\partial}{\partial x_1}$ and $\theta_2 = x_2\frac{\partial}{\partial x_2}$ it is easy to see that the *hypergeometric functions* F_1, F_2, F_3, F_4 are respectively solutions of the following systems of partial differential equations:

$$(2.9.1)\quad \begin{aligned} D_1^1(z) &= \theta_1\left(\theta_1+\theta_2+\gamma-1\right)z - x_1\left(\theta_1+\theta_2+\alpha\right)\left(\theta_1+\beta\right)z = 0,\\ D_2^1(z) &= \theta_2\left(\theta_1+\theta_2+\gamma-1\right)z - x_2\left(\theta_1+\theta_2+\alpha\right)\left(\theta_2+\beta'\right)z = 0;\end{aligned}$$

$$(2.9.2)\quad \begin{aligned} D_1^2(z) &= \theta_1(\theta_1+\gamma-1)z - x_1(\theta_1+\theta_2+\alpha)(\theta_1+\beta)z = 0,\\ D_2^2(z) &= \theta_2(\theta_2+\gamma'-1)z - x_2(\theta_1+\theta_2+\alpha)(\theta_2+\beta')z = 0; \end{aligned}$$

$$(2.9.3)\quad \begin{aligned} D_1^3(z) &= \theta_1(\theta_1+\theta_2+\gamma-1)z - x_1(\theta_1+\alpha)(\theta_1+\beta)z = 0,\\ D_2^3(z) &= \theta_2(\theta_1+\theta_2+\gamma-1)z - x_2(\theta_2+\alpha')(\theta_2+\beta')z = 0; \end{aligned}$$

$$(2.9.4)\quad \begin{aligned} D_1^4(z) &= \theta_1(\theta_1+\gamma-1)z - x_1(\theta_1+\theta_2+\alpha)(\theta_1+\theta_2+\beta)z = 0,\\ D_2^4(z) &= \theta_2(\theta_2+\gamma'-1)z - x_2(\theta_1+\theta_2+\alpha)(\theta_1+\theta_2+\beta)z = 0. \end{aligned}$$

Theorem 2.9.2.1 *The operators*

$$D^i : u \longmapsto D^i u = \begin{cases} D_1^i u, \\ D_2^i u \end{cases}$$

are regular singular.

This result means that for every pairs $(f_1(x_1,x_2), f_2(x_1,x_2))$ of holomorphic functions at the origin of $\mathbb{C}^2$ each formal solution of

$$D^i u = \begin{cases} f_1(x_1,x_2), \\ f_2(x_1,x_2) \end{cases}$$

is convergent. It is clear that for such a system to have a formal solution it is necessary to assume that $f_1(0,0)=0$ and $f_2(0,0)=0$.

1. Consider the system (2.9.1)

$$D_1^1 u = f_1(x), \qquad D_2^1 u = f_2(x)$$

and let $\hat{u}$ be a formal power series solution of this system. We have

$$(D_1^1 + D_2^1)\,\hat{u} = f_1(x) + f_2(x).$$

The linear part of $(D_1^1 + D_2^1)$ is

$$(D_1^1 + D_2^1)_1^0\, u = (\theta_1+\theta_2)^2 u + (\gamma-1)(\theta_1+\theta_2)u.$$

As $(\theta_1+\theta_2)^2$ strictly dominates $(\theta_1+\theta_2)$ and dominates also $\theta_1, \theta_2, (\theta_1)^2$, $(\theta_2)^2$, and $\theta_1 \circ \theta_2$ theorem 2.8.1.1 implies the convergence of the formal solution $\hat{u}$. Moreover, the following system:

$$\begin{aligned} D_1^1 u + R_{1,2}^1\left(x, u, \theta_1 u, \theta_2 u, (\theta_1)^2 u, (\theta_2)^2 u, \theta_1\circ\theta_2 u\right) &= 0,\\ D_2^1 u + R_{2,2}^1\left(x, u, \theta_1 u, \theta_2 u, (\theta_1)^2 u, (\theta_2)^2 u, \theta_1\circ\theta_2 u\right) &= 0, \end{aligned}$$

where for $i = 1,2$ the functions $R^1_{i,2}(x, X_0, X_1, \ldots, X_5)$ are holomorphic in their arguments and of valuation greater than or equal to two, is regular singular. We know that the system

$$D^1_1 u = 0, \qquad D^1_2 u = 0$$

has the following convergent power series as a solution:

$$F_1(\alpha, \beta, \beta', \gamma, x_1, x_2) = \sum_{m_1 \geq 0, m_2 \geq 0} \frac{(\alpha, m_1 + m_2)(\beta, m_1)(\beta', m_2)}{(\gamma, m_1 + m_2)(1, m_1)(1, m_2)} (x_1)^{m_1} (x_2)^{m_2}$$

where

$$(r, p) = r(r+1)\cdots(r+p-1) \quad \text{for all } r \in \mathbb{C} \text{ and } p \in \mathbb{N}.$$

This convergent series is also a solution of the equation

$$(x_1 - x_2)\,\theta_1 \circ \theta_2 z - \beta' x_2 \theta_1 z + \beta x_1 \theta_2 z = 0.$$

It is easy to see that this equation has a formal solution of the form

$$\sum_{m_1 \geq 0, m_2 \geq 0} \frac{(\beta, m_1)(\beta', m_2)}{(1, m_1)(1, m_2)} \Psi(m_1 + m_2)(x_1)^{m_1}(x_2)^{m_2}$$

where Ψ is an arbitrary function. If we take $\Psi(p) = \frac{(\alpha,p)}{(\gamma,p)}$ we get the convergent solution $F_1(\alpha, \beta, \beta', \gamma, x_1, x_2)$. But if we take $\Psi(p) = p!$ we get a divergent formal solution. This means that the operator

$$D : u \longmapsto Du = (x_1 - x_2)\,\theta_1 \circ \theta_2 u - \beta' x_2 \theta_1 u + \beta x_1 \theta_2 u$$

is irregular singular.

2. Consider the system (2.9.2)

$$D^2_1 u = f_1(x), \qquad D^2_2 u = f_2(x)$$

and let $\hat{u}$ be a formal power series solution of this system. We have

$$\left(D^2_1 + D^2_2\right)\hat{u} = f_1(x) + f_2(x).$$

The linear part of $\left(D^2_1 + D^2_2\right)$ is

$$\left(D^2_1 + D^2_2\right)^0_1 u = \left((\theta_1)^2 + (\theta_2)^2\right) u + (\gamma - 1)\theta_1 u + (\gamma' - 1)\theta_2 u.$$

As $(\theta_1)^2 + (\theta_2)^2$ strictly dominates θ_1 and θ_2 and dominates also $(\theta_1)^2, (\theta_2)^2$, $\theta_1 \circ \theta_2$ theorem 2.8.1.1 implies the convergence of the formal solution $\hat{u}$. And the system associated to (2.9.2) is regular singular. Moreover, the following system:

$$
\begin{aligned}
D_1^2 u + R_{1,2}^2\left(x, u, \theta_1 u, \theta_2 u, (\theta_1)^2 u, (\theta_2)^2 u, \theta_1 \circ \theta_2 u\right) &= 0,\\
D_2^2 u + R_{2,2}^2\left(x, u, \theta_1 u, \theta_2 u, (\theta_1)^2 u, (\theta_2)^2 u, \theta_1 \circ \theta_2 u\right) &= 0,
\end{aligned}
$$

where for $i = 1, 2$ the functions $R_{i,2}^2(x, X_0, X_1, \ldots, X_5)$ are holomorphic in their arguments and of valuation greater than or equal to two, is regular singular.

3. The same kind of proofs shows that the systems (2.9.3) and (2.9.4) are also regular singular. See also [1].

Remark 2.9.2.2 *We have studied the systems (2.9.1),..., (2.9.4) near the origin of $\mathbb{C}^2$. We have the same kinds of results near the other singular points of the hypergeometric systems.*

2.9.3 The systems of partial differential equations of the confluent hypergeometric functions in two variables

The *confluent hypergeometric functions* are solutions of the following systems:

$$
\begin{aligned}
(2.9.5)\quad & (1-x_1)\left((\theta_1)^2-\theta_1\right)z + (1-x_1)(\theta_1\circ\theta_2)z + \\
&\quad + [\gamma-(\alpha+\beta+1)x_1]\,\theta_1 z - x_1\theta_2 z - \alpha\beta x_1 z = 0,\\
&\left((\theta_2)^2-\theta_2\right)z + (\theta_1\circ\theta_2)z + (\gamma-x_2)\,\theta_2 z - x_2\theta_1 z - \alpha x_2 z = 0;
\end{aligned}
$$

$$
\begin{aligned}
(2.9.6)\quad & \left((\theta_1)^2-\theta_1\right)z + (\theta_1\circ\theta_2)z + (\gamma-x_1)\,\theta_1 z - \beta x_1 z = 0,\\
& \left((\theta_2)^2-\theta_2\right)z + (\theta_1\circ\theta_2)z + (\gamma-x_2)\,\theta_2 z - \beta' x_2 z = 0;
\end{aligned}
$$

$$
\begin{aligned}
(2.9.7)\quad & \left((\theta_1)^2-\theta_1\right)z + (\theta_1\circ\theta_2)z + (\gamma-x_1)\,\theta_1 z - \beta x_1 z = 0,\\
& \left((\theta_2)^2-\theta_2\right)z + (\theta_1\circ\theta_2)z + \gamma\theta_2 z - \alpha x_2 z = 0;
\end{aligned}
$$

$$
\begin{aligned}
(2.9.8)\quad & (1-x_1)\left((\theta_1)^2-\theta_1\right)z - x_1(\theta_1\circ\theta_2)z + \\
&\quad + [\gamma-(\alpha+\beta+1)x_1]\,\theta_1 z - \beta x_1\theta_2 z - \alpha\beta x_1 z = 0,\\
&\left((\theta_2)^2-\theta_2\right)z + (\theta_1\circ\theta_2)z + (\gamma'-x_2)\,\theta_2 z - x_2\theta_1 z - \alpha x_2 z = 0;
\end{aligned}
$$

$$
\begin{aligned}
(2.9.9)\quad & \left((\theta_1)^2-\theta_1\right)z + (\gamma-x_1)\,\theta_1 z - x_1\theta_2 z - \alpha x_1 z = 0,\\
& \left((\theta_2)^2-\theta_2\right)z + (\gamma'-x_2)\,\theta_2 z - x_2\theta_1 z - \alpha x_2 z = 0;
\end{aligned}
$$

$$\begin{aligned}(2.9.10)\quad (1-x_1)\left((\theta_1)^2-\theta_1\right)z+(\theta_1\circ\theta_2)\,z+&\\ +[\gamma-(\alpha+\beta+1)x_1]\,\theta_1 z-\alpha\beta x_1 z &= 0,\\ \left((\theta_2)^2-\theta_2\right)z+(\theta_1\circ\theta_2)\,z+(\gamma-x_2)\,\theta_2 z-\alpha' x_2 z &= 0;\end{aligned}$$

$$\begin{aligned}(2.9.11)\quad (1-x_1)\left((\theta_1)^2-\theta_1\right)z+(\theta_1\circ\theta_2)\,z+&\\ +[\gamma-(\alpha+\beta+1)x_1]\,\theta_1 z-\alpha\beta x_1 z &= 0,\\ \left((\theta_2)^2-\theta_2\right)z+(\theta_1\circ\theta_2)\,z+\theta_2 z-\alpha x_2 z &= 0.\end{aligned}$$

As in the section concerning hypergeometric systems we can see that *all these systems are regular singular near the origin of* $\mathbb{C}^2$. The same result is not true at infinity. In fact, the singular point at infinity is obtained by the confluence of two regular singularities and, in general, such a singular point is irregular singular. We can see that a solution of (2.9.5) is also a solution of

$$x_1\,(\theta_1\circ\theta_2)\,z-x_2\theta_1 z+x_1\theta_2 z=0$$

and a solution of (2.9.6) is a solution of

$$(x_1-x_2)\,(\theta_1\circ\theta_2)\,z-\beta' x_2\theta_1 z+\beta x_1\theta_2 z=0$$

and a solution of (2.9.7) is a solution of

$$x_1\,(\theta_1\circ\theta_2)\,z-x_2\theta_1 z+\beta x_1\theta_2 z=0$$

and all these equations are irregular singular in the same way as we did for the system (2.9.1).

2.9.4 The systems of partial differential equations of the hypergeometric functions in n variables of G. Lauricella

Each of the four functions of Lauricella satisfies a system of n partial linear differential equations of second order

$$\begin{aligned}(2.9.12)\quad & x_j\,(1-x_j)\,\frac{\partial^2 F}{\partial x_j^2}-x_j\sum_{k\neq j}x_k\frac{\partial^2 F}{\partial x_k\partial x_j}+\\ &+[\gamma_j-(\alpha+\beta_j+1)\,x_j]\,\frac{\partial F}{\partial x_j}-\beta_j\sum_{k\neq j}x_k\frac{\partial F}{\partial x_k}-\alpha\beta_j F=0\quad\text{for } j=1,\ldots,n;\end{aligned}$$

$$\begin{aligned}(2.9.13)\quad & x_j\,(1-x_j)\,\frac{\partial^2 F}{\partial x_j^2}+\sum_{k\neq j}x_k\frac{\partial^2 F}{\partial x_k\partial x_j}+\\ &+[\gamma-(\alpha_j+\beta_j+1)\,x_j]\,\frac{\partial F}{\partial x_j}-\alpha_j\beta_j F=0\quad\text{for } j=1,\ldots,n;\end{aligned}$$

$$(2.9.14)\quad x_j(1-x_j)\frac{\partial^2 F}{\partial x_j^2}-\sum_{s\neq j \text{ when } r=j} x_r x_s\frac{\partial^2 F}{\partial x_r\partial x_s}+$$
$$+[\gamma_j-(\alpha+\beta+1)x_j]\frac{\partial F}{\partial x_j}-(\alpha+\beta+1)\sum_{k\neq j}x_k\frac{\partial F}{\partial x_k}-\alpha\beta F=0 \quad \text{for } j=1,\ldots,n;$$

$$(2.9.15)\quad x_j(1-x_j)\frac{\partial^2 F}{\partial x_j^2}+(1-x_j)\sum_{k\neq j}x_k\frac{\partial^2 F}{\partial x_k\partial x_j}+$$
$$+[\gamma-(\alpha+\beta_j+1)x_j]\frac{\partial F}{\partial x_j}-\beta_j\sum_{k\neq j}x_k\frac{\partial F}{\partial x_k}-\alpha\beta_j F=0 \quad \text{for } j=1,\ldots,n.$$

These systems of partial differential equations are regular singular at the origin of $\mathbb{C}^n$. Let us prove this for the system (2.9.12). The proofs are the same for the other systems.

To the system (2.9.12) let us associate the operator

$$D_A u=(1-x_j)\left((\theta_j)^2-\theta_j\right)u-x_j\sum_{k\neq j}\theta_k\circ\theta_j u+[\gamma_j-(\alpha+\beta_j+1)x_j]\theta_j u-$$
$$-\beta_j x_j\sum_{k\neq j}\theta_k u-\alpha\beta_j x_j u=0 \quad (j=1,\ldots,n)$$

where $\theta_j=x_j\frac{\partial}{\partial x_j}$.

Let $f(x)\in(\mathbb{C}\{x\})^n$ and consider the system

$$D_A u=f(x).$$

Assume that this system has a formal solution $\hat{u}\in\mathbb{C}[[x]]$. By addition of all the equations we obtain

$$\left(\sum_{1\leq j\leq n}(\theta_j)^2\right)u-\left(\sum_{1\leq j\leq n}\theta_j\right)u-\left(\sum_{1\leq j\leq n}x_j(\theta_j)^2\right)u+\left(\sum_{1\leq j\leq n}x_j\theta_j\right)u-$$
$$-\left(\sum_{1\leq j\leq n}x_j\sum_{k\neq j}\theta_j\circ\theta_k\right)u+\left(\sum_{1\leq j\leq n}\gamma_j\theta_j\right)u-\left(\sum_{1\leq j\leq n}(\alpha+\beta_j+1)x_j\theta_j\right)u-$$
$$-\left(\sum_{1\leq j\leq n}\beta_j x_j\sum_{k\neq j}\theta_k\right)u-\left(\sum_{1\leq j\leq n}x_j\alpha\beta_j\right)u=0.$$

The linear part of D_A is

$$(D_A)_1^0 u=\left(\sum_{1\leq j\leq n}(\theta_j)^2\right)u-\left(\sum_{1\leq j\leq n}(1-\gamma_j)\theta_j\right)u.$$

As $\sum_{1\leq j\leq n} (\theta_j)^2$ strictly dominates θ_j for all $j = 1, \ldots, n$ and dominates $\theta_j \circ \theta_k$ for all $j = 1, \ldots, n$ and $k = 1, \ldots, n$ the operator D_A is regular singular. Moreover, for any function $R_2(x_1, \ldots, x_n, X_0, X_1, \ldots, X_n)$ with values in $\mathbb{C}^n$ and of valuation greater than or equal to two with respect to all its arguments the operator

$$u \longmapsto D_A u + R_2(x_1, \ldots, x_n, u, \theta_1 u, \ldots, \theta_n u)$$

is regular singular.

2.9.5 A theorem of M. Kashiwara, T. Kawai, and J. Sjöstrand

Consider the differential operator

$$Pu = \sum_{|\alpha|=|\beta|\leq m} a_{\alpha,\beta}(x) x^\alpha \frac{\partial^{|\beta|}}{\partial x^\beta} u.$$

P belongs to $(\boldsymbol{Op.l.f})$. Let us associate to it the operator

$$P_0 = \sum_{|\alpha|=|\beta|=m} a_{\alpha,\beta}(0) x^\alpha \frac{\partial^{|\beta|}}{\partial x^\beta}.$$

In [32], Kashiwara, Kawai, and Sjöstrand are showing that if

$$\sum_{|\alpha|=|\beta|=m} a_{\alpha,\beta}(0) x^\alpha (x^*)^\beta \neq 0 \quad \text{on } \mathbb{C}^n \setminus \{0\}$$

where x^* denotes the complex conjugate number of x then P is regular singular.

The method used by the authors are technics of the theory of partial differential equations. The above condition says that the spheres centered at the origin are not characteristics. We can see easily that the operator P_0 dominates the operators

$$x^\alpha \frac{\partial^{|\beta|}}{\partial x^\beta} \quad \text{for } |\alpha| = |\beta| < m$$

and the result is a corollary of theorem 2.8.2.1.

With our theory we have a more general result. In fact, we have the following non linear result that for the equation

$$Pu = R_2\left(x, x^\alpha \frac{\partial^{|\beta|}}{\partial x^\beta} u\right)$$

where R_2 is of order at least two in its arguments each formal power series solution is convergent and, in particular, the operator

$$Du = Pu - R_2\left(x, x^\alpha \frac{\partial^{|\beta|}}{\partial x^\beta} u\right)$$

is regular singular.

3 Formal and convergent solutions of singular partial differential equations

Introduction

In this chapter, we are studying operator equations having a linear part in a different way. This allows us to use the classical method of iteration which is different from the method of majorant series. This study allows us to give new results and new simple proofs of some known results such that the classical Poincaré theorem and also a more clear presentation of a theorem due to Kaplan. Finally, we are giving a proof of the classical Siegel's theorem (case of small denominators).

3.1 Notations and definitions

For a multi–index $p = (p_1, \dots, p_n) \in \mathbb{Z}^n$, we set as usual

$$|p| = p_1 + \cdots + p_n \quad \text{and} \quad x^p = (x_1)^{p_1} \cdots (x_n)^{p_n} .$$

We introduce the order $p \leq m$ if $p_j \leq m_j$ for all $j = 1, \dots, n$. We denote by 0 the multi–index $(0, \dots, 0)$.

Let A be a valuated commutative local ring with unit. In the sequel it will be the field of complex numbers or a ring of formal or convergent power series over $\mathbb{C}$ or $\mathbb{R}$.

Let

- $\mathcal{O}_f = A\,[[x_1, \dots, x_n]] = A[[x]]$ the ring of formal power series over A in the variables $x_1, \dots, x_n$;
- $\mathcal{O} = A\,\{x_1, \dots, x_n\} = A\{x\}$ the ring of convergent power series over A in the variables $x_1, \dots, x_n$;
- $\mathfrak{M}_f$ and $\mathfrak{M}$ the maximal ideals in $\mathcal{O}_f$ respectively in $\mathcal{O}$.

The elements of $\mathcal{O}_f$ or $\mathcal{O}$ are written in the form $u = \sum\limits_p u_p x^p$ where $u_p \in A$ for all p.

We consider linear operators

$$P : (\mathcal{O}_f)^q \longrightarrow (\mathcal{O}_f)^q$$

which can be written in the form

(3.1.1) $$P(u)=\sum_{|m|\geq 0}\left(\sum_{k_0\leq k\leq m,|k|\geq 0} P_k(m-k)u_{m-k}\right)x^m$$

where

- $k_0 \in \mathbb{Z}^n$ may have some negative components;
- k_0 depends on P but not on u;
- the sum is taken only over those $k \geq k_0$ with $|k| \geq 0$;
- $P_k(m)$ is a matrix valued function of m.

The set of all these operators is denoted by $\mathcal{P}_f$ and the set of all $P \in \mathcal{P}_f$ with $P(\mathcal{O}^q) \subset \mathcal{O}^q$ is $\mathcal{P}$. $\mathcal{P}_+$ and $\mathcal{P}_{f+}$ are the sets of operators with $k_0 = 0$.

Examples 3.1.1 *1. Let $P(\xi) = \sum_{|p|\leq d} a_p \xi^p$ be a polynomial with complex coefficients and consider the operator*

$$P\left(x\frac{\partial}{\partial x}\right)=\sum_{|p|\leq d} a_p\left(x\frac{\partial}{\partial x}\right)^p$$

where $\left(x\frac{\partial}{\partial x}\right)^p = \left(x_1\frac{\partial}{\partial x_1}\right)^{p_1}\cdots\left(x_n\frac{\partial}{\partial x_n}\right)^{p_n}$. We have

(3.1.2) $$P\left(x\frac{\partial}{\partial x}\right)u=\sum_{|p|\geq 0} P(p)u_p x^p.$$

So $P_k(p) \equiv 0$ if $k \neq 0$ and $P_0(p) = P(p)$ and P belongs to $\mathcal{P}_+$. An operator of this form will be called a singular partial differential operator with constant coefficients. *Note that P is not an operator with constant coefficients in the usual sense.*

2. Let P be a polynomial as above and, moreover, $P(p) \neq 0$ for all $p \in \mathbb{N}^n$. Then the operator given by (3.1.2) is invertible in $\mathcal{P}_{f+}$ and $P^{-1} : \mathcal{O}_f \longrightarrow \mathcal{O}_f$ is given by

$$P^{-1}u=\sum_{|p|\geq 0}\frac{u_p}{P(p)}x^p.$$

One of our problems is to give conditions on $P(p)$ such that $P^{-1} \in \mathcal{P}_+$. Remark that P^{-1} is not a differential operator.

3. As a special case, we consider vector fields of the form

$$\tau=\sum_{1\leq j\leq n}\lambda_j x_j\frac{\partial}{\partial x_j}$$

which define operators of the form (3.1.2) given by

(3.1.3)
$$\tau u = \sum_{|p|\geq 0} \langle \lambda ; p\rangle\, u_p x^p$$

where $\langle \lambda ; p\rangle = \sum_{1\leq j\leq n} \lambda_j p_j$.

4. *More generally, let* $P(x,\xi) = \sum_{|p|\leq d} a_p(x)\xi^p$ *be a polynomial with coefficients* $a_p(x) \in \mathcal{O}$ *and consider the operator*

$$P\left(x, x\frac{\partial}{\partial x}\right) = \sum_{|p|\leq d} a_p(x) \left(x\frac{\partial}{\partial x}\right)^p .$$

Using the Taylor expansion $a_p(x) = \sum_k a_{p,k}x^k$ *we get*

$$\begin{aligned} P\left(x, x\frac{\partial}{\partial x}\right) u &= \sum_{|p|\leq d}\sum_{k,j} a_{p,k}x^k \left(x\frac{\partial}{\partial x}\right)^p u_j x^j \\ &= \sum_{|p|\leq d}\sum_{k,j} a_{p,k} j^p u_j x^{k+j} \\ &= \sum_{k,j} P_k(j) u_j x^{k+j} \\ &= \sum_{|m|\geq 0} \left(\sum_{0\leq k\leq m} P_k(m-k) u_{m-k}\right) x^m \end{aligned}$$

where $P_k(j) = \sum_{|p|\leq d} a_{p,k} j^p$ *and* $P\left(x, x\frac{\partial}{\partial x}\right)$ *belongs to* $\mathcal{P}_+$.

5. *Consider the vector field*

$$\tau = \sum_{1\leq j\leq n} \lambda_j x_j \frac{\partial}{\partial x_j} + \sum_{1\leq j\leq n-1} a_j x_{j+1} \frac{\partial}{\partial x_j} = {}^s\tau + {}^n\tau.$$

We have

$$[{}^s\tau, {}^n\tau] = \sum_{1\leq j\leq n-1} (\lambda_{j+1} - \lambda_j)\, a_j x_{j+1} \frac{\partial}{\partial x_j}.$$

If $(\lambda_{j+1} - \lambda_j)\, a_j = 0$ *for all* $j = 1, \dots, n-1$ *then* $[{}^s\tau, {}^n\tau] = 0$ *and the decomposition* $\tau = {}^s\tau + {}^n\tau$ *is the Jordan normal form of* τ *([15]) where* ${}^s\tau$ *is the semi-simple part and* ${}^n\tau$ *the nilpotent part of the vector field* τ*. The operator* ${}^s\tau$ *belongs to* $\mathcal{P}_{f+}$ *and*

$$\begin{aligned} {}^n\tau u &= \sum_{|m|\geq 0}\sum_{1\leq j\leq n-1} a_j m_j u_m x^{m+e_j} \\ &= \sum_{|m|\geq 0} \left(\sideset{}{'}\sum_{1\leq j\leq n-1} a_j\,(m_j + 1)\, u_{m-e_j}\right) x^m \end{aligned}$$

with $e_j = (0,\dots,0,-1,1,0,\dots,0)$ *the* -1 *being at the* j*-th place and* $\sum'$ *for the inner summation means that terms where* u_{m-e_j} *is not defined have to be ommitted. If we set* $k_0 = (-1,\dots,-1,1)$ *then* $P_k(p) = 0$ *when* $k \neq e_j$ *for* $j = 1,\dots,n-1$ *and when* $k = e_j$ *we have*

$$P_{e_j}(p) = a_j p_j.$$

We get

$$(3.1.4) \qquad {}^n\tau u = \sum_{|m|\geq 0} \left(\sum_{k_0 \leq k \leq m} P_k(m-k) u_{m-k} \right) x^m.$$

Then we see that ${}^n\tau$ *is in* $\mathcal{P}_f$ *but not in* $\mathcal{P}_{f+}$*. The sum* $\tau = {}^s\tau + {}^n\tau$ *is given by the same formula if we set in addition*

$$P_0(p) = \langle \lambda; p \rangle = \sum_{1 \leq j \leq n} \lambda_j p_j.$$

6. *In an application in section 3.4, we have to consider more general vector fields of the form*

$$\tau = \sum_{1\leq j \leq d} \left[x_j \left(\lambda_j + R_j \left(x_{d+1}, \dots, x_n \right) \right) + P_j \left(x_1, \dots, x_{j-1}, x_{j+1}, \dots, x_n \right) \right] \frac{\partial}{\partial x_j}$$

where $d < n$*,* R_j *is a formal power series in* $x_{d+1},\dots,x_n$ *without constant term and* P_j *is a polynomial in* $x_1,\dots,x_{j-1},x_{j+1},\dots,x_d$ *without constant term and with formal power series in* $x_{d+1},\dots,x_n$ *as coefficients.*

Considered as an operator

$$\tau : A\left[\left[x_1,\dots,x_d\right]\right] \longrightarrow A\left[\left[x_1,\dots,x_d\right]\right]$$

with $A = \mathbb{C}\left[\left[x_{d+1},\dots,x_n\right]\right]$ *this operator is in* $\mathcal{P}_f$*.*

7. *As an example of an operator which is neither a differential operator nor the inverse of a differential operator, let* $\tau = \sum_{1\leq j\leq n} \lambda_j x_j \frac{\partial}{\partial x_j}$ *be a vector field and consider*

$$P = \exp(\tau) = \sum_{k\geq 0} \frac{\tau^k}{k!}.$$

P is given by the formula

$$Pu = \sum_{|p|\geq 0} \exp\left(\langle \lambda; p \rangle\right) u_p x^p$$

and $P \in \mathcal{P}_{f+}$*.*

Remark 3.1.2 *If we consider the formal power series* $\sum u_p x^p$ *as given by its sequence of coefficients* $(u_p)_{p\in\mathbb{N}^n}$ *an operator* $P \in \mathcal{P}_{f+}$ *can be considered as a lower triangular infinite matrix with diagonal elements* $P_0(p)$. *The operators* $P \in \mathcal{P}_f$ *are then given by infinite matrices.*

As an easy consequence, we have

Proposition 3.1.3 *If* $P \in \mathcal{P}_{f+}$ *is such that* $P_0(m)$ *is invertible in* A *for all* $m \in \mathbb{N}^n$ *then* P *is invertible in* $\mathcal{P}_{f+}$.

This is obviously not true in $\mathcal{P}_f$.

For later use let us introduce the following decomposition of an operator $P \in \mathcal{P}_f$

$$P = P_- + P_0 + P_+ \tag{3.1.5}$$

where

$$\begin{aligned} P_0 u &= \sum_m P_0(m) u_m x^m, \\ P_+ u &= \sum_m \left(\sum_{k\le m, |k|>0} P_k(m-k) u_{m-k} \right) x^m, \\ P_- u &= \sum_m \left(\sum_{|k|=0, k\neq 0} P_k(m-k) u_{m-k} \right) x^m. \end{aligned}$$

3.2 Holomorphic solutions of certain equations

In this section, we are looking for holomorphic solutions of equations

$$Pu = F(x,u)$$

where $P \in \mathcal{P}$ and $F(x,u)$ is holomorphic in a neighborhood of $(0,0)$ in $\mathbb{C}^n \times \mathbb{C}^q$. Let us be more precise. For $u \in \mathcal{O}^q$ with $u = \sum_p u_p x^p$ there exist a polydisk given by the poly–radius $t = (t_1, \dots, t_n) \in \mathbb{R}^n_+$ and ρ with $0 < \rho < 1$ such that

$$\|u\|_\rho = \sum_{|p|\ge 0} |u_p|\, \rho^{|p|} t^p < \infty \tag{3.2.1}$$

where $|\cdot|$ is any norm on $\mathbb{C}^q$. Given t and ρ we denote by $B^q(\rho)$ the space of the $u \in \mathcal{O}^q$ for which $\|u\|_\rho$ is finite. $\|\cdot\|_\rho$ is a norm on $B^q(\rho)$ and $B^q(\rho)$ is a Banach space. For $q=1$ $B^q(\rho)$ is a Banach algebra. We do not indicate the dependance on t since in general t will be kept fixed.

We set $\mathfrak{M}(\rho) = \mathfrak{M} \cap B^q(\rho)$. Sometimes we write $B_\rho = B(\rho)$ and $\mathfrak{M}_\rho = \mathfrak{M}(\rho)$.

3.2.1 Solution of the linear equation

For $P \in \mathcal{P}_f$ we denote by $|P_k(m)|$ a matrix norm of $P_k(m)$ compatible with the norm chosen in $\mathbb{C}^q$.

Theorem 3.2.1.1 *Let $P \in \mathcal{P}_f$ be an operator satisfying the following condition: there exist a positive function φ and positive constants c_k for $k \geq k_0$ such that*

$$|P_k(m)| \leq c_k \varphi(m) \quad \textit{for all } m \in \mathbb{N}^n \tag{3.2.2}$$

and such that

$$c(\rho) = \sum_{k \geq k_0} c_k \rho^{|k|} t^k < \infty \tag{3.2.3}$$

for some $t \in \mathbb{R}^n_+$ and ρ with $0 < \rho < 1$. Assume moreover that there exist a μ with $0 < \mu \leq 1$ and a positive constant $d(\mu)$ such that

$$\varphi(m)\mu^{|m|} \leq d(\mu) \quad \textit{for all } m \in \mathbb{N}^n. \tag{3.2.4}$$

Then

$$P : B^q(\rho) \longrightarrow B^q(\sigma)$$

is continuous where $\sigma = \mu\rho$ and

$$\|Pu\|_\sigma \leq c(\sigma) d(\mu) \|u\|_\rho \tag{3.2.5}$$

Proof : For $u \in B^q(\rho)$ and $\sigma = \mu\rho$ we estimate $\|Pu\|_\sigma$ by

$$\begin{aligned}
\|Pu\|_\sigma &\leq \sum_{|m| \geq 0} \left(\left| \sum_{k_0 \leq k \leq m, |k| \geq 0} P_k(m-k) u_{m-k} \right| \right) \sigma^{|m|} t^m \\
&\leq \sum_{|m| \geq 0} \left(\sum_k c_k \varphi(m-k) \, |u_{m-k}| \right) \sigma^{|m|} t^m \\
&\leq \left(\sum_{k_0 \leq k, |k| \geq 0} c_k \sigma^{|k|} t^k \right) \left(\sum_p \varphi(p) \, |u_p| \, \sigma^{|p|} t^p \right) \\
&\leq c(\sigma) \sum_p \varphi(p) \mu^{|p|} \, |u_p| \, \rho^{|p|} t^p \\
&\leq c(\sigma) d(\mu) \, \|u\|_\rho \, .
\end{aligned}$$

■

We apply this theorem to the inverse of singular operators with constant coefficients of the form (3.1.2) and give three conditions on $P(m)$ which imply the conditions (3.2.2) to (3.2.4).

Corollary 3.2.1.2 *Let P be an operator of the form (3.1.2) and let c, δ be positive constants.*

a) Assume that P satisfies the Poincaré condition

$$|P(m)| \geq c\,|m|^{\delta} \quad \text{for all } m \in \mathbb{N}^n \setminus \{0\} \tag{3.2.6}$$

then

$$P^{-1} : \mathfrak{M}_\rho \longrightarrow \mathfrak{M}_\rho$$

is continuous.

b) Assume that P satisfies the Siegel condition

$$|P(m)| \geq c\,|m|^{-\delta} \quad \text{for all } m \in \mathbb{N}^n \setminus \{0\} \tag{3.2.7}$$

then

$$P^{-1} : \mathfrak{M}_\rho \longrightarrow \mathfrak{M}_\sigma$$

is continuous for any $\sigma < \rho < 1$ and we have the estimate

$$\left\|P^{-1}u\right\|_\sigma \leq \frac{C}{(\rho-\sigma)^\delta}\left\|u\right\|_\rho \quad \text{for some } C. \tag{3.2.8}$$

c) Assume that P satisfies the exponential condition

$$|P(m)| \geq c\delta^{-|m|} \quad \text{for all } m \in \mathbb{N}^n \setminus \{0\} \tag{3.2.9}$$

then

$$P^{-1} : \mathfrak{M}_\rho \longrightarrow \mathfrak{M}_\sigma$$

is continuous for $\sigma = \frac{\rho}{\delta}$.

Proof : In the cases a) and c), we have the estimates

$$\left|P(m)^{-1}\right| \leq \frac{1}{c\,|m|^\delta} \qquad \text{resp.} \qquad \left|P(m)^{-1}\right| \leq \frac{\delta^{|m|}}{c}.$$

We set $c_0 = \frac{1}{c}, c_k = 0$ for $k \neq 0$ and $\varphi(m) = \frac{1}{|m|^\delta}$ resp. $\varphi(m) = \delta^{|m|}$. The assertion now follows from theorem 3.2.1.1 with $\mu = 1$ in case a) and $\mu = \delta^{-1}$ in case c).

In case b), we have c_k as above and $\varphi(m) = |m|^\delta, c(\rho) = \frac{1}{c}$ and to prove (3.2.4) we have to find an upper bound of $|m|^\delta \mu^{|m|}$ with $0 < \mu < 1$. By elementary calculus the maximum of the function $f(x) = x^\delta \mu^x$ is $\left(\frac{\delta}{e}\right)^\delta \frac{1}{(-\log\mu)^\delta}$ (where $\log e = 1$) and since $\mu = \frac{\sigma}{\rho}$ we have by the mean value theorem and the fact that $\sigma < \rho < 1$

$$|m|^\delta \mu^{|m|} \leq \frac{\left(\frac{\delta}{e}\right)^\delta}{(\log\rho - \log\sigma)^\delta} \leq \frac{\left(\frac{\delta}{e}\right)^\delta}{(\rho-\sigma)^\delta}.$$

And (3.2.5) gives the estimate (3.2.8) with $C = \frac{1}{c}\left(\frac{\delta}{e}\right)^\delta$. ∎

Remarks 3.2.1.3 *1. If we make the additional assumption $\left|P(0)^{-1}\right| \leq \frac{1}{c}$ we have the same results with B_ρ instead of $\mathfrak{M}_\rho$.*

2. The same results hold for matrix valued operators if we require the corresponding bounds for the inverse matrices $P(m)^{-1}$.

3. A special case are vector fields as in (3.1.3). The estimates (3.2.6), (3.2.7), (3.2.9) are required for $P(m) = \langle \lambda; m \rangle$.

4. If $P(m) \neq 0$ for all m but $P(m_k)$ decreases faster than exponentially for some sequence m_k it is possible that the formal power series $P^{-1}u$ which exists by proposition 3.1.3 does not converge for some u. In fact, $\tau = x_1 \frac{\partial}{\partial x_1} - \lambda x_2 \frac{\partial}{\partial x_2}$ where λ is a transcendental number which is well approximated by rationals is such an example (see [6], [31]).

Corollary 3.2.1.4 *If $P = P\left(x, x\frac{\partial}{\partial x}\right)$ is a singular differential operator with variable coefficients (example 3.1.1, 4.) there exist ρ_0 and t_0 such that*

$$P : B_\rho \longrightarrow B_\sigma$$

is continuous for all $\sigma < \rho \leq \rho_0, t \leq t_0$.

Proof :

$$P(u) = \sum_{m \geq 0} \left(\sum_{0 \leq k \leq m} P_k(m-k) u_{m-k} \right) x^m$$

where

$$P_k(\xi) = \sum_{|p| \leq d} a_{p,k} \xi^p .$$

Since P_k is a polynomial of degree d we have the estimate

$$|P_k(m)| \leq c_k \, |m|^d \quad \text{with } c_k = \sum_{|p| \leq d} |a_{p,k}|$$

and since the series $\sum_k a_{p,k} x^k$ converges in some polydisk there are ρ_0, t_0 such that $\sum_k c_k \rho^{|k|} t^k = c(\rho) < \infty$. For $\varphi(m) = |m|^d$ the same argument as in corollary 3.2.1.2 b) shows that $\mu^{|m|}\varphi(m)$ is bounded for every $\mu < 1$. Theorem 3.2.1.1 gives then the conclusion. ∎

For the solution of $Pu = f$ where $P \in \mathcal{P}$ we consider P as a perturbation of P_0 by $P_- + P_+$ where P_0, P_-, P_+ are given by (3.1.5). In order that the Neumann series converges we have to impose conditions which guarantee that the norm of the operator $(P_0)^{-1}(P_- + P_+)$ is strictly less than 1.

Theorem 3.2.1.5 *Let $P \in \mathcal{P}_f$ be decomposed as in (3.1.5), $P = P_0 + P_- + P_+$.*

a) Suppose that $P_0(m)$ is invertible for $m \in \mathbb{N}^n$ and that the formal inverse P_0^{-1} (which exists by proposition 3.1.3) is continuous from

$$B^q(\rho) \longrightarrow B^q(\rho) \quad \text{for all } \rho \le \rho_0, t \le t_0.$$

b) Suppose that there exist a positive function φ and for all $k \ge k_0, k \ne 0$ positive constants c_k and d_k such that

$$|P_k(m)| \le c_k \varphi(m) \qquad \text{and} \qquad \varphi(m)\left|P_0(m+k)^{-1}\right| \le d_k \quad \text{for all } m \in \mathbb{N}^n$$

and such that

$$c(\rho) = \sum_{k \ge k_0, k \ne 0} c_k d_k \rho^{|k|} t^k < \infty \quad \text{for some } \rho \text{ and } t.$$

c) Suppose that the finite sum

$$c_-(\rho) = \sum_{|k|=0, k \ne 0} c_k d_k t^k$$

corresponding to P_- can be made arbitrarily small by a suitable choice of t.

Then P is invertible and for all sufficiently small ρ and suitably chosen t

$$P^{-1} : B^q(\rho) \longrightarrow B^q(\rho)$$

is continuous.

Proof : $P_0(m)$ is invertible. We estimate

$$\begin{aligned}
\left\|(P_0)^{-1}(P_- + P_+)u\right\|_\rho &\le \sum_{m \ge 0} \left| \sum_{k_0 \le k \le m, k \ne 0} P_0(m)^{-1} P_k(m-k) u_{m-k} \right| \rho^{|m|} t^m \\
&\le \sum_{m \ge 0} \sum_{k_0 \le k \le m, k \ne 0} \left|P_0(m)^{-1}\right| c_k \varphi(m-k) \left|u_{m-k}\right| \rho^{|m|} t^m \\
&\le \sum_{m \ge 0} \sum_{k_0 \le k \le m, k \ne 0} c_k d_k \rho^{|k|} t^k \left|u_{m-k}\right| \rho^{|m-k|} t^{m-k} \\
&\le c(\rho) \left\|u\right\|_\rho .
\end{aligned}$$

Now $c(\rho) = c_-(\rho) + c_+(\rho)$ with $c_+(\rho) = \sum_{|k|>0} c_k d_k \rho^{|k|} t^k$. The series $c_+(\rho)$ has no constant term and tends to zero as ρ goes to zero. Together with hypothesis c) this implies that there exist ρ and t such that $c(\rho) < 1$. Now $P^{-1} = (1 + P_- + P_+)^{-1} P_0^{-1}$ where the first factor is given by the Neumann series and

$$\left\|P^{-1}u\right\|_\rho \le \frac{C \left\|u\right\|_\rho}{1 - c(\rho)}$$

where C is the norm of P_0^{-1}. ■

Remarks 3.2.1.6 *1. The condition c) seems to be rather technical to make the proof work. But for the applications that we have in mind it is easily verified.*

2. The theorem means that the equation $Pu = v$ has a unique solution for every $v \in B_\rho$.

Corollary 3.2.1.7 *Let $P = P\left(x, x\frac{\partial}{\partial x}\right)$ be a singular differential operator with variable coefficients and assume that P_0 satisfies the Poincaré condition*

$$|P_0(m)| \geq c\,|m|^d \quad \text{for } m \in \mathbb{N}^n \setminus \{0\}$$

where d is the degree of the operator. Then P is invertible and

$$P^{-1} : \mathfrak{M}_\rho \longrightarrow \mathfrak{M}_\rho$$

is continuous for small ρ.

Proof : As in the proof of corollary 3.2.1.4 we have

$$|P_k(m)| \leq c_k\,|m|^d = c_k\varphi(m)$$

and

$$\frac{\varphi(m)}{|P_0(m+k)|} \leq \frac{|m|^d}{c\,|m+k|^d} \leq \frac{1}{c}.$$

So $d_k = \frac{1}{c}$ and $c(\rho) = \frac{1}{c}\sum_{k\neq 0} c_k\rho^{|k|}t^k$ is finite for some ρ and t. The condition c) is void since $P_- = 0$. ■

This corollary is not true if we replace the Poincaré condition by the Siegel condition. Even an exponent $\delta < d$ in the Poincaré condition is not sufficient as the following counter–example shows:

$$\begin{aligned}
P &= \left(x\frac{\partial}{\partial x}+1\right)^2 - \left(y\frac{\partial}{\partial y}+1\right)^2 + 1 - xy\left(x\frac{\partial}{\partial x}+y\frac{\partial}{\partial y}\right),\\
P_0 &= \left(x\frac{\partial}{\partial x}+1\right)^2 - \left(y\frac{\partial}{\partial y}+1\right)^2 + 1,\\
P_{1,1} &= -\left(x\frac{\partial}{\partial x}+y\frac{\partial}{\partial y}\right).
\end{aligned}$$

We have

$$|P_0(n,m)| = \left|(n+1)^2 - (m+1)^2 + 1\right| \geq 1$$

and

$$P_{1,1}(n,m) = -(n+m).$$

Let us consider the equation $Pu = v$ where $v = (1-xy)^{-1} = \sum_{n\geq 0}(xy)^n$. Comparing coefficients we get

$$P_0(n,m)u_{n,m} + P_{1,1}(n-1,m-1)u_{n-1,m-1} = \delta_{n,m} \quad \text{for } n,m \geq 1$$

and

$$\begin{aligned} P_0(n,0)u_{n,0} &= 0 \quad \text{for } n \neq 0, \\ P_0(0,m)u_{0,m} &= 0 \quad \text{for } m \neq 0, \\ P_0(0,0)u_{0,0} &= 1. \end{aligned}$$

These equations can be solved inductively and we get $u_{n,m} = 0$ for $n \neq m$ and

$$u_{n,n} = 1 + 2(n-1) + 2^2(n-1)(n-2) + \cdots + 2^{n-1}(n-1)!$$

and the series $\sum u_{n,m}x^n y^m$ is not convergent.

As an example where $P_- \neq 0$ we consider vector fields as in example 3.1.1, 5.

Corollary 3.2.1.8 *Consider the vector field*

$$\tau = \sum_{1\leq j\leq n} \lambda_j x_j \frac{\partial}{\partial x_j} + \sum_{1\leq j\leq n-1} a_j x_{j+1} \frac{\partial}{\partial x_j}$$

and let λ satisfy the Poincaré condition $|\langle\lambda; p\rangle| \geq c\,|p|$. Then there exist ρ and t such that $\tau^{-1} : \mathfrak{M}_\rho \longrightarrow \mathfrak{M}_\rho$ is continuous.

Proof : With the notations of example 3.1.1, 5. we have $P_0 = {}^s\tau, P_- = {}^n\tau$ and $P_+ = 0$. The Poincaré condition implies condition a) of the theorem. With $\varphi(m) = |m|$ we have

$$|P_k(m)| = |a_j m_j| \leq |a_j|\,\varphi(m) \quad \text{for } k = e_j.$$

So $c_k = |a_j|$ for $k = e_j$ and so $c_k = 0$ for $k \neq e_j$. Since $|k| = 0$ for $k = e_j$ we have

$$\left|\varphi(m)\langle\lambda; m+k\rangle^{-1}\right| \leq \frac{|m|}{c\,|m+k|} = \frac{|m|}{c\,|m|} \leq \frac{1}{c} = d_k.$$

Finally, we get

$$c(\rho) = c_-(\rho) = \sum_k c_k d_k t^k = \frac{1}{c} \sum_{1\leq j\leq n-1} |a_j|\,(t_j)^{-1}\,t_{j+1}$$

and all the $(t_j)^{-1}\,t_{j+1}$ can be made small simultaneously. ■

3.2.2 Solution of the non linear equation

In this section, we solve the equation $Pu = F(x,u)$ where P is an operator of the form (3.1.1). We suppose that $F : \mathbb{C}^n \times \mathbb{C}^q \longrightarrow \mathbb{C}^q$ is a vector valued holomorphic function in a neighborhood of the origin.

Theorem 3.2.2.1 *Assume $F(0,0) = 0$, $\left(\frac{\partial F}{\partial u}\right)(0,0) = 0$ where $\left(\frac{\partial F}{\partial u}\right)$ is the Jacobian matrix of F with respect to the variables u. If P satisfies the conditions a), b), and c) of the theorem 3.2.1.5 the equation*

$$Pu = F(x,u) \tag{3.2.10}$$

has a unique holomorphic solution in a neighborhood of the origin satisfying

$$u(0) = 0.$$

Proof : We use Newton's method to solve the equation

$$\mathfrak{F}(x,u) = Pu - F(x,u) = 0.$$

Let u_0 be an approximate solution. We try to give a better approximation $u_0 + \Delta_0 = u_1$. The Taylor formula gives

$$\mathfrak{F}(x,u_1) = \mathfrak{F}(x,u_0) + \left(P - \left(\frac{\partial F}{\partial u}\right)(x,u_0)\right)\Delta_0 + O\left((\Delta_0)^2\right).$$

We see that if Δ_0 is a solution of

$$\left(P - \left(\frac{\partial F}{\partial u}\right)(x,u_0)\right)\Delta_0 = -\mathfrak{F}(x,u_0) \tag{3.2.11}$$

we have

$$\mathfrak{F}(x,u_1) = O\left((\Delta_0)^2\right).$$

Before starting the iteration let us study the linearized equation (3.2.11).

Lemma 3.2.2.2 *There exist ρ, t and $r > 0$ such that for any holomorphic function $v : \mathbb{C}^n \longrightarrow \mathbb{C}^q$ in a neighborhood of the origin with*

$$v(0) = 0 \qquad \textit{and} \qquad \|v\|_\rho \le r$$

the operator

$$P - \left(\frac{\partial F}{\partial u}\right)(x, v(x)) : B^q(\rho) \longrightarrow B^q(\rho)$$

is invertible. The norm of the inverse can be estimated by a constant K depending only on ρ, r and on the norm of P^{-1} but not on v.

Proof : We make use again of a perturbation argument, so we have to estimate $\left\|P^{-1}\left(\frac{\partial F}{\partial u}\right)(x,v(x))\right\|_\rho$. By theorem 3.2.1.5

$$\|P^{-1}\|_\rho \leq \frac{C}{1-c(\rho)}$$

and this is decreasing to C with ρ. We get

$$\left\|P^{-1}\left(\frac{\partial F}{\partial u}\right)(x,v(x))\,u\right\|_\rho \leq \frac{C}{1-c(\rho)}\left\|\left(\frac{\partial F}{\partial u}\right)(x,v(x))\right\|_\rho \|u\|_\rho .$$

Let

$$\left(\frac{\partial F}{\partial u}\right)(x,u) = \sum_j g_j(x)u^j = \sum_{|j|+|k|\geq 1} g_{j,k}k^k u^j$$

be the Taylor expansion of $\frac{\partial F}{\partial u}$. This double series converges absolutely in some polydisk centered at 0. We have

$$\left\|\left(\frac{\partial F}{\partial u}\right)(x,v(x))\right\|_\rho \leq \sum_j \|g_j(x)\|_\rho \left(\|v(x)\|_\rho\right)^{|j|} \leq \sum_{|j|+|k|\geq 1} |g_{j,k}|\,\rho^{|k|}t^k r^{|j|}.$$

This series converges for some ρ, t, r and since there is no term with $(j,k)=(0,0)$ it can be made as small as we like by choosing ρ and r small enough independently of $v(x)$. ∎

Now let us come back to the proof of theorem 3.2.2.1. We define inductively the sequence u_k by

$$u_0 = 0, \tag{3.2.12}$$

$$u_{k+1} = u_k + \Delta_k, \tag{3.2.13}$$

$$\left(P - \left(\frac{\partial F}{\partial u}\right)(x,u_k)\right)\Delta_k = -\mathfrak{F}(x,u_k). \tag{3.2.14}$$

Since $F(0,0)=0$ and $\mathfrak{F}(x,u_0) = -F(x,0)$ the norm $\|\mathfrak{F}(x,u_0)\|_\rho = \varepsilon$ can be made arbitrarily small by choosing a sufficiently small ρ.

As in the proof of lemma 3.2.2.2, we see that for some r and ρ we have

$$\|F''(x,v(x))\|_\rho \leq C \quad \text{for } \|v\|_\rho \leq r$$

where F'' is the second derivative of F and C is a constant independent of v. By (3.2.14) we get the estimates

$$\|\Delta_k\|_\rho \leq K\,\|\mathfrak{F}(x,u_k)\|_\rho \tag{3.2.15}$$

and

$$(3.2.16)\quad \begin{aligned} \|\mathfrak{F}(x,u_{k+1})\|_\rho &= \left\|O\left((\Delta_k)^2\right)\right\|_\rho \\ &= \left\|\left\langle \int_0^1 (1-s)F''(u_k+s\Delta_k)\,ds\Delta_k;\Delta_k\right\rangle\right\|_\rho \\ &\le \frac{C}{2}\|\Delta_k\|_\rho^2 \\ &\le \frac{CK^2}{2}\|\mathfrak{F}(x,u_k)\|_\rho^2 \qquad \text{if } \|u_k+s\Delta_k\|_\rho \le r. \end{aligned}$$

Now assume ε chosen so small that

$$\frac{2}{CK}\sum_{k\ge 0}\left(\frac{CK^2\varepsilon}{2}\right)^{2^k} < r.$$

Then we have

$$(3.2.17)\qquad \|u_k\|_\rho < r,$$

$$(3.2.18)\qquad \|\Delta_k\|_\rho \le \frac{2}{CK}\left(\frac{CK^2\varepsilon}{2}\right)^{2^k},$$

$$(3.2.19)\qquad \|\mathfrak{F}(x,u_k)\|_\rho \le \frac{2}{CK^2}\left(\frac{CK^2\varepsilon}{2}\right)^{2^k}.$$

This is proved by induction. In fact, if (3.2.17), (3.2.18) and (3.2.19) are true for k then $u_{k+1} = \sum_{0\le m\le k}\Delta_m$ satisfies

$$\|u_{k+1}\|_\rho \le \sum_{m\ge 0}\|\Delta_m\|_\rho < r$$

and the estimate for the second derivative holds

$$\|F''(u_k+\Delta_k)\|_\rho \le C.$$

(3.2.15) implies (3.2.18) and (3.2.16) implies (3.2.19) for $k+1$. This proves that $u = \sum_{k\ge 0}\Delta_k$ converges in B_ρ, $\mathfrak{F}(x,u_k)$ tends to zero, and $\mathfrak{F}(x,u) = 0$. The uniqueness follows from the fact that the formal solution of (3.2.10) is unique. ∎

Remark 3.2.2.3 *The condition* $\left(\frac{\partial F}{\partial u}\right)(0,0) = 0$ *is not essential. If* $\left(\frac{\partial F}{\partial u}\right)(0,0) \ne 0$ *we replace* P *by* $P - \left(\frac{\partial F}{\partial u}\right)(0,0)$ *and* F *by* $F - \left(\frac{\partial F}{\partial u}\right)(0,0)u$. *Of course the conditions of theorem 3.2.1.5 are now to be required for the operator* $P - \left(\frac{\partial F}{\partial u}\right)(0,0)$.

This theorem applies in particular to singular partial differential equations of the form

$$P\left(x, x\frac{\partial}{\partial x}\right) u = F(x,u)$$

where P satisfies the Poincaré condition $|P_0(m)| \geq c\,|m|^d$. Another application is Poincaré's classical theorem about reduction of singular vector fields to normal form.

Corollary 3.2.2.4 *Let* $\tau = \sum_{1\leq j\leq n} a_j(x)\frac{\partial}{\partial x_j}$ *be a vector field where* $a_j(x) \in \mathfrak{M}$ *for all* $j = 1,\dots,n$ *such that the Jacobian matrix* $J(a_1,\dots,a_n)$ *is semi-simple with eigenvalues* $\lambda_1,\dots,\lambda_n$ *satisfying the Poincaré condition*

$$\left|\sum_{1\leq j\leq n} \lambda_j m_j - \lambda_k\right| \geq c\,|m| \quad \textit{for } k = 1,\dots,n.$$

Then there exists a holomorphic transformation tangent to the identity

$$x_j = y_j + \varphi_j(y_1,\dots,y_n) \quad \textit{with } \varphi_j \in \mathfrak{M}^2$$

which transforms the vector field τ *to its linear part*

$$\sum_{1\leq j\leq n} \lambda_j y_j \frac{\partial}{\partial y_j}.$$

Proof : Such a transformation exists formally. By a linear change of coordinates we can assume that $J(a_1,\dots,a_n)$ is diagonal and then that τ can be written in the form

$$\tau = \sum_{1\leq j\leq n} (\lambda_j x_j + f_j(x_1,\dots,x_n))\frac{\partial}{\partial x_j}.$$

An easy computation shows that φ_j satisfies the system of partial differential equations

$$\sum_{1\leq j\leq n} \lambda_j y_j \frac{\partial \varphi_k}{\partial y_j} - \lambda_k \varphi_k = f_k(y+\varphi)$$

written in the shorter form

(3.2.20) $$(\tau - \Lambda)\varphi = F(y+\varphi)$$

where $\varphi = (\varphi_1,\dots,\varphi_n)$, Λ is the diagonal matrix with entries $\lambda_1,\dots,\lambda_n$, and $F = (f_1,\dots,f_n)$. The Poincaré condition implies by corollary 3.2.1.7 that the conditions of theorem 3.2.1.5 are satisfied for the operator $\tau - \Lambda$. ■

We have a slightly more general version of theorem 3.2.1.5.

Theorem 3.2.2.5 *Let P be an operator of the form (3.1.1) satisfying the conditions a), b), and c) of theorem 3.2.1.5 for all m with $|m| \geq \mu > 0$ and all $k \geq k_0$ and let F be as in theorem 3.2.2.1. If the equation*

$$Pu = F(x, u)$$

has a formal power series solution then this series converges.

Proof : If some of the $P_0(m)$ with $|m| < \mu$ are not invertible the equations giving the coefficients of a formal power series solution are solvable only under additional compatibility conditions and the existence of a formal solution $\hat{u}$ means that these conditions are satisfied. But since $P_0(m)$ is invertible for all $|m| \geq \mu$ the coefficients u_m with $|m| \geq \mu$ are uniquely determined if the u_m for $|m| < \mu$ are given.

Now set $w(x) = \sum_{|m|<\mu} u_m x^m$ and $v = u - w$. We get a new equation

$$Pv = F_1(x, v)$$

which we can solve as in theorem 3.2.2.1 but in the space $(\mathfrak{M}^\mu(\rho))^q$ the subspace of $B^q(\rho)$ consisting of power series beginning with terms of order at least μ. ■

This applies e.g. to vector fields

$$\tau = \sum_{1 \leq j \leq n} \lambda_j x_j \frac{\partial}{\partial x_j}$$

satisfying the Poincaré condition $|\langle \lambda; p \rangle| \geq c\,|p|$ which implies

$$\left| \sum_{1 \leq j \leq n} \lambda_j p_j - p_k \right| \geq c_1\,|p| \quad \text{for } k = 1, \ldots, n$$

for sufficiently large p and some c_1. So we have the following result:

> *if the equation $\tau u = F(x, u)$ has a formal solution this solution converges.*

3.3 Equations with parameters

We extend now the results of section 3.2 to equations of the form

(3.3.1) $$Pu = F(x, y, u)$$

where y is considered as a parameter and P is an operator of the form (3.1.1) depending holomorphically on y.

Let $r = (r_1, \ldots, r_p)$ be a polyradius and let $\mathfrak{A}^q(r)$ be the space of convergent power series $u(y) = \sum_j u_j y^j$ with $u_j \in \mathbb{C}^q$ for which

$$|u| = |u|_r = \sum_j |u_j| r^j < \infty.$$

Let us denote by $B^q(\rho, r)$ the Banach space of power series $\sum_k u_k(y)x^k$ with $u_k(y) \in \mathfrak{A}^q(r)$ for which

$$\|u\|_\rho = \sum_k |u_k|_r \rho^{|k|} t^k < \infty. \tag{3.3.2}$$

$B^1(\rho, r) = B(\rho, r)$ is again a Banach algebra and $\mathfrak{M}^\mu(\rho, r)$ is the subspace of those series $\sum_k u_k(y)x^k$ with $u_k = 0$ for $|k| < \mu$.

We consider operators P acting on $B^q(\rho, r)$ of the form

$$Pu = \sum_{|m|\geq 0} \left(\sum_{k_0 \leq k \leq m, |k| \geq 0} P_k(m-k, y) u_{m-k}(y) \right) x^m \tag{3.3.3}$$

where $P_k(m, y)$ is a $q \times q$–matrix with entries in $\mathfrak{A}(r)$. P can be written as in (3.1.5) in the form $P = P_0 + P_- + P_+$. With the same proof as in section 3.2 we have results analogous to theorems 3.2.1.1 and 3.2.1.5 and their corollaries.

For the non linear equation (3.3.1) we have

Theorem 3.3.1 *Let P be an operator of the form (3.3.3) and $F = (F_1, \ldots, F_q)$ be a holomorphic function near the origin of $\mathbb{C}^{n+p+q}$. Assume that $F(0,0,0) = 0$ and $\left(\frac{\partial F}{\partial u}\right)(0,0,0) = 0$. Let $P = P_0 + P_- + P_+$.*

a) Assume that $P_0(m, y)$ is invertible for all $m \in \mathbb{N}^n$ and that the formal inverse P_0^{-1} is continuous

$$P_0^{-1} : B^q(\rho, r) \longrightarrow B^q(\rho, r)$$

for $\rho \leq \rho_0, t \leq t_0, r \leq r_0$.

b) Assume that there exist a positive function φ and for all $k \geq k_0, k \neq 0$ positive constants c_k and d_k such that

$$|P_k(m,y)|_r \leq c_k \varphi(m) \quad \textit{and} \quad \varphi(m) \left|P_0(m+k, y)^{-1}\right|_r \leq d_k \quad \textit{for all } m \in \mathbb{N}^n$$

and such that

$$c(\rho) = \sum_{k \geq k_0, k \neq 0} c_k d_k \rho^{|k|} t^k < \infty \quad \textit{for some } \rho, t, r.$$

c) Assume that the finite sum

$$c_-(\rho) = \sum_{|k|=0,k\neq 0} c_k d_k t^k$$

corresponding to P_- can be made arbitrarily small by a suitable choice of t.

Then the equation $Pu = F(x,y,u)$ has a unique holomorphic solution u near the origin satisfying $u(0,0) = 0$.

The proof is the same as the proof of theoem 3.2.2.1 using instead of lemma 3.2.2.2 the following:

Lemma 3.3.2 *Let P and F be as above. There exist $\rho, s, r > 0$ such that for any $v \in B^q(\rho, r)$ with $v(0,0) = 0$ and $\|v\|_\rho \leq s$ the operator*

$$P - \left(\frac{\partial F}{\partial u}\right)(x,y,v(x,y)) : B^q(\rho,r) \longrightarrow B^q(\rho,r)$$

is invertible with bounded inverse.

The proof is the same as for lemma 3.2.2.2 if we notice the following fact: $\left(\frac{\partial F}{\partial u}\right)(x,y,v(x,y))$ can be made arbitrarily small by choosing ρ, s, and r small enough since $\left(\frac{\partial F}{\partial u}\right)(0,0,0) = 0$.

Remark 3.3.3 *As in theorem 3.2.2.1 the condition $\left(\frac{\partial F}{\partial u}\right)(0,0,0) = 0$ is not essential. If $\left(\frac{\partial F}{\partial u}\right)(0,0,0) \neq 0$ replace P by $P - \left(\frac{\partial F}{\partial u}\right)(0,0,0)$.*

We give now two theorems where the conditions of theorem 3.3.1 are required only for $|m| \geq \mu > 0$.

Theorem 3.3.4 *Assume that F satisfies the conditions of theorem 3.3.1 and let P satisfy the conditions:*

a) $P_0(m,y)$ is invertible for $|m| \geq \mu$ and

$$P_0^{-1} : (\mathfrak{M}^\mu(\rho,r))^q \longrightarrow (\mathfrak{M}^\mu(\rho,r))^q$$

is continuous for $\rho \leq \rho_0, t \leq t_0, r \leq r_0$;

b) condition b) of theorem 3.3.1 but only for $|m| \geq \mu$;

c) condition c) of theorem 3.3.1.

Then if the equation $Pu = F(x,y,u)$ has a formal power series solution $u(x,y) = \sum_k u_k(y)x^k$ where $u_k \in \mathfrak{A}^q(r)$ for some r and $u_0(0) = 0$ this series converges.

Proof : Since $P_0(m,y)$ is invertible for $|m| \geq \mu$ the coefficients $u_k(y)$ are uniquely determined for $|k| \geq \mu$ if the u_k are given for $|k| < \mu$. As in the proof of theorem 3.2.2.5 we set $w = \sum_{|m|<\mu} u_m(y)x^m$ and $v = u - w$ and get a new equation $Pv = F_1(x,y,v)$ which we solve in $(\mathfrak{M}^\mu(\rho,r))^q$. ■

Theorem 3.3.5 *Let P and F be as in theorem 3.3.4. If the equation $Pu = F(x,y,u)$ has a formal solution*

$$\hat{u} = \sum_{k\geq 0} u_k(y)x^k$$

where $u_k \in \mathbb{C}[[y_1,\dots,y_p]]$ for all k and $u_0(0) = 0$ then the equation has a formal solution $v = \sum_{k\geq 0} v_k(y)x^k$ with $v_k \in \mathfrak{A}(r)$ for some r and by the preceding theorem this series converges. Moreover, the v_k can be chosen such that $v_k \equiv u_k \mod (\mathfrak{M}_f)^\nu$ for an arbitrary ν.

Proof : Expand F in a series with respect to x and u

$$F(x,y,u) = \sum_{k,m} F_{k,m}x^k u^m = \sum_{k,m} F_{k,m}(y)x^k \left(\sum_p u_p(y)x^p\right)^m$$

and rearrange the series with respect to the powers of x. In the coefficient of x^m only occur the $u_p(y)$ with $p \leq m$ and this coefficient is an analytic function in y and the u_p for $p \leq m$. By comparing coefficients in the equation $Pu = F(x,y,u)$, we get

$$\sum_{k_0\leq k\leq m, |k|\geq 0} P_k(m-k,y)u_{m-k}(y) = \text{ coefficient of } x^m.$$

This is a finite system of analytic equations for the unknown u_m with $|m| < \mu$ which by hypothesis has a formal solution. So by Artin's theorem [2] there exists a convergent solution v_m with $|m| < \mu$ and for some r $v_m(y) \in \mathfrak{A}(r)$ for all $|m| < \mu$. This solution can be chosen to coincide with u_m up to any order ν. The other coefficients v_m for $|m| \geq \mu$ are now uniquely determined and can be found as in the proof of the preceding theorem by setting $w = \sum_{|m|<\mu} v_m(y)x^m$ and solving the equation

$$P(u-w) = F_1(x,y,u-w) \quad \text{in } (\mathfrak{M}^\mu(\rho,r))^q .$$

Moreover, the coefficients $v_{m,k}$ of $v_m = \sum_{k\geq 0} v_{m,k}y^k$ for $|k| < \nu, |m| \geq \mu$ are determined uniquely by the coefficients $v_{m,k}$ with $|m| < \mu, |k| < \nu$ and for these $v_{m,k} = u_{m,k}$ and hence we get $v_{m,k} = u_{m,k}$ for $|k| < \nu$ and all m. ■

3.4 An application: A theorem of S. Kaplan

In [31], Kaplan considers vector fields $\tau = \sum_{1\le j\le n} X_j(z)\frac{\partial}{\partial z_j}$ where $X_j(z)$ is holomorphic in a neighborhood of the origin in $\mathbb{C}^n$ and $X_j(0) = 0$ for $j = 1, \dots, n$. Let S be the germ of subvariety defined by

$$X_j(z) = 0 \quad \text{for } j = 1, \dots, n \qquad \text{near the origin of } \mathbb{C}^n.$$

Assume that S satisfies the following condition:

(H1) S is the germ at 0 of a complex analytic submanifold of $\mathbb{C}^n$ of dimension s with $0 \le s \le n-1$. We denote $d = n - s$.

By a local holomorphic change of variables $\xi = \xi(z)$, we may suppose that

$$S = \{z \in \mathbb{C}^n : \xi_1(z) = \cdots = \xi_d(z) = 0\}$$

near the origin and τ is transformed into

$$\sum_{1\le j\le n} X_j^*(\xi)\frac{\partial}{\partial \xi_j} \quad \text{where } X^*(\xi(z)) = \left(\frac{\partial \xi}{\partial z}\right)(z)X(z)$$
$$\text{and } X_j^*(0, \dots, 0, \xi_{d+1}, \dots, \xi_n) = 0.$$

This implies $\left(\frac{\partial X_j^*}{\partial \xi_k}\right)(0) = 0$ for $k = d+1, \dots, n$ and since S is regular of dimension s the Jacobian matrix $\left(\frac{\partial X^*}{\partial \xi}\right)(0)$ has rank $d = n - s$.

Let $\lambda_1, \dots, \lambda_d$ be the non zero eigenvalues of $\left(\frac{\partial X^*}{\partial \xi}\right)(0)$. By a linear change of coordinates, we transform τ into a vector field such that the Jacobian $\left(\frac{\partial X^*}{\partial \xi}\right)(0)$ is in Jordan normal form and, since the Jacobian has a block of zeros of size $s \times s$ in the lower right-hand corner, this Jordan normal form looks like

$$\text{(3.4.1)} \qquad \begin{pmatrix} \lambda_1 & a_1 & 0 & \cdots & \cdots & \cdots & \cdots & 0 \\ 0 & \lambda_2 & a_2 & 0 & \cdots & \cdots & \cdots & 0 \\ \cdots & \cdots & \cdots & \cdots & \cdots & \cdots & \cdots & \cdots \\ 0 & \cdots & \lambda_{d-1} & a_{d-1} & 0 & \cdots & & 0 \\ 0 & \cdots & 0 & \lambda_d & 0 & \cdots & & 0 \\ 0 & \cdots & 0 & 0 & 0 & \cdots & & 0 \\ \cdots & \cdots & \cdots & \cdots & \cdots & \cdots & \cdots & \cdots \\ 0 & \cdots & 0 & 0 & 0 & \cdots & & 0 \end{pmatrix}.$$

Thus, the linear part of the vector field has the form of example 3.1.1, 5. of section 3.1.

We make now the hypothesis that $(\lambda_1, \dots, \lambda_d)$ satisfies a Poincaré condition.

(H2) Let $\lambda_1, \ldots, \lambda_d$ be the non zero eigenvalues of $\left(\frac{\partial X^*}{\partial \xi}\right)(0)$. There exists $c > 0$ such that

$$|\langle \lambda; m\rangle| = \left|\sum_{1\le j\le d} \lambda_j m_j\right| \ge c\,|m| = c \sum_{1\le j\le d} m_j.$$

The idea is now to apply the theorems of section 3.3 treating $\xi_{d+1}, \ldots, \xi_n$ as parameters. In order to do this we set $\xi_j = \zeta_j$ for $j = 1, \ldots, d$ and $\xi_{d+j} = \eta_j$ for $j = 1, \ldots, s$ and $X_j^*(\xi) = X_j(\zeta, \eta)$ for $j = 1, \ldots, d$ and $X_{d+j}^*(\xi) = Y_j(\zeta, \eta)$ for $j = 1, \ldots, s$. The vector field τ is now written in the form

(3.4.2)

$$\tau = \sum_{1\le j\le d} X_j(\zeta,\eta)\frac{\partial}{\partial \zeta_j} + \sum_{1\le j\le s} Y_j(\zeta,\eta)\frac{\partial}{\partial \eta_j} \qquad \text{with } X_j(0,\eta) = Y_j(0,\eta) = 0$$

and the Jacobian matrix $\left(\frac{\partial(X,Y)}{\partial(\zeta,\eta)}\right)(0)$ has the form (3.4.1).

It is shown in [15] that any singular vector field $\tau = \sum_{1\le j\le n} X_j \frac{\partial}{\partial x_j}$ can be reduced by a formal change of variables to a Jordan form

$$\tau = {}^s\tau + {}^n\tau$$

where ${}^s\tau = \sum_{1\le j\le n} \lambda_j z_j \frac{\partial}{\partial z_j}$ is the semi–simple part and ${}^n\tau = \sum_{1\le j\le n} g_j(z)\frac{\partial}{\partial z_j}$ is the nilpotent part of τ. Here λ_j are the eigenvalues of the Jacobian $\left(\frac{\partial X}{\partial x}\right)(0)$ and the Jacobian matrix $\left(\frac{\partial g}{\partial z}\right)(0)$ is nilpotent and has Jordan normal form. Moreover, ${}^s\tau$ and ${}^n\tau$ commute $[{}^s\tau, {}^n\tau] = 0$ or equivalently ${}^s\tau g_j = \lambda_j g_j$. If the Jacobian $\left(\frac{\partial X}{\partial x}\right)(0)$ is already in Jordan normal form the formal transformation transforming the vector field to normal form can be chosen tangent to the identity.

If we apply this to our case we get

Lemma 3.4.1 *Let τ satisfy* **(H1)** *and* **(H2)** *and assume that the Jacobian matrix $\left(\frac{\partial X}{\partial x}\right)(0)$ has already the Jordan normal form (3.4.1). Then there exists a formal change of variables tangent to the identity $z = z(\zeta,\eta)$ which reduces τ to*

$$\tau = {}^s\tau + {}^n\tau$$

where

$$(3.4.3)\qquad \begin{aligned} {}^s\tau &= \sum_{1\le j\le d} \lambda_j z_j \frac{\partial}{\partial z_j}, \\ {}^n\tau &= \sum_{1\le j\le d} (z_j R_j + P_j)\frac{\partial}{\partial z_j} + \sum_{d+1\le j\le n} G_j^0(z_{d+1}, \ldots, z_n)\frac{\partial}{\partial z_j} \end{aligned}$$

where $R_j = R_j(z_{d+1}, \ldots, z_n)$ and $G_j^0(z_{d+1}, \ldots, z_n)$ are formal power series in $z_{d+1}, \ldots, z_n$ and $P_j = P_j(z_1, \ldots, z_{j-1}, z_{j+1}, \ldots, z_n)$ is a polynomial in $z_1, \ldots, z_{j-1}$, $z_{j+1}, \ldots, z_d$ with coefficients in $\mathbb{C}[[z_{d+1}, \ldots, z_n]]$ and the Jacobian matrix at 0 is just the nilpotent part of (3.4.1).

Proof : Let ${}^n\tau = \sum_{1\le j\le n} g_j(z)\frac{\partial}{\partial z_j}$ be the nilpotent part of τ. Now divide the variables into the two groups $z_j = x_j$ for $j = 1,\dots,d$ and $z_{j+d} = y_j$ for $j = 1,\dots,s$ and write $g_k(x,y) = \sum G_k^m(y)x^m$. The commutation $[{}^s\tau, {}^n\tau] = 0$ gives us the equation

$$\sum_{1\le j\le d} \lambda_j x_j \frac{\partial g_k}{\partial x_j} = 0 \quad \text{for } k = d+1,\dots,n$$

which implies $\langle\lambda; m\rangle\, G_k^m = 0$ for $k = d+1,\dots,n$ and since $|\langle\lambda; m\rangle| \ge c\,|m|$ we have $G_k^m = 0$ for $|m| \ne 0$ and $g_k(x,y) = G_k^0(y)$.

For $k = 1,\dots,d$ we have

$$\sum_{1\le j\le d} \lambda_j x_j \frac{\partial g_k}{\partial x_j} = \lambda_k g_k$$

and this gives

$$(\langle\lambda; m\rangle - \lambda_k)\, G_k^m = 0.$$

But **(H2)** implies $(\langle\lambda; m\rangle - \lambda_k) \ne 0$ so $G_k^m = 0$ for $m_k > 1$.

If $m_k = 1$ $(\langle\lambda; m\rangle - \lambda_k) = 0$ only if $m = (0,\dots,0,1,0,\dots,0)$ (the 1 is at the k–th place); so possibly for this m $G_k^m(y) = R_k(y) \ne 0$.

For $m_k = 0$ the equation $\sum_{j\ne k} \lambda_j m_j = \lambda_k$ has by **(H2)** only a finite number of solutions m. So there are only a finite number of $G_k^m(y) \ne 0$ and $m_k = 0$. It follows that $\sum_{m_k=0} G_k^m(y)x^m = P_k$ is a polynomial which does not contain x_k. ■

Lemma 3.4.2 *In the normal form (3.4.3), we have* $G_j^0(y) = 0$.

Proof : Let $\xi = \xi(x,y), \eta = \eta(x,y)$ be the formal change of variables which transforms (3.4.2) to (3.4.3) and write

$$\begin{aligned}
\xi_j &= \xi_j(x,y) = \xi_j^0(y) + \sum_{1\le k\le d} x_k \xi_{j,k}(x,y) \quad \text{for } j = 1,\dots,d,\\
\eta_j &= \eta_j(x,y) = \eta_j^0(y) + \sum_{1\le k\le d} x_k \eta_{j,k}(x,y) \quad \text{for } j = 1,\dots,s.
\end{aligned}$$

Writing down the change of variables formula (3.4.1) for this case and setting $x = 0$ we get

$$X_j\left(\xi^0,\eta^0\right) = \sum_{1\le k\le s} \frac{\partial \xi_j^0}{\partial y_k} G_{d+k}^0 \qquad \text{and} \qquad Y_j\left(\xi^0,\eta^0\right) = \sum_{1\le k\le s} \frac{\partial \eta_j^0}{\partial y_k} G_{d+k}^0.$$

Since $\left(\frac{\partial X}{\partial \xi}\right)(0)$ has eigenvalues $\lambda_1,\dots,\lambda_d$ different from zero this Jacobian is invertible and since G_{d+k}^0 begins with terms of order at least two the first equation shows that $\xi_j^0 \in (\mathfrak{M}_f)^\nu$ implies $\xi_j^0 \in (\mathfrak{M}_f)^{\nu+1}$. This is possible only if $\xi_j^0 = 0$.

Looking now at the second equation, we get

$$\sum_{1\le k\le s} \frac{\partial \eta_j^0}{\partial y_k} G_{d+k}^0 = Y_j\left(0,\eta^0\right) = 0$$

by the definition of S. The transformation being tangent to the identity, the matrix $\left(\frac{\partial \eta^0}{\partial y}\right)$ is invertible and we get $G_{d+k}^0 = 0$. ■

Finally, the vector field τ can be written in the form

$$\sum_{1\le j\le d} \lambda_j x_j \frac{\partial}{\partial x_j} + \sum_{1\le j\le d} \left(x_j R_j(y) + P_j(x,y)\right) \frac{\partial}{\partial x_j}$$

and we have the equation

$${}^s\tau\left(x_j R_j(y) + P_j(x,y)\right) = \lambda_j\left(x_j R_j(y) + P_j(x,y)\right).$$

We apply theorem 3.3.5 to this equation where $P_0 = {}^s\tau - \lambda_j$ and **(H2)** implies that the conditions a),b) of this theorem are satisfied; c) is void since $P_- = 0$. We get

Theorem 3.4.3 *If the vector field τ satisties the conditions* **(H1)** *and* **(H2)** *it can by a formal change of variables be reduced to the normal form*

$$\tau^* = {}^s\tau + {}^n\tau$$

where

(3.4.4)

$${}^s\tau = \sum_{1\le j\le d} \lambda_j x_j \frac{\partial}{\partial x_j} \qquad \textit{and} \qquad {}^n\tau = \sum_{1\le j\le d} \left(x_j R_j(y) + P_j(x,y)\right) \frac{\partial}{\partial x_j}$$

where R_j is holomorphic in a neighborhood of 0 *and P_j is a polynomial in $x_1, \ldots, x_{j-1}, x_{j+1}, \ldots, x_d$ with holomorphic coefficients in y.*

The formal transformation $\xi = \xi(x,y), \eta = \eta(x,y)$ transforms the vector field (3.4.2) to the convergent normal form (3.4.4) if and only if ξ, η satisfy the equations $\left({}^s\tau + {}^n\tau\right)\xi_j = X_j(\xi,\eta)$ and $\left({}^s\tau + {}^n\tau\right)\eta_j = Y_j(\xi,\eta)$. Now we apply theorem 3.3.5 to the operator

$${}^s\tau + {}^n\tau = \sum_{1\le j\le d} \left(x_j\left(\lambda_j + R_j(y)\right) + P_j(x,y)\right) \frac{\partial}{\partial x_j}.$$

Here $P_0 = {}^s\tau, P_- = \sum_{1\le j\le d-1} a_j x_{j+1} \frac{\partial}{\partial x_j}$.

As in corollary 3.2.1.8, we see that ${}^s\tau + {}^n\tau - \left(\frac{\partial(x,y)}{\partial(\xi,\eta)}\right)(0)$ satisfies the conditions of theorem 3.3.5. Since by [15] there exists a formal solution ξ, η we get

Theorem 3.4.4 *There exists a change of coordinates tangent to the identity given by convergent power series $\xi = \xi(x,y), \eta = \eta(x,y)$ transforming the vector field (3.4.2) to the normal form (3.4.4).*

As a corollary we get the following Kaplan's result:

Theorem 3.4.5 *If the vector field τ satisfies conditions* **(H1)** *and* **(H2)** *and if the equation*

$$\tau u = F(x,u) \quad \text{with } F(0,0) = 0$$

has a formal solution u then there exists a convergent solution v with $v \equiv u$ mod $(\mathfrak{M}_f)^\nu$ for any given ν.

Proof: By a holomorphic change of variables transform τ to the normal form (3.4.4). The transformed equation $\tau^* u = F_1(\xi, u)$ has a formal solution. By theorem 3.3.5 it has a convergent solution v^* and we get v by inverting this transformation. ■

3.5 The case of small denominators

The example in section 3.2 shows that our theorems do not hold in general when the Poincaré condition is not satisfied. This is due to the fact that the linearized equation $\left(P - \frac{\partial F}{\partial u}\right) v = f$ cannot be solved. Here we give an example with Siegel condition (3.2.7) where the linearized equation is approximately solvable by a special trick due to Rüssmann [43] (cf. also [50]).

Let $\tau = \sum_{1 \leq j \leq n} X_j(x) \frac{\partial}{\partial x_j}$ be a vector field with $X_j(0) = 0$ such that the Jacobian matrix $\left(\frac{\partial X_j}{\partial x_k}\right)(0)$ can be diagonalized. So after a linear change of coordinates we assume that $\left(\frac{\partial X_j}{\partial x_k}\right)(0) = \Lambda$ is diagonal with eigenvalues $\lambda_1, \ldots, \lambda_n$ and suppose that the eigenvalues satisfy the Siegel condition

$$\text{(3.5.1)} \qquad |\langle \lambda; m\rangle - \lambda_k| = \left| \sum_{1 \leq j \leq n} \lambda_j m_j - \lambda_k \right| \geq c\,|m|^{-\gamma} \quad \text{for } k = 1, \ldots, n$$

for some positive constants c, γ.

Since $\langle \lambda; m\rangle - \lambda_k \neq 0$ it is easy to see that there is a formal change of variables tangent to the identity $x_j = y_j + \varphi_j(y)$ with $\varphi_j(y) \in (\mathfrak{M}_f)^2$ reducing τ to the normal form ${}^s\tau = \sum_{1 \leq j \leq n} \lambda_j y_j \frac{\partial}{\partial y_j}$. The vector field τ is formally semi-simple and the normal form is trivially convergent. It was proved by Siegel [44] that the transformation converges. We will give a proof of this fact using singular partial differential equations.

Theorem 3.5.1 *If the vector field* $\tau = \sum\limits_{1\le j\le n} X_j(x)\frac{\partial}{\partial x_j}$ *with* $X_j(0)=0$ *is formally semi-simple with normal form* ${}^s\tau = \sum\limits_{1\le j\le n} \lambda_j y_j \frac{\partial}{\partial y_j}$ *and if the* λ_j *satisfy the Siegel condition (3.5.1) then the transformation* $x_j = y_j + \varphi_j(y)$ *transforming* τ *to its normal form converges.*

Proof : An easy calculation shows that $u = (\varphi_1, \ldots, \varphi_n)$ satisfies

$$ {}^s\tau u - \Lambda u = F(y+u) \tag{3.5.2}$$

where $\Lambda = \left(\frac{\partial X_j}{\partial x_k}\right)(0)$ is diagonal and

$$F = (F_1, \ldots, F_n) \qquad \text{with } F_k(u) = X_k(u) - \frac{\partial X_k}{\partial x_k}(0)u.$$

Now set

$$\mathfrak{F}(x,u) = ({}^s\tau - \Lambda)\, u - F(x+u). \tag{3.5.3}$$

To solve the equation by Newton's method (cf. [40], [43], [45], [50]) we set as in the proof of theorem 3.2.2.1

$$\begin{aligned} u_0 &= 0, \\ u_{k+1} &= u_k + \Delta_k, \qquad (3.5.4)\\ ({}^s\tau - \Lambda)\,\Delta_k - \left(\frac{\partial F}{\partial u}\right)(x+u_k)\,\Delta_k &= -\mathfrak{F}\,(x, u_k)\,. \qquad (3.5.5) \end{aligned}$$

Since $({}^s\tau - \Lambda)^{-1}$ is only continuous $B_\rho \longrightarrow B_{\rho'}$ with $\rho' < \rho$ (corollary 3.2.1.2, b)) we cannot use the perturbation argument to solve (3.5.5).

Following Rüssmann [43] we set

$$\Delta_k = J\,(x+u_k)\cdot E_k \tag{3.5.6}$$

where $J(v)$ is the Jacobian of v with respect to x and E_k is a yet unknown column. Let us forget for a moment the index k. We get the commutation relation

$$J\,({}^s\tau v) = {}^s\tau J(v) + J(v)\cdot\Lambda. \tag{3.5.7}$$

With (3.5.6) and (3.5.7) the equation (3.5.5) becomes

$$\begin{aligned} -\mathfrak{F}(x,u) &= ({}^s\tau - \Lambda)\,(J(x+u)E) - \left(\frac{\partial F}{\partial u}\right)(x+u)J(x+u)\cdot E \\ &= ({}^s\tau J(x+u))\cdot E + J(x+u)^s\tau E - \Lambda J(x+u)\cdot E - JF(x+u)\cdot E \end{aligned}$$

where JF is the total Jacobian of $F(x+u(x))$ with respect to x. Using (3.5.7) we continue

$$\begin{aligned} -\mathfrak{F}(x,u) &= \left(J^s\tau(x+u)\right)\cdot E - J(x+u)\Lambda E - J(\Lambda(x+u))E + \\ &\qquad + J(x+u)^s\tau E - J(F(x+u))\cdot E \\ &= J\left(\left({}^s\tau - \Lambda\right)(x+u) - F(x+u)\right)\cdot E + J(x+u)\cdot\left({}^s\tau-\Lambda\right)E. \end{aligned}$$

Finally, we get

$$-\mathfrak{F}(x,u) = J\left(\mathfrak{F}(x,u)\right)\cdot E + J(x+u)\cdot\left({}^s\tau-\Lambda\right)E. \tag{3.5.8}$$

Now if u is an approximate solution $J(\mathfrak{F}(x,u))$ is small and E is small so the first term $J(\mathfrak{F}(x,u))\cdot E$ should be small of the second order and if we forget this term we can solve the remaining equation

$$J(x+u)\cdot\left({}^s\tau-\Lambda\right)E = -\mathfrak{F}(x,u)$$

since $J(x+u) = I + J(u)$ is invertible if u is small enough and by corollary 3.2.1.2 b) $\left({}^s\tau-\Lambda\right)$ is invertible.

We get the new iteration schema

$$\begin{aligned} u_0 &= 0, \\ u_{k+1} &= u_k + \Delta_k, \\ \Delta_k &= J\left(x+u_k\right)E_k, \\ E_k &= -\left({}^s\tau-\Lambda\right)^{-1} J\left(x+u_k\right)^{-1}\mathfrak{F}\left(x,u_k\right). \end{aligned} \tag{3.5.9}$$

Then the calculations leading to (3.5.8) give now

$$\begin{aligned} \mathfrak{F}\left(x,u_{k+1}\right) &= \mathfrak{F}\left(x,u_k\right) + \left(\left({}^s\tau-\Lambda\right) - \left(\frac{\partial F}{\partial u}\right)\left(x+u_k\right)\right) J\left(x+u_k\right)E_k + \\ &\qquad + O\left(\left(\Delta_k\right)^2\right) \\ &= \mathfrak{F}\left(x,u_k\right) + J\left(\mathfrak{F}\left(x,u_k\right)\right)E_k + J\left(x+u_k\right)\left({}^s\tau-\Lambda\right)E_k + O\left(\left(\Delta_k\right)^2\right) \qquad (3.5.10) \\ &= J\left(\mathfrak{F}\left(x,u_k\right)\right)E_k + O\left(\left(\Delta_k\right)^2\right). \end{aligned}$$

Since $u_0 \equiv 0$ we have $\mathfrak{F}\left(x,u_0\right) = \mathfrak{F}(x,0) = -F(x+0) = -F(x) \in \left(\mathfrak{M}^2\right)^n$ and, therefore, $E_0, \Delta_0 \in \left(\mathfrak{M}^2\right)^n$ are well-defined.

If u_k, E_k, Δ_k are well-defined and satisfy

$$u_k \in \left(\mathfrak{M}^2\right)^n \qquad \text{and} \qquad \mathfrak{F}\left(x,u_k\right), E_k, \Delta_k \in \left(\mathfrak{M}^{2^k+1}\right)^n$$

then we have $u_{k+1} = u_k + \Delta_k \in \left(\mathfrak{M}^2\right)^n$,

$$\mathfrak{F}\left(x,u_{k+1}\right) = J\left(\mathfrak{F}\left(x,u_k\right)\right)E_k + O\left(\left(\Delta_k\right)^2\right) \in \left(\mathfrak{M}^{2^k+(2^k+1)}\right)^n = \left(\mathfrak{M}^{2^{k+1}+1}\right)^n$$

and, therefore, $E_{k+1}, \Delta_{k+1} \in \left(\mathfrak{M}^{2^{k+1}+1}\right)^n$ are also well-defined.

Thus, by induction we see that u_k, E_k, Δ_k for $k = 0,1,2,\ldots$ are well-defined in the germ sense and satisfy

(3.5.11)
$$u_k \in \left(\mathfrak{M}^2\right)^n, \qquad \mathfrak{F}(x, u_k), E_k, \Delta_k \in \left(\mathfrak{M}^{2^k+1}\right)^n \quad \text{for } k = 0, 1, 2, \ldots.$$

Since u_k is given by

$$u_k = \sum_{0 \le m \le k} \Delta_m$$

to have theorem 3.5.1 it is enough to prove the following **(S1)** and **(S2)** for some $\rho > 0$:

(S1) $u = \sum_{0 \le m < \infty} \Delta_m$ converges in $(B_\rho)^n$;

(S2) $\mathfrak{F}(x, u_k)$ tends to zero in $(B_\rho)^n$ (as k tends to ∞).

We note the following. By (3.2.8) we have

$$\left\| \left({}^s\tau - \Lambda\right)^{-1} v \right\|_\sigma \le \frac{C}{(\rho - \sigma)^\gamma} \|v\|_\rho$$

where we suppose without restriction $\gamma \ge 1$. The arguments leading to (3.2.8) give also

$$\|J(v)\|_\sigma \le \frac{C}{\rho - \sigma} \|v\|_\rho .$$

Suppose that $F(x + u)$ is holomorphic in a neighborhood of the polydisc $|x_j| \le t_j, |u_j| \le r_j$. As in the proof of theorem 3.2.2.1 we get

$$\|F''(x + v(x))\|_\rho \le C \quad \text{for } \|v\|_\rho \le r.$$

In all the inequalities, all the constants are denoted by C and we suppose moreover $C > 1$.

Let $p \in \mathbb{N}$ be sufficiently large and fix it. Choose R so that $0 < R < 1$ and

(3.5.12)
$$\sum_{0 \le m \le p-1} \|\Delta_m\|_R \le \frac{r}{2},$$
$$\sum_{0 \le m \le p-1} \|J(\Delta_m)\|_R \le \frac{1}{2}\left(1 - \frac{1}{C}\right).$$

This is possible since p is fixed and $\Delta_m(0) = 0$ and $J(\Delta_m)(0) = 0$ hold.

Note the following lemma.

Lemma 3.5.2 *There are $\rho_0 > 0$ and two sequences $a_k > 0, \delta_k > 0$ for $k = 0, 1, 2, \ldots$ such that the following conditions (c1) to (c7) are satisfied:*

(c1) $0 < \rho_0 < R$;

(c2) $\sum_{0 \le k < \infty} \delta_k \le \frac{\rho_0}{2}$;

(c3) $\|\mathfrak{F}(x, u_p)\|_{\rho_0} \leq a_0$;

(c4) $a_k \longrightarrow 0$ *(as* $k \longrightarrow \infty$*)*;

(c5) $a_{k+1} \geq C^7 \cdot 2^{2\gamma} \frac{a_k^2}{\delta_k^{2\gamma}}$;

(c6) $\sum_{0 \leq k < \infty} \frac{a_k}{\delta_k^\gamma} \leq \frac{r}{2^{\gamma+1} C^3}$;

(c7) $\sum_{0 \leq k < \infty} \frac{a_k}{\delta_k^\gamma \delta_{k+1}} \leq \frac{1}{2^{\gamma+2} C^4} \left(1 - \frac{1}{C}\right)$.

The proof of this lemma will be given later.
If we put

$$b_k = 2^\gamma C^2 \frac{a_k}{\delta_k^\gamma} \tag{3.5.13}$$

then (c6) implies

$$\sum_{0 \leq k < \infty} b_k \leq \frac{r}{2C} \tag{3.5.14}$$

and (c7) implies

$$\sum_{0 \leq k < \infty} \frac{b_k}{\delta_{k+1}} \leq \frac{1}{4C^2} \left(1 - \frac{1}{C}\right). \tag{3.5.15}$$

By using lemma 3.5.2, let us prove the assertions **(S1)** and **(S2)** from now. Set

$$\begin{aligned} \rho_{k+1} &= \rho_0 - \sum_{0 \leq m \leq k} \delta_m \quad \text{for } k = 0, 1, 2, \ldots, \\ \eta_k &= \rho_k - \frac{\delta_k}{2} \qquad\qquad \text{for } k = 0, 1, 2, \ldots. \end{aligned}$$

Then $\rho_{k+1} = \rho_k - \delta_k$ and $\frac{\rho_0}{2} < \rho_{k+1} < \eta_k < \rho_k < R$ hold. Since $u_p = \sum_{0 \leq m \leq p-1} \Delta_m$ by (3.5.12) we see

$$\|J(x + u_p)\|_{\rho_1} \leq 1 + \sum_{0 \leq m \leq p-1} \|J(\Delta_m)\|_R \leq 1 + \frac{1}{2}\left(1 - \frac{1}{C}\right) \leq C$$

and

$$\|J(u_p)\|_{\eta_0} \leq \sum_{0 \leq m \leq p-1} \|J(\Delta_m)\|_R \leq \frac{1}{2}\left(1 - \frac{1}{C}\right) < 1.$$

Hence, by using the Neumann series we see

$$\left\|J(x + u_p)^{-1}\right\|_{\eta_0} \leq \frac{1}{1 - \frac{1}{2}\left(1 - \frac{1}{C}\right)} = \frac{2C}{1 + C} \leq C.$$

Let $k \geq 0$ and suppose that for all $m \leq k$ u_{p+m} satisfies the following estimates:

$$\|\mathfrak{F}(x, u_{p+m})\|_{\rho_m} \leq a_m, \tag{3.5.16}$$

$$\|J(x + u_{p+m})\|_{\rho_{m+1}} \leq C, \tag{3.5.17}$$

$$\left\|J(x + u_{p+m})^{-1}\right\|_{\eta_m} \leq C. \tag{3.5.18}$$

Then by (3.5.9), (3.5.18), (3.5.16), and (3.5.13) we see

$$\begin{aligned}
\|E_{p+m}\|_{\rho_{m+1}} &= \left\|({}^s\tau - \Lambda)^{-1} J(x + u_{p+m})^{-1} \mathfrak{F}(x, u_{p+m})\right\|_{\rho_{m+1}} \\
&\leq \frac{C}{(\eta_m - \rho_{m+1})^\gamma} \left\|J(x + u_{p+m})^{-1}\right\|_{\eta_m} \|\mathfrak{F}(x, u_{p+m})\|_{\rho_m} \\
&\leq 2^\gamma C^2 \frac{a_m}{\delta_m^\gamma} \\
&= b_m
\end{aligned} \tag{3.5.19}$$

which yields

$$\|\Delta_{p+m}\|_{\rho_{m+1}} = \|J(x + u_{p+m}) E_{p+m}\|_{\rho_{m+1}} \leq C b_m \tag{3.5.20}$$

for all $m \leq k$. Hence, for $0 \leq s \leq 1$

$$\begin{aligned}
\|u_{p+k} + s\Delta_{p+k}\|_{\rho_{k+1}} &\leq \sum_{0 \leq m \leq p-1} \|\Delta_m\|_R + \sum_{0 \leq m \leq k} \|\Delta_{p+m}\|_{\rho_{m+1}} \\
&\leq \frac{r}{2} + \sum_{0 \leq m < \infty} C b_m \leq r
\end{aligned}$$

and as in the proof of theorem 3.2.2.1 we get

$$\left\|O\left((\Delta_{p+k})^2\right)\right\|_{\rho_{k+1}} \leq \frac{C}{2} \left\|(\Delta_{p+k})^2\right\|_{\rho_{k+1}} \leq C^3 \frac{b_k^2}{2}. \tag{3.5.21}$$

Thus, by (3.5.10), (3.5.21), and (3.5.16), (3.5.19) for $m = k$, and by the conditions $\delta_k \leq 1, \gamma \geq 1$, and (c5) in lemma 3.5.2 we obtain

$$\begin{aligned}
\|\mathfrak{F}(x, u_{p+k+1})\|_{\rho_{k+1}} &\leq \|J(\mathfrak{F}(x, u_{p+k})) E_{p+k}\|_{\rho_{k+1}} + \left\|O\left((\Delta_{p+k})^2\right)\right\|_{\rho_{k+1}} \\
&\leq \frac{C}{\rho_k - \rho_{k+1}} \|\mathfrak{F}(x, u_{p+k})\|_{\rho_k} \cdot \|E_{p+k}\|_{\rho_{k+1}} + \frac{C^3}{2} b_k^2 \\
&\leq \frac{C a_k b_k}{\delta_k} + \frac{C^3}{2} b_k^2 \\
&= \left(\frac{C^3 2^\gamma}{\delta_k^{\gamma+1}} + \frac{C^7 2^{2\gamma}}{2\delta_k^{2\gamma}}\right) a_k^2
\end{aligned}$$

$$
\begin{aligned}
&= \frac{C^7 2^{2\gamma}}{\delta_k^{2\gamma}} a_k^2 \left(\frac{1}{2} + \frac{\delta_k^{\gamma-1}}{C^4 2^\gamma} \right) \\
&\le \frac{C^7 2^{2\gamma}}{\delta_k^{2\gamma}} a_k^2 \\
&\le a_{k+1}
\end{aligned}
$$

which implies that (3.5.16) holds also for $m = k+1$.

By (3.5.12), (3.5.15), (3.5.20) (for $m \le k$), and the condition $C > 1$ the Jacobian $J(x + u_{p+k+1})$ is estimated as

$$
\begin{aligned}
\|J(x + u_{p+k+1})\|_{\rho_{k+2}} &= \left\| I + \sum_{0 \le m \le p+k} J(\Delta_m) \right\|_{\rho_{k+2}} \\
&\le 1 + \sum_{0 \le m \le p-1} \|J(\Delta_m)\|_R + \sum_{0 \le m \le k} \|J(\Delta_{p+m})\|_{\rho_{m+2}} \\
&\le 1 + \frac{1}{2}\left(1 - \frac{1}{C}\right) + \sum_{0 \le m \le k} \frac{C}{\delta_{m+1}} \|\Delta_{p+m}\|_{\rho_{m+1}} \\
&\le 1 + \frac{1}{2}\left(1 - \frac{1}{C}\right) + C^2 \sum_{0 \le m \le k} \frac{b_m}{\delta_{m+1}} \\
&\le 1 + \frac{1}{2}\left(1 - \frac{1}{C}\right) + C^2 \left(1 - \frac{1}{C}\right) \frac{1}{4C^2} \\
&= 1 + \frac{3}{4}\left(1 - \frac{1}{C}\right) \\
&\le 1 + \left(1 - \frac{1}{C}\right) \\
&\le 1 + (C - 1) \\
&= C
\end{aligned}
$$

which implies that (3.5.17) holds for $m = k+1$.

Since (3.5.12), (3.5.15), and (3.5.20) (for $m \le k$) give

$$
\begin{aligned}
\|J(u_{p+k+1})\|_{\eta_{k+1}} &= \left\| \sum_{0 \le m \le p+k} J(\Delta_m) \right\|_{\eta_{k+1}} \\
&\le \sum_{0 \le m \le p-1} \|J(\Delta_m)\|_R + \sum_{0 \le m \le k} \|J(\Delta_{p+m})\|_{\eta_{m+1}} \\
&\le \frac{1}{2}\left(1 - \frac{1}{C}\right) + \sum_{0 \le m \le k} \frac{2C}{\delta_{m+1}} \|\Delta_{p+m}\|_{\rho_{m+1}} \\
&\le \frac{1}{2}\left(1 - \frac{1}{C}\right) + 2C^2 \sum_{0 \le m \le k} \frac{b_m}{\delta_{m+1}}
\end{aligned}
$$

$$\begin{aligned} &\le \frac{1}{2}\left(1-\frac{1}{C}\right)+2C^2\frac{1-\frac{1}{C}}{4C^2} \\ &= \left(1-\frac{1}{C}\right) \\ &< 1 \end{aligned}$$

by the Neumann series we get

$$\left\| J\left(x+u_{p+k+1}\right)^{-1}\right\|_{\eta_{k+1}} \le \frac{1}{1-\left(1-\frac{1}{C}\right)} = C$$

which implies that (3.5.18) holds also for $m = k+1$.

Summing up, we have seen by induction that (3.5.16), (3.5.17), and (3.5.18) are valid for all m. In particular, we have obtained

$$\begin{aligned} \sum_{0\le m<\infty} \|\Delta_m\|_{\frac{\rho_0}{2}} &\le \sum_{0\le m\le p-1} \|\Delta_m\|_R + \sum_{0\le m<\infty} \|\Delta_{p+m}\|_{\rho_{m+1}} \\ &\le \frac{r}{2} + C\sum_{0\le m<\infty} b_m \\ &\le \frac{r}{2}+\frac{Cr}{2C} \\ &= r \end{aligned}$$

and

$$\|\mathfrak{F}(x,u_{p+k})\|_{\rho_k} \le a_k \longrightarrow 0 \quad (\text{as } k\longrightarrow\infty)$$

from the condition (c4) in lemma 3.5.2. Thus, the assertions **(S1)** and **(S2)** are proved. ■

Proof of lemma 3.5.2: We set

$$\begin{aligned} a_k &= \varepsilon^{\alpha^k} && \text{for } k=0,1,2,\ldots, \\ \delta_k &= K\varepsilon^{\beta\alpha^k} && \text{for } k=0,1,2,\ldots \end{aligned}$$

and try to detremine $\rho_0, \varepsilon, \alpha, \beta, K$ with $0<\rho_0<R, 0<\varepsilon<1, 1<\alpha<2$ so that the conditions (c2) to (c7) of lemma 3.5.2 are satisfied.

Since $0<\varepsilon<1$ and $1<\alpha$ the condition (c4) is clear. The condition (c5) is equivalent to

$$\varepsilon^{\alpha^{k+1}} \ge \frac{C^7\cdot 2^{2\gamma}\varepsilon^{2\alpha^k}}{K^{2\gamma}\varepsilon^{2\beta\gamma\alpha^k}}. \tag{3.5.22}$$

If we set $K^{2\gamma}=C^7\cdot 2^{2\gamma}$ (3.5.22) is written as

$$\varepsilon^{\alpha^{k+1}} \ge \varepsilon^{(2-2\beta\gamma)\alpha^k}$$

which is equivalent to $\alpha \leq (2 - 2\beta\gamma)$. Therefore, by putting

$$K^{2\gamma} = C^7 \cdot 2^{2\gamma} \qquad \text{and} \qquad \beta = \frac{2-\alpha}{2\gamma} > 0$$

we obtain (c5). The condition (c6) is equivalent to

$$\frac{1}{K^\gamma} \sum_{0 \leq k < \infty} \varepsilon^{(1-\beta\gamma)\alpha^k} \leq \frac{r}{2^{\gamma+1}C^3}$$

and the condition (c7) is equivalent to

$$\frac{1}{K^{\gamma+1}} \sum_{0 \leq k < \infty} \varepsilon^{(1-\alpha\beta-\beta\gamma)\alpha^k} \leq \frac{1}{2^{\gamma+2}C^4}\left(1 - \frac{1}{C}\right).$$

For the exponent of ε we have

$$\begin{aligned} (1-\beta\gamma)\alpha^k &= \left(1 - \frac{(2-\alpha)\gamma}{2\gamma}\right)\alpha^k = \frac{1}{2}\alpha^{k+1}, \\ (1-\alpha\beta-\beta\gamma)\alpha^k &= \left(1 - \frac{\alpha(2-\alpha)}{2\gamma} - \frac{(2-\alpha)\gamma}{2\gamma}\right)\alpha^k = \frac{(\alpha+\gamma-2)\alpha^{k+1}}{2\gamma}. \end{aligned}$$

Since $\alpha > 1$ and $\gamma \geq 1$ we have $(\alpha+\gamma-2) > 0$. Thus, by choosing $\varepsilon > 0$ sufficiently small we get the condition (c6) and (c7).

Let us investigate (c2) and (c3). The condition (c2) is equivalent to

$$K \sum_{0 \leq k < \infty} \varepsilon^{\beta\alpha^k} \leq \frac{\rho_0}{2}. \tag{3.5.23}$$

Since $\mathfrak{F}(x, u_p) \in \left(\mathfrak{M}^{2^p+1}\right)^n$ (by (3.5.11)) we have

$$\|\mathfrak{F}(x, u_p)\|_{\rho_0} \leq M\rho_0^{2^p+1}$$

for some $M > 0$. Therefore, the condition (c3) is obtained from

$$M\rho_0^{2^p+1} \leq \varepsilon. \tag{3.5.24}$$

To get (c2) and (c3) we must choose $\varepsilon > 0, \rho_0 > 0$ so that (3.5.23) and (3.5.24) are satisfied. To do so we recall the fact that p is chosen sufficiently large and, therefore, we may assume

$$2^p\beta > 1.$$

Set

$$\varepsilon = \rho_0^{2^p} \tag{3.5.25}$$

and note the following:

$$\sum_{0\le k<\infty} \varepsilon^{\beta\alpha^k} = O\left(\varepsilon^{\beta}\right) = O\left(\rho_0^{\beta 2^p}\right).$$

Hence, if we choose $\rho_0 > 0$ sufficiently small under (3.5.25) we easily get (3.5.23) and (3.5.24). Thus, lemma 3.5.2 is proved. ■

Lastly, let us give a variation of the above discussion. Let $(x,t) \in \mathbb{C}^2$ and let us consider

$$x\frac{\partial u}{\partial x} + bt\frac{\partial u}{\partial t} = ax + bu + R_2(x,u) \tag{3.5.26}$$

where $a, b \in \mathbb{C}$ and $R_2(x,u)$ is a holomorphic function defined near the origin of $\mathbb{C}^2$ with a valuation greater than or equal to two with respect to the variables x and u.

Theorem 3.5.3 *Suppose that the Siegel condition*

$$|m + bn - b| \ge c(m+n)^{-\gamma} \quad \text{for any } (m,n) \in \mathbb{N}^2 \setminus \{(0,0),(0,1)\} \tag{3.5.27}$$

is satisfied for some $c > 0$ *and* $\gamma > 0$. *Then the equation (3.5.26) has a formal solution* $u = \sum_{m+n\ge 1} u_{m,n}x^m t^n$ *with* $u_{0,1}$ *being arbitrary and non zero. Moreover, this formal solution is convergent.*

Proof : Put

$$u = \alpha x + \beta t + w \quad \text{with } w \in \left(\mathfrak{M}_f\right)^2$$

and substitute it into (3.5.26). By the formal identification modulo $\left(\mathfrak{M}_f\right)^2$ we have

$$\alpha = a + b\alpha, \qquad b\beta = b\beta$$

and, hence, we see that α is determined uniquely and β can be chosen arbitrarily. Choose $\beta \ne 0$ and fix it. Then the equation (3.5.26) can be rewritten into an equation with respect to w

$$(\tau - b)w = R_2(x, \alpha x + \beta t + w) \tag{3.5.28}$$

where $\tau = x\frac{\partial}{\partial x} + bt\frac{\partial}{\partial t}$. Since (3.5.27) implies $m + bn - b \ne 0$ for any $(m+n) \ge 2$ it is easy to see that (3.5.28) has a unique formal solution w in $\left(\mathfrak{M}_f\right)^2$. Thus, to have theorem 3.5.3 it is enough to prove that (3.5.28) has a holomorphic solution in $\mathfrak{M}^2$. Let us show this now.

Set

$$\mathfrak{F}(x,w) = (\tau - b)w - R_2(x, \alpha x + \beta t + w)$$

and let us find a holomorphic solution $w \in \mathfrak{M}^2$ of $\mathfrak{F}(x,w) = 0$. By the Taylor expansion we have

$$\mathfrak{F}(x, w+\Delta) = \mathfrak{F}(x,w) + (\tau - b)\Delta - \left(\left(\frac{\partial R_2}{\partial u}\right)(x, \alpha x + \beta t + w)\right)\Delta + O\left(\Delta^2\right).$$

If we set $\Delta(x,t) = \left(\beta + \frac{\partial w}{\partial t}\right) \cdot E(x,t)$, as (3.5.8) we get

$$\mathfrak{F}(x, w+\Delta) = \mathfrak{F}(x,w) + \left(\beta + \frac{\partial w}{\partial t}\right)(\tau - b)E + \frac{\partial}{\partial t}\left(\mathfrak{F}(x,w)\right) \cdot E + O\left(\Delta^2\right).$$

Hence, if we use the scheme

$$\begin{aligned} w_0 &= 0, \\ w_{k+1} &= w_k + \Delta_k, \\ \Delta_k &= \left(\beta + \frac{\partial w_k}{\partial t}\right) E_k, \\ E_k &= -(\tau - b)^{-1}\left(\beta + \frac{\partial w_k}{\partial t}\right)^{-1} \mathfrak{F}\left(x, w_k\right) \end{aligned}$$

and if we apply the same argument as in the proof of theorem 3.5.1 we can prove that $\sum_{0\le k<\infty} \Delta_k$ converges in B_ρ for some $\rho > 0$ and that $\mathfrak{F}\left(x, w_k\right)$ tends to zero in B_ρ (as k tends to infinity).

Thus, by putting $w = \sum_{0\le k<\infty} \Delta_k$ we obtain a solution $w \in \mathfrak{M}^2$ of

$$\mathfrak{F}(x,w) = 0.$$

■

4 Local study of differential equations of the form $xy' = f(x,y)$ near $x = 0$

Introduction

In this chapter, we are studying the singularity of the solutions near the origin of a singular differential equation of the form

$$xy' = f(x,y) \tag{4.0.1}$$

where $f(x,y)$ is holomorphic near the origin of $\mathbb{C}^2$. In particular, we are looking for a normal form of (4.0.1). We have a transformation $z = z(x,y)$ which is reducing (4.0.1) to a normal form and this transformation is given as a solution of a partial differential equation of the form

$$\tau z = F(x,y,z) \tag{4.0.2}$$

where $F(x,y,z)$ is holomorphic near the origin of $\mathbb{C}^3$ and τ is a holomorphic vector field having a singular point at the origin of $\mathbb{C}^2$.

There exist many results on this theory: for example [5], [16], [26], [28], and [33]. In all these papers, the equation (4.0.2) plays no important role and sometimes it is completely forgotten.

In our study, the equation (4.0.2) is playing the central role. Using the results of the preceding chapters on partial differential equations we are able to describe more completely the singularities of the solutions of (4.0.1) near the origin and, moreover, in a more simple way than in the classical theory given in [5].

4.1 Coupling of two differential equations

Consider two differential equations

$$\frac{dy}{dx} = f(x,y), \tag{4.1.1}$$

$$\frac{dz}{dx} = g(x,z) \tag{4.1.2}$$

where f and g are meromorphic near the origin of $\mathbb{C}^2$.

The coupling of these two equations consists in introducing the system

$$\frac{dx}{1} = \frac{dy}{f(x,y)} = \frac{dz}{g(x,z)} \tag{4.1.3}$$

and considering this system as the system of characteristics of one of the quasilinear partial differential equations:

$$\frac{\partial z}{\partial x} + f(x,y)\frac{\partial z}{\partial y} = g(x,z), \tag{4.1.4}$$

$$\frac{\partial y}{\partial x} + g(x,z)\frac{\partial y}{\partial z} = f(x,y), \tag{4.1.5}$$

$$f(x,y)\frac{\partial x}{\partial y} + g(x,z)\frac{\partial x}{\partial z} = 1. \tag{4.1.6}$$

Instead of the system (4.1.3) we can also write

$$\frac{dx}{1} = \frac{dy}{f(x,y)} = \frac{dz}{g(x,z)} = \frac{du}{k(x,y,z)u} \tag{4.1.7}$$

and consider this system as being the system of characteristics of the partial differential equation

$$\frac{\partial u}{\partial x} + f(x,y)\frac{\partial u}{\partial y} + g(x,z)\frac{\partial u}{\partial z} = k(x,y,z)u.$$

The interest of this notion of coupling is showed by the following remarks.

Remark 4.1.1 *Let $z = \psi(x,y)$ be a solution of (4.1.4) and $y = h(x)$ a solution of (4.1.1). Then $z = \psi(x, h(x))$ is a solution of (4.1.2). In fact,*

$$\frac{dz}{dx} = \frac{\partial \psi}{\partial x} + h'(x)\frac{\partial \psi}{\partial y} = \frac{\partial \psi}{\partial x} + f(x,h(x))\frac{\partial \psi}{\partial y} = g(x, \psi(x, h(x))).$$

It is important to remark that if $\psi(x,y)$ is a particular solution of (4.1.4) and $h(x)$ the general solution of (4.1.1) then $\psi(x,h(x))$ is the general solution of (4.1.2). In the same way, if $\psi(x,y)$ is a solution of (4.1.4) depending on an arbitrary constant and $h(x)$ a particular solution of (4.1.1) then $\psi(x,h(x))$ is the general solution of (4.1.2).

Here the meaning of a solution is to be taken in the sense of a local solution.

Remark 4.1.2 *If $y = \psi(x,z)$ is a solution of (4.1.5) and $z = k(x)$ a solution of (4.1.2) then $y = \psi(x,k(x))$ is a solution of (4.1.1). We have also the following results: If $\psi(x,z)$ depends on an arbitrary constant and if $k(x)$ is a particular solution of (4.1.2) then $\psi(x,k(x))$ is the general solution of (4.1.1). In the same way, if $\psi(x,z)$ is a solution of (4.1.5) and $k(x)$ the general solution of (4.1.2) then $\psi(x,k(x))$ is the general solution of (4.1.1).*

Remark 4.1.3 *If $z = \psi(x,y)$ is a solution of (4.1.4) which is locally invertible in y; this means that locally we can write*

$$y = \varphi(x,z) \quad \textit{with } z = \psi(x, \varphi(x,z))$$

then $\varphi(x,z)$ is a solution of (4.1.5) and $z = \psi(x,y)$ defines a transformation which transforms (4.1.2) into (4.1.1) and the inverse transformation is $y = \varphi(x,z)$. In this case, equations (4.1.1) and (4.1.2) are said equivalent.

Remark 4.1.4 *Consider the partial differential equation*

$$\frac{\partial u}{\partial x} + f(x,y)\frac{\partial u}{\partial y} + g(x,z)\frac{\partial u}{\partial z} = k(x,y,z)u$$

where $k(x,y,z) = 0$. *Let* $\psi(x,y,z) = C$ *be a prime integral of the vector field*

$$\frac{\partial}{\partial x} + f(x,y)\frac{\partial}{\partial y} + g(x,z)\frac{\partial}{\partial z}.$$

Assume that ψ *is invertible in* y. *We can then write this integral locally in the form*

$$y = \varphi(x,z,C).$$

If $h(x)$ *is a solution of (4.1.2) then* $y = \varphi(x, h(x), C)$ *is the local general solution of (4.1.1). Also if* ψ *is locally invertible in* z *we can write locally in the form*

$$z = \varphi(x,y,C)$$

and if $k(x)$ *is a solution of (4.1.1) then* $z = \varphi(x, k(x), C)$ *gives us all local solutions of (4.1.2).*

Remark 4.1.5 *If* $\psi(x,y,z)$ *is a formal power series in* x, y, z *satisfying formally*

$$\frac{\partial \psi}{\partial x} + f(x,y)\frac{\partial \psi}{\partial y} + g(x,z)\frac{\partial \psi}{\partial z} = 0$$

we shall say that $\psi(x,y,z)$ *defines near the origin a* formal coupling *between the equations (4.1.1) and (4.1.2). If the formal power series* $\psi(x,y,z)$ *is convergent we shall say that* $\psi(x,y,z)$ *gives a* holomorphic coupling *between the two equations (4.1.1) and (4.1.2). If this coupling can be written in the form* $y = \varphi(x,z)$ *where* φ *is invertible in* z *we say that this two equations are* formally *(resp.* holomorphically*)* equivalent.

In general, a formal or holomorphic coupling does not define an equivalence.

4.2 Behavior of solutions of a differential equation near a regular point

We can assume without loosing generality that the point near which we are studying our equation is the origin $(0,0)$ of $\mathbb{C}^2$. Consider

$$\frac{dy}{dx} = f(x,y) \tag{4.2.1}$$

where $f(x,y)$ is holomorphic at the origin of $\mathbb{C}^2$ that is in a polydisk centered at the origin. We have

$$f(x,y) = \sum_{m+n\geq 0} f_{m,n} x^m y^n.$$

Let us make a coupling of (4.2.1) with an equation

$$\frac{dz}{dx} = g(x,z) \tag{4.2.2}$$

which is also holomorphic near the origin of $\mathbb{C}^2$.

Consider the coupling partial differential equation

$$\frac{\partial y}{\partial x} + g(x,z)\frac{\partial y}{\partial z} = f(x,y) \tag{4.2.3}$$

and look for a formal power series solution of the form

$$y = \sum_{m+n\geq 1} y_{m,n} x^m z^n.$$

$$g(x,z) = \sum g_{m,n} x^m z^n, \qquad\qquad f(x,y) = \sum f_{m,n} x^m y^n.$$

By identification we obtain

$$y_{1,0} + g_{0,0} y_{0,1} = f_{0,0}$$

and

$$A_{p+1}\begin{pmatrix} y_{p+1,0} \\ y_{p,1} \\ \vdots \\ y_{0,p+1}\end{pmatrix}$$

$$= -\left(\sum_{1\leq q\leq p}\ \sum_{r+s=q} g_{m-r,p-m-s+1}\, s\, y_{r,s}\right)_{m=p,p-1,\dots,0} + \left(f_{p,0}\delta_{p,m}\right)_{m=p,p-1,\dots,0}$$

$$+\left(\sum_{r+s\leq p, s\neq 0} f_{r,s} \sum_{i_1+\cdots+i_s=m-r,\, j_1+\cdots+j_s=p-m} y_{i_1,j_1}\cdots y_{i_s,j_s}\right)_{m=p,p-1,\dots,0}$$

where $\delta_{p,m}$ is the Kronecker symbol and

$$A_{p+1} = \begin{pmatrix} p+1 & g_{0,0} & 0 & 0 & \dots & 0 \\ 0 & p & 2g_{0,0} & 0 & \dots & 0 \\ \dots & \dots & \dots & \dots & \dots & \dots \\ 0 & \dots & \dots & 0 & 1 & (p+1)g_{0,0} \end{pmatrix}.$$

The matrix A_{p+1} has $(p+2)$ columns and $(p+1)$ lines but for all p it is of maximal rank $(p+1)$. We deduce that if $y_{0,p}$ is given for all $p = 1, 2, \dots$ equation (4.2.3) has a unique formal solution.

In particular, we can take $y_{0,1} \neq 0$ which proves that two differential equations regular near the origin are always formally equivalent. The transformation is given by

$$y = \sum_{m+n\geq 1} y_{m,n} x^m z^n \quad \text{with } y_{0,1} \neq 0.$$

Consequently, every equation

$$\frac{dy}{dx} = f(x,y)$$

where $f(x,y)$ is holomorphic at the origin of $\mathbb{C}^2$ is formally equivalent to

$$\frac{dz}{dx} = 0.$$

Remark 4.2.1 *This result remains true if $f(x,y)$ is only a formal power series in the variables x and y.*

In this case, the coupling equation is

$$\frac{\partial y}{\partial x} = f(x,y)$$

with y being considered as a function of x and z. This equation has a formal solution of the form

$$y = \sum_{n\geq 1} a_{0,n} z^n + \sum_{m\neq 0, m+n\geq 1} a_{m,n} x^m z^n$$

where for $m \neq 0$ the coefficient $a_{m,n}$ is uniquely determined by the coefficients $a_{0,n}$. By Cauchy–Kowalewski theorem if the series $\sum\limits_{n\geq 1} a_{0,n} z^n$ is convergent at $z = 0$ then the series

$$y = \sum_{n\geq 1} a_{0,n} z^n + \sum_{m\neq 0, m+n\geq 1} a_{m,n} x^m z^n$$

is convergent near the origin of $\mathbb{C}^2$. It is the solution of the Cauchy problem with data on $x = 0$.

Let $\varphi(z) = \sum\limits_{m\geq 1} \varphi_m z^m$ be holomorphic near $z = 0$, for example holomorphic in $|z| < R$. To this function is associated a solution

$$y\,(x, \varphi(z)) = \sum_{m\geq 1} \varphi_m z^m + \sum_{m+n\geq 1, m\neq 0} a_{m,n} x^m z^n$$

of the coupling equation

$$\frac{\partial y}{\partial x} = f(x,y).$$

For all k such that $|k| < R$ $y(x, \varphi(k))$ is a solution of (4.2.1). If $\varphi(z) \equiv 0$ the solution is $\sum_{m \geq 1} a_{m,0} x^m$ which is the solution of (4.2.1) taking the value 0 at the origin of $\mathbb{C}$. All the solutions of (4.2.1) near the origin are given by $y(x, \varphi(k))$. We can assume that $\varphi_m = 0$ for all $m \neq 1$ and $\varphi_1 \neq 0$.

All these considerations show us that the solutions of (4.2.1) depend holomorphically on the initial conditions and, moreover, the local image of the solutions of (4.2.1) can be represented after a holomorphic transformation by the simple picture:

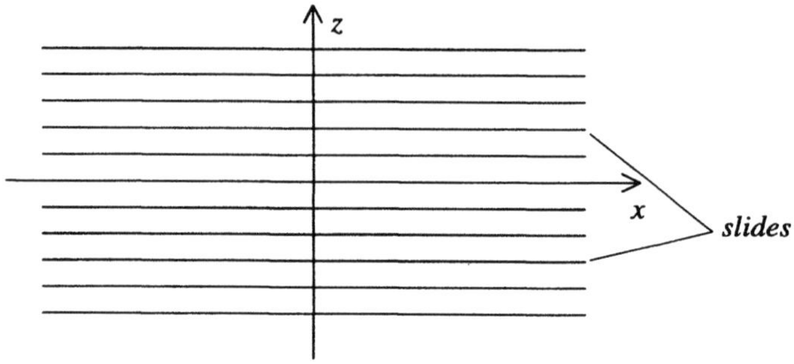

This simple picture is at the origin of the theory of foliations and was first remarked by Painlevé (Oeuvres T.I). Then we have locally in the plane (x, y) the following picture:

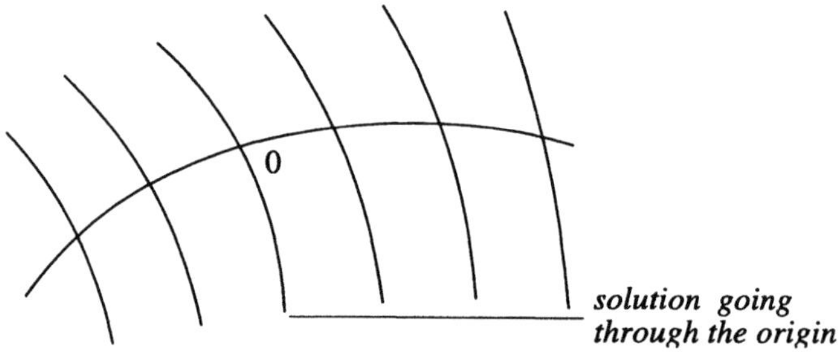

4.3 Local study of a differential equation near a singular point of regular type

4.3.1 Singular point of regular type

Consider a differential equation of the form

$$\frac{dy}{dx} = \frac{P(x,y)}{Q(x,y)} \tag{4.3.1}$$

where P and Q are holomorphic at the origin of $\mathbb{C}^2$ with $Q(0,0)=0$, $Q(0,z)\equiv 0$, and $P(0,0)\neq 0$. Under these conditions we shall say that the origin is a *singular point of regular type* for the equation (4.3.1) if, moreover,

$$Q(x,y)=xR(x,y) \quad \text{with } R(0,0)\neq 0.$$

Equation (4.3.1) can be written

$$x\frac{dy}{dx}=\frac{P(x,y)}{R(x,y)} \quad \text{with } R(0,0)\neq 0 \text{ and } P(0,0)\neq 0$$

or

$$x\frac{dy}{dx}=f(x,y) \tag{4.3.2}$$

where $f(x,y)$ is holomorphic at the origin of $\mathbb{C}^2$ satisfying $f(0,0)\neq 0$. This equation does not have a formal solution of the form $\sum a_n x^n$ and there does not exist an algebraic extension of the form $x=t^p$ having the property that the equation obtained after this extension has a formal solution of the form $\sum b_n t^n$.

Let us couple this equation with an other equation of the same form

$$x\frac{dz}{dx}=g(x,z) \quad \text{where } g(0,0)\neq 0$$

and let us try to find a function $g(x,z)$ holomorphic at the origin of $\mathbb{C}^2$ as simple as possible, for example polynomial. Consider the coupling partial differential equation

$$x\frac{\partial y}{\partial x}+g(x,z)\frac{\partial y}{\partial z}=f(x,y). \tag{4.3.3}$$

The vector field $x\frac{\partial}{\partial x}+g(x,z)\frac{\partial}{\partial z}$ is not singular at the origin because $g(0,0)\neq 0$. This implies that all formal solutions of the equation (4.3.3) are convergent.

Look for a formal solution of (4.3.3) of the form $\sum y_{m,n}x^m z^n$. By identification we obtain $g_{0,0}y_{0,1}=f_{0,0}$. If $g_{0,0}$ is taken arbitrary but non zero then $y_{0,1}$ is well determined and non zero. It is easy to see by more identifications that it is possible to determine the coefficients $y_{p,q}$. Moreover, the coefficients $y_{1,0}, y_{2,0},\ldots,y_{p,0}$ can be chosen arbitrarily, for example $y_{p,0}=0$ for all $p=1,2,\ldots$. This means that the equation

$$x\frac{dy}{dx}=f(x,y)$$

is formally equivalent to the equation

$$x\frac{dz}{dx}=g(0,0)$$

by a formal transformation $\sum y_{m,n}x^m z^n$ where $y_{p,0}=0$ for all p.

Setting $y=zu$ in the coupling equation we get

$$xz\frac{\partial u}{\partial x}+g(0,0)z\frac{\partial u}{\partial z}=f(x,zu)-g(0,0)u. \tag{4.3.4}$$

This equation has a formal solution. The vector field

$$\tau = xz\frac{\partial}{\partial x} + g(0,0)z\frac{\partial}{\partial z}$$

satisfies the following conditions:

A) its singular set is $z = 0$ which near the origin of $\mathbb{C}^2$ defines a regular germ of variety;

B) the Jacobian of the vector field τ at the origin is given by

$$J(\tau)(0,0) = \begin{pmatrix} 0 & \vdots & 0 \\ \cdots & \cdots & \cdots \\ 0 & \vdots & g(0,0) \end{pmatrix}.$$

Kaplan's theorem (section 3.4) tells us that if (4.3.4) has a formal solution it has a convergent one which coincides with the formal one at any order we want.

Consequence: *There exists a holomorphic transformation* $\sum\limits_{m+n\geq 1} y_{m,n}x^m z^n$ *which transforms the equation (4.3.1) into the equation*

$$x\frac{dz}{dx} = 1 \quad (g(0,0) = 1).$$

Moreover,

$$\sum y_{m,n}x^m(\log x + k)^n$$

is for all $k \in \mathbb{C}$ a formal solution of (4.3.1). If R_1 and R_2 are a pair of radius of convergence of the series $\sum y_{m,n}x^m z^n$ we obtain an actual solution of (4.3.1) so far as $|\log x + k| < R_2$ for $|x| < R_1$.

We are coming to the problem of the validity of such a solution. Let us be more precise. Assume that $y(x,z) = \sum\limits_{m+n\geq 1} y_{m,n}x^m z^n$ is convergent in $|x| < R_1$ and $|z| < R_2$. Let x_0 be a point near the origin of $\mathbb{C}$ and $y(x, x_0, y_0)$ the solution of (4.3.1) satisfying $y(x_0, x_0, y_0) = y_0$ and let z_0 be such that

$$y_0 = \sum y_{m,n}(x_0)^m (z_0)^n.$$

Such a z_0 exists and is also small. The solution of $x\frac{dz}{dx} = 1$ taking the value z_0 at the point x_0 is $z = z_0 + \log\left(\frac{x}{x_0}\right)$. For x near x_0 z is near z_0 and

$$y(x) = \sum y_{m,n}x^m\left(z_0 + \log\left(\frac{x}{x_0}\right)\right)^n$$

has a meaning as x is near x_0. It is then possible to give the domain of validity of a solution.

4.3.2 The singularities of solutions of the Briot–Bouquet equation

Consider now an equation of the form

$$x\frac{dy}{dx} = ax + by + R_2(x,y). \tag{4.3.5}$$

This is an equation of the type that we have considered above with

$$f(0,0) = 0.$$

For such an equation we know that each formal solution is convergent. Moreover,

1. *if $b \notin \mathbb{N}^*$ this equation has a unique formal solution which is convergent;*

2. *if $b \in \mathbb{N}^*$ and if this equation has a formal solution then all the solutions are holomorphic at the origin.*

1. is proved in subsection 1.4.2 on page 24.

2. is verified as follows. Assume $b = k \in \mathbb{N}^*$ and let $\varphi(x)$ be a formal solution. We know that $\varphi(x)$ is convergent. Put

$$y = \varphi(x) + x^k z$$

and let us rewrite the equation (4.3.5) into an equation with respect to z. Then we have

$$x^{k+1}\frac{dz}{dx} = x^{k+1}h(x,z)z$$

for some holomorphic function $h(x,z)$ and by a simplification we get

$$\frac{dz}{dx} = h(x,z)z$$

which is studied in section 4.2 and all the solutions of which are holomorphic. This concludes 2.

It remains to study the following cases:

1. when $b \notin \mathbb{N}^*$ the nature of the other solutions,

2. if $b \in \mathbb{N}^*$ and equation (4.3.5) has no formal solutions.

This type of equations was studied completely in the book of Hukuhara, Kimura, and Matuda [28] by a different method without using the theory of partial differential equations as we are doing.

Study of the first case

We have $b \notin \mathbb{N}^*$. In this case, the equation

$$x\frac{dy}{dx} = ax + by + R_2(x,y) = f(x,y) \tag{4.3.6}$$

has a unique holomorphic solution at the origin of $\mathbb{C}$. Let us study the other solutions.

Let us find a coupling of the equation (4.3.6) with an equation of the form

$$x\frac{dz}{dx} = g(x,z) = g_{1,0}x + g_{0,1}z + H_2(x,z)$$

using the coupling partial differential equation

$$x\frac{\partial y}{\partial x} + g(x,z)\frac{\partial y}{\partial z} = ax + by + R_2(x,y);$$

this means

$$x\frac{\partial y}{\partial x} + (g_{1,0}x + g_{0,1}z)\,\frac{\partial y}{\partial z} = ax + by + R_2(x,y) - H_2(x,z)\frac{\partial y}{\partial x}.$$

Look for the coupling equation for a formal solution of the form $\sum\limits_{m+n\geq 1} y_{m,n}x^m z^n$. By formal identification we obtain

$$\begin{aligned} g_{0,1}y_{0,1} &= by_{0,1}, \\ y_{1,0} + g_{1,0}y_{0,1} &= a + by_{1,0}. \end{aligned}$$

The first equation gives us $g_{0,1} = b$ or $y_{0,1} = 0$. The coefficients $(y_{j,k})_{j+k=n}$ must satisfy the system

$$\begin{pmatrix} ng_{0,1} - b & 0 & \cdots\cdots & & \cdots\cdots & 0 \\ ng_{1,0} & 1 + (n-1)g_{0,1} - b & 0 & \cdots & \cdots\cdots & 0 \\ 0 & \cdots\cdots & & & \cdots\cdots & \\ \cdots\cdots & & & & \cdots\cdots & \\ 0 & \cdots & 0 & (n-p+1)g_{1,0} & p + (n-p)g_{0,1} - b \quad 0 \quad \cdots & 0 \\ \cdots\cdots & & & & \cdots\cdots & \\ 0 & \cdots\cdots & & \cdots & 0 \quad g_{1,0} & n - b \end{pmatrix} \begin{pmatrix} y_{0,n} \\ y_{1,n-1} \\ \cdots \\ \cdots \\ y_{p,n-p} \\ \cdots \\ y_{n,0} \end{pmatrix}$$

$$= F_n\left((y_{i,k})_{i+k<n}, f_{i,k}, g_{k,l}\right) = \begin{pmatrix} F_{0,n} \\ F_{1,n-1} \\ \vdots \\ F_{n-1,1} \\ F_{0,n} \end{pmatrix} \left((y_{i,k})_{i+k<n}, f_{i,k}, g_{k,l}\right).$$

This system shows us that if $g_{0,1} = b$ and if the matrix

$$A_n = \begin{pmatrix} ng_{0,1} - b & 0 & \cdots & & & & & 0 \\ ng_{1,0} & 1+(n-1)g_{0,1} - b & 0 & \cdots & & & & 0 \\ 0 & \cdots & & & & & & \cdots \\ \cdots & & & & & & & \cdots \\ 0 & \cdots & 0 & (n-p+1)g_{1,0} & p+(n-p)g_{0,1} - b & 0 & \cdots & 0 \\ \cdots & & & & & & & \cdots \\ 0 & \cdots & & & & 0 & g_{1,0} & n-b \end{pmatrix}$$

is invertible the choice of $y_{0,1}$ arbitrary but non zero implies that equations of the form

$$x\frac{dy}{dx} = a_{1,0}x + by + R_2(x,y)$$

are formally equivalent. We are under the hypothesis $b \notin \mathbb{N}^*$. If in the matrix A_n we are replacing $g_{0,1}$ by b we see immediately that if $b \notin \mathbb{Q}^-$ and $b \neq 0$ the matrix A_n is invertible for all values of n $(n = 2, 3, \ldots)$.

Conclusion: *If $b \notin \mathbb{N}^*$ and, moreover, $b \notin \mathbb{Q}^-$ and $b \neq 0$ the equation (4.3.6) is formally equivalent to the equation*

$$x\frac{dz}{dx} = bz. \tag{4.3.7}$$

This equivalence is given by the transformation

$$y = \sum_{m+n\geq 1} y_{m,n}(y_{0,1})\, x^m z^n$$

the coefficients $(y_{m,n}(y_{0,1}))_{m+n\geq 1}$ of which are given by the infinite sytem introduced above. The formal power series $\sum\limits_{m+n\geq 1} y_{m,n}x^m z^n$ is a formal solution of

(4.3.8)
$$x\frac{\partial y}{\partial x} + bz\frac{\partial y}{\partial z} = ax + by + R_2(x,y).$$

This implies that

(4.3.9)
$$\sum_{m+n\geq 1} y_{m,n}x^m \left(x^b\right)^n$$

is a formal solution of (4.3.6).

Remark 4.3.1 *We have taken $y_{0,1} \neq 0$ (arbitrary). We see easily that if $b \notin \mathbb{N}^*$, $b \notin \mathbb{Q}^-$, $b \neq 0$ and if we take $y_{0,1} = 0$ then we obtain as a solution of the equation (4.3.8) the power series $\sum_{m\geq 1} y_{m,0}x^m$ which is the holomorphic solution of (4.3.6) the existence of which follows from the fact that $b \notin \mathbb{N}^*$.*

Convergence of the formal solution:
If the vector field

$$x\frac{\partial}{\partial x} + bz\frac{\partial}{\partial z}$$

satisfies the Poincaré condition

$$|m + bn| \geq c(m+n) \quad \text{for any } (m,n) \in \mathbb{N}^2 \setminus \{0,0\}$$

or the Siegel condition

$$|m + bn| \geq \frac{c}{(m+n)^\gamma} \quad \text{for any } (m,n) \in \mathbb{N}^2 \setminus \{0,0\}$$

the formal power series $\sum_{m+n\geq 1} y_{m,n}\left(y_{0,1}\right)x^m z^n$ is convergent at the origin of $\mathbb{C}^2$ (see theorem 3.2.2.1 and theorem 3.5.3).

Let us recall that the Poincaré condition is equivalent to the fact that 0 does not belong to the segment joining the point 1 to the point b.

Assume that this power series is convergent in

$$D = \{|x| < R_1\} \times \{|z| < R_2\}$$

and denote by $F(x,z)$ the holomorphic function defined in D by this power series.

If $F(x,z)|_{z=x^b}$ (the restriction of $F(x,z)$ to the analytic curve $\Gamma : z = x^b$) is well defined that is if $\Gamma \cap D \neq \emptyset$ then the function $\Psi(x) = F(x,z)|_{z=x^b}$ defined on $\Gamma \cap D$ is a solution of (4.3.6). As it depends on an arbitrary constant $y_{0,1}$ we have all solutions of (4.3.6) on $\Gamma \cap D$. The complex number b is given. It follows that in every domain having the origin in its closure (sector, spiralling sector, ...) in which x^b tends to zero as x tends to zero we get the form and the behavior of the solution of (4.3.6) near the origin.

In the case when $F(x,z)$ is an entire function, then $F\left(x,x^b\right)$ gives us all the solutions of (4.3.6).

Examples of domains in which x^b tends to zero when x tends to zero

1. *Position of b*

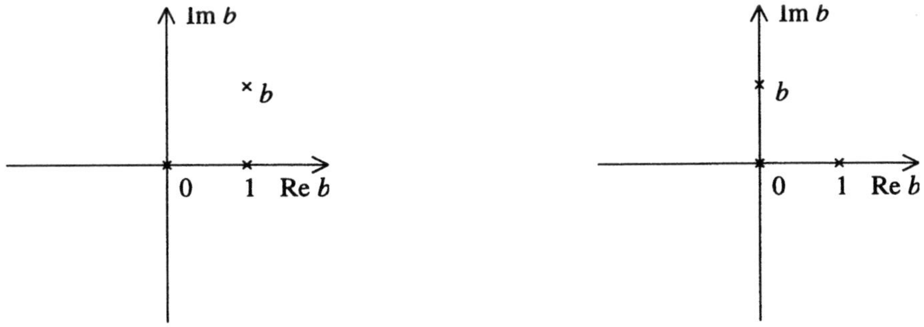

cases (a) and (b)

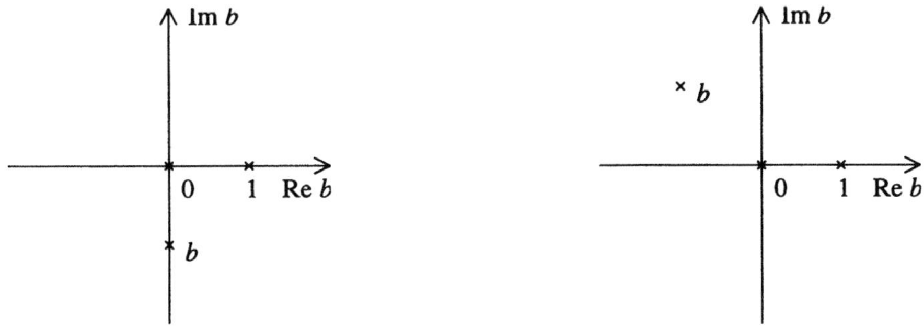

cases (c) and (d)

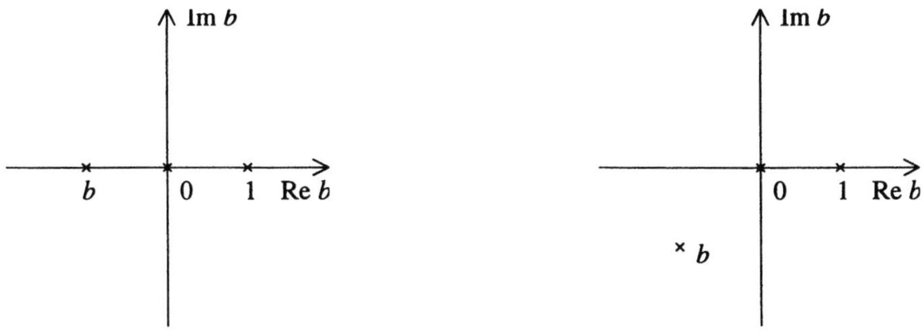

cases (e) and (f)

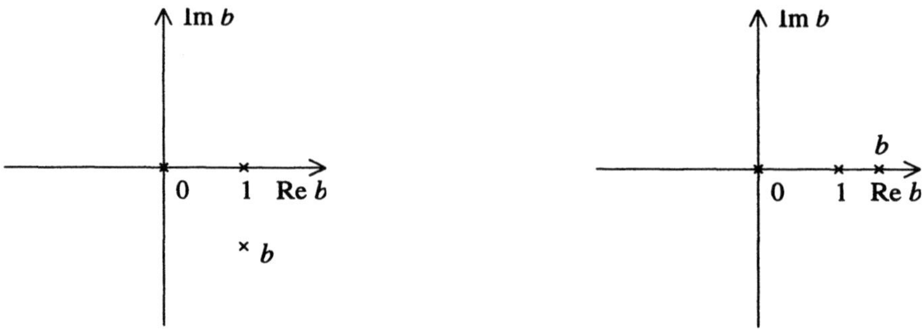

cases (g) and (h)

2. *Domain in the plane* $(\log|x|, \arg x)$
Put $b = c + id$.

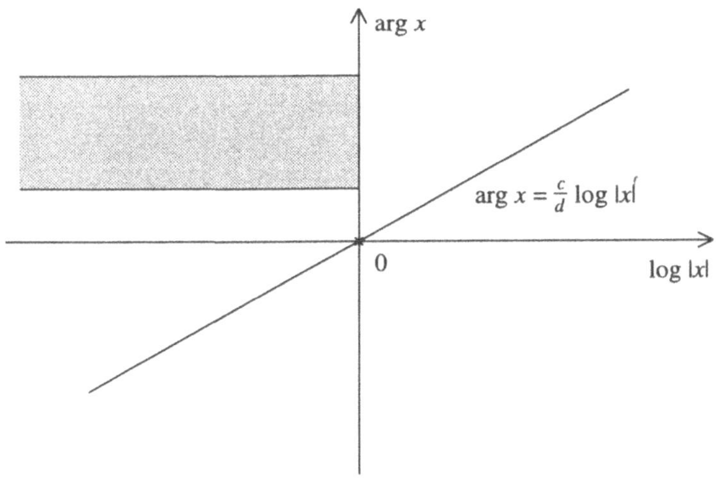

case (a)

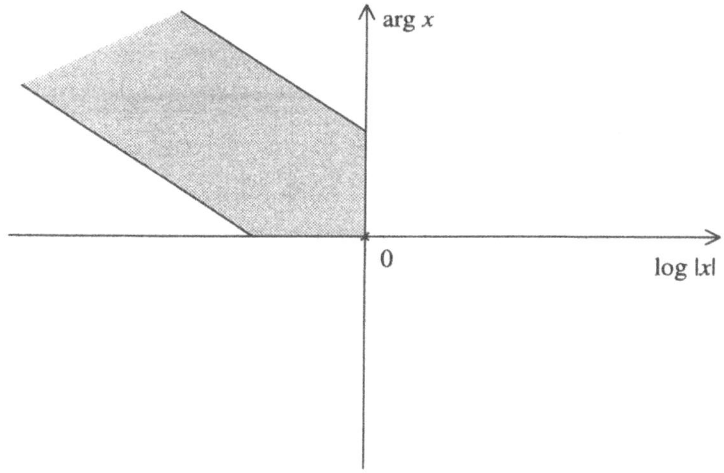

case (b)

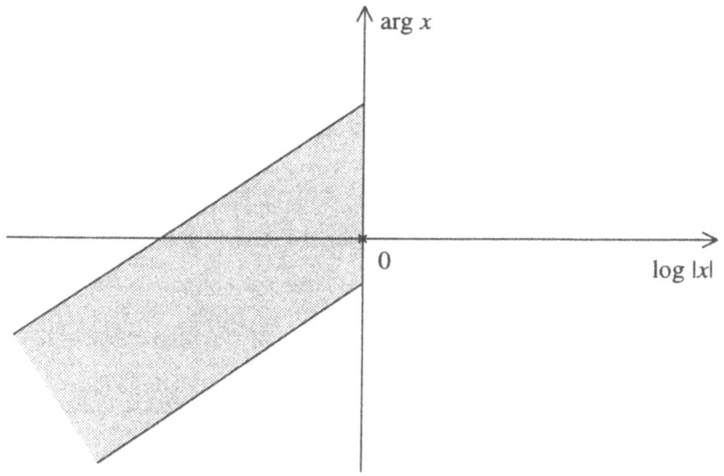

case (c)

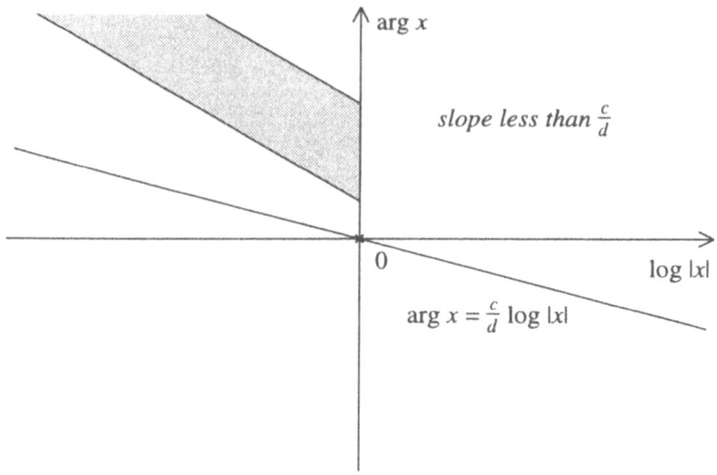

case (d)

For the case (e) we cannot give a picture of the domain because $|x^b|$ tends to infinity. There does not exist a domain in the x–plane having the origin in its closure such that when x tends to zero $|x^b|$ tends to zero. In this case, we will have a solution of (4.3.6) when $F(x, z)$ is an entire function. We think that in this case an asymptotic study will be very interesting and useful.

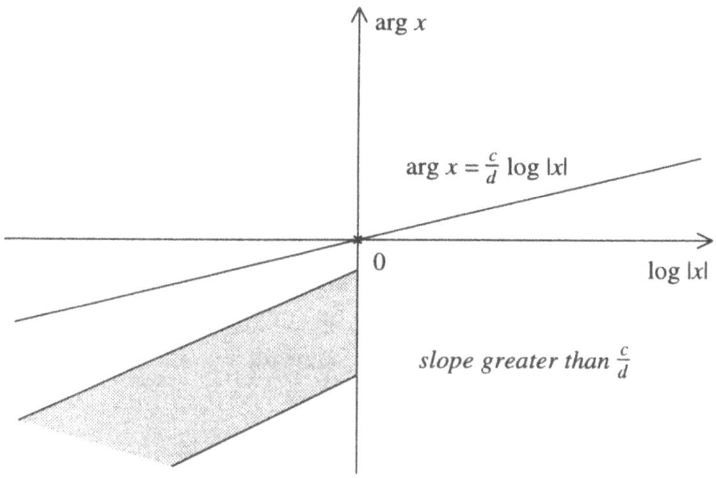

case (f)

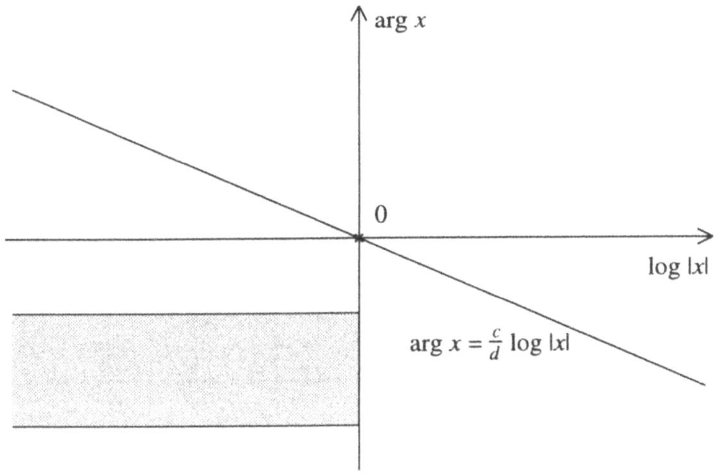

case (g)

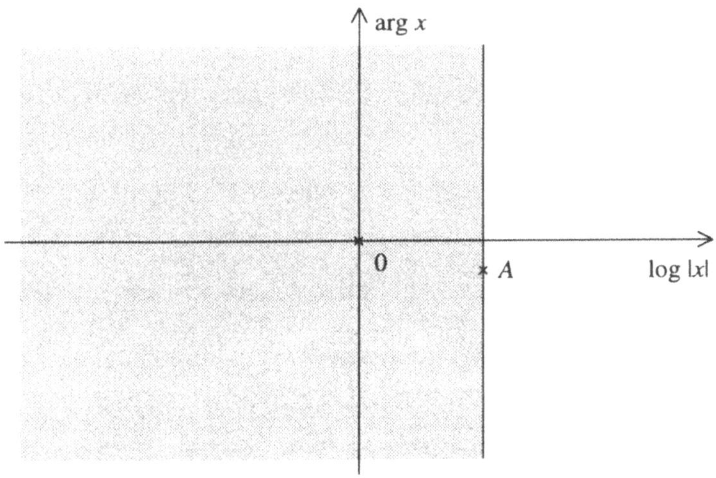

case (h)

3. *Domain in the x–plane*

The cases (a) and (g) give in the complex plane of the variable x sectors centered at the origin.

The cases (b),(c),(d),(f) give in the plane of the variable x spiralling sectors of one of the form

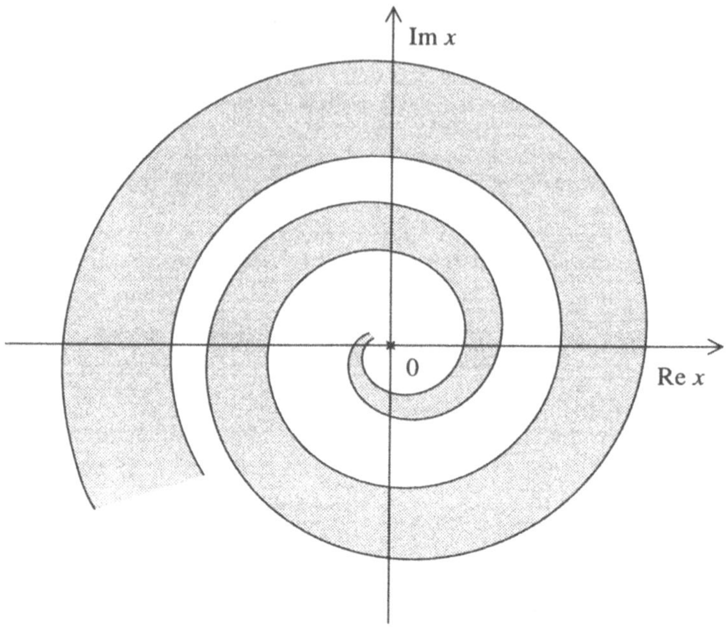

or

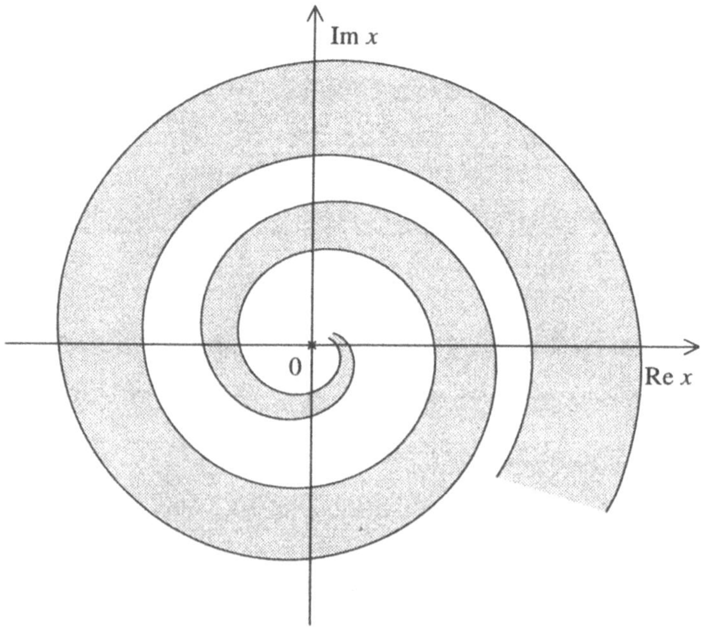

The general solution of the equation $x\frac{dz}{dx} = bz$ is $z = cx^b$. Assume to be under the conditions which imply the convergence of the power series $\sum_{m+n\geq 1} y_{m,n}(y_{0,1})\, x^m z^n$ in $\{|x| < R_1\} \times \{|z| < R_2\}$. Let (x_0, y_0) be initial conditions and determine z_0 by

$$y_0 = \sum_{m+n\geq 1} y_{m,n}\,(x_0)^m\,(z_0)^n\,.$$

This is possible because $y_{0,1}$ is non zero.

The solution of (4.3.6) which for $x = x_0$ takes the value y_0 is then given by

$$y\,(x, x_0, y_0) = \sum_{m+n\geq 1} y_{m,n}x^m \left(z_0\left(\frac{x}{x_0}\right)^b\right)^n$$

and this expression has a meaning as long as

$$\left|\left(\frac{x}{x_0}\right)^b\right| \leq 1.$$

It is easy to give the domain of validity of this solution for the different values of b. We obtain domains of validity which are sectors in the Riemann surface of the logarithm if for example $b_1 > 0$ and $b_2 > 0$ $(b = b_1 + ib_2)$, or spiralling sectors in this surface if for example $b_1 < 0$ and $b_2 \neq 0$. In the case where the convergence is obtained by a Siegel type condition [4], that is, in particular, when $b_2 = 0$ and $b_1 < 0$, the domain of validity of the solution $y\,(x, x_0, y_0)$ is a crown $|x_0| \leq |x| < R_1$.

The case $b \in \mathbb{Q}^-$ and $b \neq 0$

Assume that $b = -\frac{p}{q}$ with $p, q \in \mathbb{N}^*$ being relatively prime. The equation

$$x\frac{dy}{dx} = f(x, y)$$

has a unique holomorphic solution $y = \varphi(x)$ such that $\varphi(0) = 0$. By a simple translation we can assume that this solution is $y = 0$ and $f(x, 0) = 0$. Now our equation has the form

$$x\frac{dy}{dx} = -\frac{p}{q}y + \sum_{m+n\geq 2} f_{m,n}x^m y^n = f(x, y) \tag{4.3.10}$$

with $f_{m,0} = 0$ for all m.

Consider the coupling equation

$$x\frac{\partial y}{\partial x} + g(x, z)\frac{\partial y}{\partial z} = -\frac{p}{q}y + \sum_{m+n\geq 2} f_{m,n}x^m y^n.$$

By formal identification we get

$$\begin{aligned} g_{0,1}y_{0,1} &= -\frac{p}{q}y_{0,1}, \\ y_{1,0} + g_{1,0}y_{0,1} &= -\frac{p}{q}y_{1,0}. \end{aligned}$$

We know that the choice $y_{0,1} = 0$ gives us the holomorphic solution of the equation (4.3.10). We have to take $y_{0,1} \neq 0$ (arbitrary) and $g_{0,1} = -\frac{p}{q}$. The second equation gives us $y_{1,0}$.

Considering the matrix A_m introduced before we see that in the computation by induction of the coefficients $(y_{j,k})$ of a formal solution we have an indetermination in the formula giving $y_{p,q+1}$ the corresponding equation being

$$py_{p,q+1} - \frac{p}{q}(q+1)y_{p,q+1} + \frac{p}{q}y_{p,q+1} + g_{p,q+1}y_{0,1} = G_{p,q+1}\left(\left(y_{j,k}\right)_{j+k\leq p+q}, (g_{j,k})\right)$$

where $G_{p,q+1}\left(\left(y_{j,k}\right)_{j+k\leq p+q}, (g_{j,k})\right)$ is known by induction. This equation gives $g_{p,q+1}$.

The next indetermination or better the next *resonance* comes for the computation of $y_{2p,2q+1}$; that is in the identification of the terms in $x^{2p}z^{2q+1}$. The corresponding equation is

$$g_{p,q+1}(q+1)y_{p,q+1} + g_{2p,2q+1}y_{0,1} = G_{2p,2q+1}\left(\left(y_{j,k}\right)_{j+k\leq 2p+2q}, (g_{j,k})\right).$$

As $y_{p,q+1}$ does not occur in the right member of this equation it is of the form

$$H_{p,q+1}y_{p,q+1} + g_{2p,2q+1}y_{0,1} = G^*_{2p,2q+1}\left(\left(y_{j,k}\right)_{j+k\leq 2p+2q}, (g_{j,k})\right)$$

where $y_{p,q+1}$ does not occur in G^*. If $H_{p,q+1} \neq 0$ this equation gives us $y_{p,q+1}$ and $g_{2p,2q+1}$ is arbitrary. If $H_{p,q+1} = 0$ this equation gives us $g_{2p,2q+1}$ and $y_{p,q+1}$ is arbitrary. All the precedent equations are solvable without problems.

The next resonance occurs in the identification of the coefficients in $x^{3p}z^{3q+1}$. The corresponding equation is

$$\begin{aligned} g_{p,q+1}(2q+1)y_{2p,2q+1} + g_{2p,2q+1}(q+1)y_{p,q+1} + g_{3p,3q+1}y_{0,1} \\ = G_{3p,3q+1}\left(\left(y_{j,k}\right)_{j+k\leq 3p+3q}, (g_{j,k})\right) \end{aligned}$$

which takes the form

$$H_{2p,2q+1}y_{2p,2q+1} + g_{3p,3q+1}y_{0,1} = G^*_{3p,3q+1}\left(\left(y_{j,k}\right)_{j+k\leq 3p+3q}, (g_{j,k})\right)$$

where G^* does not contain $y_{2p,2q+1}$. If $H_{2p,2q+1} \neq 0$ this equation gives us $y_{2p,2q+1}$ and if $H_{2p,2q+1} = 0$ it gives us $g_{3p,3q+1}$.

It is now easy to see by induction that the given equation is formally equivalent to an equation of the form

$$x\frac{dz}{dx} = -\frac{p}{q}z + zh\left(x^p z^q\right)$$

where h is a formal power series in one variable satisfying $h(0) = 0$.

By the change of function $z = u^{\frac{1}{q}} x^{-\frac{p}{q}}$ we obtain the equation

$$x\frac{du}{dx} = quh(u). \tag{4.3.11}$$

If in the coupling equation

$$x\frac{\partial y}{\partial x} + \left(-\frac{p}{q}z + zh\left(x^p z^q\right)\right)\frac{\partial y}{\partial z} = f(x,y) \tag{4.3.12}$$

we introduce the transformation $z = u^{\frac{1}{q}} x^{-\frac{p}{q}}$ we obtain

$$x\frac{\partial y}{\partial x} + (quh(u))\frac{\partial y}{\partial u} = f(x,y)$$

which defines a formal coupling between the two equations (4.3.10) and (4.3.11). Moreover, this coupling gives us the formal transformation which transforms (4.3.10) into (4.3.11).

We are now back to the study of an equation (4.3.6) with $b = 0$. This is the case that we study now.

The case $b = 0$

Once more, if we are setting $y_{0,1} = 0$ we get the holomorphic solution of the equation (4.3.6). Let m $(m > 1)$ be the integer such that $f_{0,s} = 0$ for all $s \le m$ and $f_{0,m+1} \neq 0$. This integer does exist. In fact, $f_{0,s} = 0$ for all s implies that $f(x,y)$ is of the form $f(x,y) = xf'(x,y)$ and we are back to the study of an equation regular at the point $x = 0$.

The matrix A_n has in this case the form

$$A_n = \begin{pmatrix} 0 & 0 & \cdots\cdots\cdots\cdots\cdots\cdots & 0 \\ ng_{1,0} & 1 & 0 \quad \cdots\cdots\cdots\cdots\cdots & 0 \\ 0 & \cdots\cdots\cdots\cdots\cdots\cdots\cdots\cdots & & \\ \cdots\cdots\cdots & \cdots\cdots\cdots\cdots\cdots\cdots\cdots\cdots & & \\ 0 & \cdots \; 0 \;\; (n-p+1)g_{1,0} \;\; p \;\; 0 & \cdots & 0 \\ \cdots\cdots\cdots & \cdots\cdots\cdots\cdots\cdots\cdots\cdots\cdots & & 0 \\ \cdots\cdots\cdots & \cdots\cdots\cdots\cdots\cdots\cdots\cdots\cdots & & \\ 0 & \cdots\cdots\cdots\cdots\cdots\cdots \; 0 & g_{1,0} & n \end{pmatrix}.$$

There is only a problem in the determination of $y_{0,n}$ for all n. The equations giving $y_{0,1}$ are

$$\begin{aligned} g_{0,i}y_{0,1} &= 0 \quad \text{for all } i \le m, \\ g_{0,m+1} &= f_{0,m+1}\left(y_{0,1}\right)^m . \end{aligned}$$

This equation gives us $g_{0,m+1}$. The next equation is

$$g_{0,m+1}2y_{0,2} + g_{0,m+2}y_{0,1} = f_{0,m+1}y_{0,2}(m+1)\left(y_{0,1}\right)^m + f_{0,m+2}\left(y_{0,1}\right)^{m+2}$$

which is

$$f_{0,m+1}y_{0,2}\left(y_{0,1}\right)^{m-1}(1-m) = f_{0,m+2}\left(y_{0,1}\right)^{m+1} - g_{0,m+2} \quad (m>1)$$

and then we have $y_{0,2}$ and we can take $g_{0,m+2} = 0$.

To calculate $y_{0,m+1}$ we have the equation

$$g_{0,m+1}(m+1)y_{0,m+1} + g_{0,2m+1}y_{0,1} - f_{0,m+1}y_{0,m+1}\left(y_{0,1}\right)^{m}(m+1)$$
$$= (\text{by induction a known expression}).$$

As $g_{0,m+1} = f_{0,m+1}\left(y_{0,1}\right)^{m}$ this equation does not give $y_{0,m+1}$ but it allows us to determine $g_{0,2m+1}$.

We see easily that the resolution of the infinite sytem of equations can be done without any new problems. The only constraint is that $g_{0,m+1} = f_{0,m+1}\left(y_{0,1}\right)^{m}$ and $g_{0,2m+1}$ is given by the above equation. $y_{0,1}$ and $y_{0,m+1}$ being arbitrary, $g_{i,k} = 0$ for all $i \neq 0$ and $g_{0,j} = 0$ for all $j \neq m+1$ and $j \neq 2m+1$.

We have now the following result:

When $b = 0$ the equation (4.3.6) is formally equivalent to an equation of the form

$$H(A,B,m): \qquad x\frac{dz}{dx} = Az^{m+1} + Bz^{2m+1}$$

where A and B are constants.

We have also the following result:

When $b = -\frac{p}{q} \in \mathbb{Q}^-$ the equation (4.3.6) is formally equivalent to an equation of the form

$$E(b,A,B,m): \qquad x\frac{dz}{dx} = z\left(b + A\left(x^p z^q\right)^m + B\left(x^p z^q\right)^{2m}\right).$$

In general, the formal transformation is not convergent. It is necessary in this case to make an asymptotic study which we are doing later.

It is also interesting to study *Hukuhara's equation*

$$x\frac{dy}{dx} = Ay^{m+1} + By^{2m+1}.$$

We are doing this in the next section.

Remark 4.3.2 *Two equations $H(A,B,m)$ and $H(A',B',m)$ (resp. $E(b,A,B,m)$ and $E(b',A',B',m)$) are formally equivalent if and only if $\frac{A^2}{B} = \frac{(A')^2}{B'}$ (resp. $b = b'$ and $\frac{A^2}{B} = \frac{(A')^2}{B'}$).*

Study of the second case

We have $b \in \mathbb{N}^*$ and our equation has no formal solutions. We are looking at the infinite system associated to the coupling equation

$$x\frac{\partial y}{\partial x} + g(x,z)\frac{\partial y}{\partial z} = f(x,y) = ax + by + \sum_{m+n\geq 2} f_{m,n}x^m y^n,$$

$$g_{0,1}y_{0,1} = by_{0,1}, \qquad y_{1,0} + g_{1,0}y_{0,1} = a + by_{1,0}.$$

We can take $y_{0,1}$ arbitrary and non zero then we have to take $g_{0,1} = b \in \mathbb{N}^*$. The coefficients of higher order of a formal solution $\sum_{m+n\geq 1} y_{m,n}x^m z^n$ are given by induction by an infinite system $(S_m)_{m\in\mathbb{N}^*}$ of the form

$$A_m Y_m = F_m\left((y_{j,k})_{j+k<m}\right)$$

where $Y_m =$ transposed of $(y_{0,m}, y_{1,m-1}, \dots, y_{m,0})$ and

$$A_m = \begin{pmatrix} mg_{0,1}-b & 0 & \cdots & & & & \cdots & 0 \\ mg_{1,0} & 1+(m-1)g_{0,1}-b & 0 & \cdots & & & \cdots & 0 \\ 0 & \cdots & & & & & & \cdots \\ \cdots & & & & & & & \cdots \\ 0 & \cdots & 0 & (m-p+1)g_{1,0} & p+(m-p)g_{0,1}-b & 0 & \cdots & 0 \\ \cdots & & & & & & & \cdots \\ 0 & \cdots & & & \cdots & 0 & g_{1,0} & m-b \end{pmatrix}.$$

The matrix A_m is invertible for all $m \neq b$. In solving the system S_b, we have a problem only to calculate $y_{b,0}$. In fact, the equation giving $y_{b,0}$ is

$$\begin{aligned} 0y_{b,0} + g_{1,0}y_{b-1,1} + g_{b,0}y_{0,1} &= B_{b,0}\left((y_{j,k})_{j+k<b}, (f_{j,k}), (g_{j,k})_{j+k<b}\right) \\ &= \text{(an expression known by induction).} \end{aligned}$$

To obtain a differential equation as simple as possible and equivalent to (4.3.6) take $g_{0,1} = b$ and $g_{j,k} = 0$ for all $(j,k) \neq (0,1)$ such that $j+k<b$. Then the above equation has the form

$$g_{b,0}y_{0,1} = B_{b,0}\left((y_{j,k})_{j+k<b}, (f_{j,k})\right).$$

Remark 4.3.3 *If $B_{b,0}\left(\left(y_{j,k}\right)_{j+k<b},\left(f_{j,k}\right)\right) = 0$ for the equation (4.3.6) we take $g_{b,0} = 0$ and $y_{0,1}$ arbitrary and different from zero. In this case, the equation (4.3.6) is equivalent to the equation $x\frac{dz}{dx} = bz$ and all the solutions of (4.3.6) are holomorphic at the origin.*

Assume now that $B_{b,0}\left(\left(y_{j,k}\right)_{j+k<b},\left(f_{j,k}\right)\right) \neq 0$. In this case, the equation (4.3.6) is equivalent to an equation of the form

$$x\frac{dz}{dx} = bz + cx^b \quad \text{with } c \in \mathbb{C}^* \tag{4.3.13}$$

and two equations of this form are equivalent. In this case, $g_{b,0}$ being fixed equal to c, $y_{0,1}$ is well determined and, moreover, different from zero. The solutions of the equation (4.3.13) are of the form

$$z = kx^b + cx^b \log x$$

where k is an arbitrary constant. As all equations of the form (4.3.13) are equivalent we can assume that $c = 1$.

Since the vector field

$$x\frac{\partial}{\partial x} + \left(bz + x^b\right)\frac{\partial}{\partial z}$$

satisfies the Poincaré condition every formal solution of the coupling equation

$$x\frac{\partial y}{\partial x} + \left(bz + x^b\right)\frac{\partial y}{\partial z} = f(x,y) \tag{4.3.14}$$

is convergent. If $\sum_{m+n\geq 1} y_{m,n}x^m z^n$ is the convergent solution of the equation (4.3.14) all solutions of the equation (4.3.6) are given by

$$\sum_{m+n\geq 1} y_{m,n}x^m \left(kx^b + x^b \log x\right)^n$$

where k is an arbitrary constant.

Let be given some initial condition (x_0, y_0) and introduce z_0 defined by

$$y_0 = \sum_{m+n\geq 1} y_{m,n}\left(x_0\right)^m \left(z_0\right)^n .$$

z_0 is well defined by the fact $y_{0,1} \neq 0$. Then

$$z = z_0\left(\frac{x}{x_0}\right)^b + x^b \log\left(\frac{x}{x_0}\right)$$

and the solution of (4.3.6) is given near the origin by

$$\sum_{m+n\geq 1} y_{m,n}x^m \left(z_0\left(\frac{x}{x_0}\right)^b + x^b \log\left(\frac{x}{x_0}\right)\right)^n .$$

As $b \in \mathbb{N}^*$ $z(x)$ tends to zero with x and this expression has a meaning for x near zero because by restriction of z to a solution of (4.3.13) it gives us an expression which is in the domain of convergence of the formal solution of (4.3.14). The expression given above shows us the nature of the singularity near the origin of the solution. They are of non–essential transcendental type. In fact, each solution tends to zero when x tends to zero.

4.4 Study of the Hukuhara equation and of the Hukuhara function

The *Hukuhara equation* has the form

(4.4.1) $$x\frac{dy}{dx} = Ay^{m+1} + By^{2m+1}.$$

The particular cases $A = 0$ or $B = 0$ are left to the reader. Let us assume $AB \neq 0$. By setting $\eta = y^{-m}$ we get

$$x\frac{d\eta}{dx} = -Am - B\frac{m}{\eta}$$

and by setting $x = e^{-\frac{B}{mA^2}X}$ and $\eta = \frac{B}{A}Y$

$$\frac{dY}{dX} = 1 + \frac{1}{Y}$$

that is

$$Y\frac{dY}{dX} = 1 + Y.$$

We have

$$Y - \log(Y+1) = X + C.$$

The solution of this equation is denoted by $\mathbb{H}(X + C)$. By solution we mean a solution in the large that is the analytic extension of a local solution. We have

$$y = \left[\frac{B}{A}\mathbb{H}\left(C - \frac{mA^2}{B}\log x\right)\right]^{-\frac{1}{m}}.$$

Let us come back to the formal integration of an equation of the form

(4.4.2) $$x\frac{dy}{dx} = ax + by + R_2(x,y)$$

when $b = 0$ or when $b = -\frac{p}{q} \in \mathbb{Q}^-$.

When $b = 0$ the equation (4.4.2) has a formal solution of the form

$$y = \sum_{p+q\geq 1} y_{p,q}x^p \left[\frac{B}{A}\mathbb{H}\left(C - \frac{mA^2}{B}\log x\right)\right]^{-\frac{q}{m}}.$$

When $b = -\frac{p}{q}$ the equation (4.4.2) has a formal solution which can be easily expressed in terms of the Hukuhara function because we know how to reduce this case to the case $b = 0$.

Study of the Hukuhara function

The *Hukuhara function* $\mathbb{H}(X)$ is defined by the equation

$$\mathbb{H}(X) - \log(\mathbb{H}(X) + 1) = X.$$

If we set $u = \mathbb{H}(X) + 1$ and $v = X + 1$ this relation is

$$u - \log u = v.$$

By introducing

$$\begin{aligned} u_1 &= u^{-1}, & u_2 &= u^{-1}\log u, \\ v_1 &= v^{-1}, & v_2 &= v^{-1}\log v, \end{aligned}$$

we get

$$\begin{aligned} v_1 &= u_1 (1 - u_2)^{-1}, \\ v_2 &= u_2 (1 - u_2)^{-1} + u_1 (1 - u_2)^{-1} \log(1 - u_2). \end{aligned}$$

The implicit function theorem shows that the functions u_1 and u_2 have convergent power series expansions in terms of v_1 and v_2. We get an expansion of the form

$$u_1 = v_1 (1 - v_2 + \cdots)$$

where the non written terms have a valuation greater than or equal to two with respect to v_1 and v_2.

We conclude that the function $\mathbb{H}(X)$ has a convergent expansion of the form

$$\mathbb{H}(X) = X\left[1 + X^{-1}\log X + \cdots\right]$$

where the non written terms have a valuation greater than or equal to two with respect to the variables X^{-1} and $X^{-1}\log X$, from which we deduce the inequality

$$|\mathbb{H}(X) - X - \log X| \leq M\frac{|\log X|^2}{|X|}$$

if the modules of X^{-1} and $X^{-1}\log X$ are small enough where M is a constant.

Consider the domain

$$R_0 < R < \infty, \qquad |\theta| < \alpha \log R \qquad (X = Re^{i\theta})$$

where α is an arbitrary non negative constant. In this domain when X tends to infinity

$$\mathbb{H}(X) - X - \log X \longrightarrow 0.$$

The relation

$$\mathbb{H}(X) - \log(\mathbb{H}(X) + 1) = X$$

can be written in the form

$$(\mathbb{H}(X) + 1)\, e^{-(\mathbb{H}(X)+1)} = e^{-(X+1)}$$

and once more the implicit function theorem tells us that $\mathbb{H}(X)$ has the expansion

$$\mathbb{H}(X) = -1 + e^{-(X+1)} + e^{-2(X+1)} + \cdots$$

which is convergent for e^{-X} small enough. The branch of $\mathbb{H}(X)$ defined by the analytic expansion of this series tends to -1 as e^{-X} tends to zero.

The function $\mathbb{H}(X)$ has no pole. Set $Y = \frac{1}{Z}$ in the equation

$$Y\frac{dY}{dX} = 1 + Y.$$

We get

$$-\frac{dZ}{dX} = Z^3 + Z^2.$$

The line at infinity $Z = 0$ is a solution of this equation and, moreover, there are no singularities on this line. This implies that the Hukuhara function has no pole.

5 Holomorphic and singular solutions of non linear singular first order partial differential equations

Introduction

In this chapter, we will discuss the following types of non linear singular first order partial differential equations

$$(5.0.1) \qquad t\frac{\partial u}{\partial t} = F\left(t, x, u, \frac{\partial u}{\partial x}\right)$$

where $(t,x) \in \mathbb{C}_t \times (\mathbb{C}_x)^n, x = (x_1, \ldots, x_n), \frac{\partial u}{\partial x} = \left(\frac{\partial u}{\partial x_1}, \ldots, \frac{\partial u}{\partial x_n}\right), F(t,x,u,v)$ is a function defined in a polydisk Δ centered at the origin of $\mathbb{C}_t \times (\mathbb{C}_x)^n \times \mathbb{C}_u \times (\mathbb{C}_v)^n$, and $v = (v_1, \ldots, v_n)$. Let us denote $\Delta_0 = \Delta \cap \{t = 0, u = 0, v = 0\}$.

The assumptions are as follows:

(A1) $F(t,x,u,v)$ is holomorphic in Δ;

(A2) $F(0,x,0,0) = 0$ in Δ_0;

(A3) $\dfrac{\partial F}{\partial v_i}(0,x,0,0) = 0$ in Δ_0 for $i = 1, \ldots, n$.

The purpose of this chapter is the following:

Problem: Investigate the structure of holomorphic and singular solutions of (5.0.1).

The most typical model of equation (5.0.1) is the following ordinary differential equation:

$$(5.0.2) \qquad t\frac{du}{dt} = f(t,u) \quad \text{with } f(0,0) = 0$$

which was first studied by Briot–Bouquet [5].

Nowadays it is called the *Briot–Bouquet equation* and the structure of solutions of (5.0.2) near the origin is well-known (see chapter 4 and also [13], [26], [28], [33], ...).

Let us recall some results when

$$\rho = \frac{\partial f}{\partial u}(0,0)$$

is in a generic position.

Theorem 5.0.1 *Assume that $f(t,u)$ is a holomorphic function defined near the origin of $\mathbb{C}_t \times \mathbb{C}_u$. Then we have*

(I) (Holomorphic solutions) *If $\rho \notin \mathbb{N}^*$ the equation (5.0.2) has a unique solution $u_0(t)$ holomorphic near the origin of $\mathbb{C}_t$ satisfying $u_0(0) = 0$.*

(II) (Singular solutions) *If $\rho \notin \mathbb{N}^* \cup \{a \in \mathbb{R} : a \leq 0\}$ the general solution $u(t)$ of (5.0.2) near the origin is given by*

$$u(t) = ct^\rho + a_{1,0}t + \sum_{i+j\geq 2} a_{i,j}t^i (ct^\rho)^j \tag{5.0.3}$$

where $c \in \mathbb{C}$ is arbitrary, the coefficients $a_{i,j} \in \mathbb{C}$ are uniquely determined by the equation (5.0.2), and the series

$$w + a_{1,0}t + \sum_{i+j\geq 2} a_{i,j}t^i w^j$$

is a convergent power series in (t, w). If we take $c = 0$ then this solution is $u_0(t)$.

Now the questions are

1. Is it possible to generalize these results (I) and (II) to the equation (5.0.1)?
2. What kinds of series appear in the expansion of singular solutions of (5.0.1)?

In this chapter, we are giving the answers to these two questions.

5.1 Notations and definitions

Let us denote by

- $\mathfrak{R}(\mathbb{C} \setminus \{0\})$ the universal covering space of $\mathbb{C} \setminus \{0\}$,
- S_θ the sector in $\mathfrak{R}(\mathbb{C} \setminus \{0\})$ defined by $\{t \in \mathfrak{R}(\mathbb{C} \setminus \{0\}) : |\arg t| < \theta\}$,
- $S(\varepsilon(s)) = \{t \in \mathfrak{R}(\mathbb{C} \setminus \{0\}) : 0 < |t| < \varepsilon(\arg t)\}$ for some positive-valued function $\varepsilon(s)$ defined and continuous on $\mathbb{R}$,
- $D(\delta) = \{x \in \mathbb{C}^n : |x_i| < \delta, i = 1, \ldots, n\}$,
- $\mathbb{C}\{x\}$ the ring of germs of holomorphic functions at the origin of $\mathbb{C}^n$.

Definition 5.1.1 *We define $\widetilde{\mathcal{O}}_+$ the set of all functions $u(t,x)$ satisfying the following conditions:*

1. *$u(t,x)$ is holomorphic in $S(\varepsilon(s)) \times D(\delta)$ for some $\varepsilon(s)$ and $\delta > 0$;*

2. *there is an* $a > 0$ *such that for any* $\theta > 0$ *and any compact subset* K *of* $D(\delta)$

$$\max_{x \in K} |u(t,x)| = O\left(|t|^a\right) \qquad \text{as } t \text{ tends to zero in } S_\theta.$$

Note that $\mu(s) = s^a$ with $a > 0$ is a particular form of a function satisfying the following properties:

(μ1) $\mu(s) \in C^0([0;\infty[)$ (the set of continuous function on $[0;\infty[$),

(μ2) $\mu(s) > 0$ and $\mu(s)$ is increasing on $]0;\infty[$,

(μ3) $\int_0^1 \frac{\mu(s)}{s} ds < +\infty$.

Definition 5.1.2 *We define* $\widetilde{\mathcal{O}}_{\text{int}}$ *the set of all functions* $u(t,x)$ *satisfying the following conditions:*

1. $u(t,x)$ *is holomorphic in* $S(\varepsilon(s)) \times D(\delta)$ *for some* $\varepsilon(s)$ *and* $\delta > 0$,
2. *there is a function* $\mu(s)$ *satisfying* (μ1), (μ2), *and* (μ3) *such that for any* $\theta > 0$ *and any compact subset* K *of* $D(\delta)$

$$\max_{x \in K} |u(t,x)| = O\left(\mu\left(|t|\right)\right) \qquad \text{as } t \text{ tends to zero in } S_\theta.$$

We will employ the space $\widetilde{\mathcal{O}}_+$ or $\widetilde{\mathcal{O}}_{\text{int}}$ as a framework of singular solutions. Clearly we have $\widetilde{\mathcal{O}}_+ \subset \widetilde{\mathcal{O}}_{\text{int}}$.

5.2 Statement of the main theorem

Definition 5.2.1 *If (5.0.1) satisfies (A1), (A2), and (A3) we say that (5.0.1) is of* Briot–Bouquet type *with respect to* t.

Definition 5.2.2 *The holomorphic function*

$$\rho(x) = \frac{\partial F}{\partial u}(0,x,0,0)$$

is called the characteristic exponent *of (5.0.1).*

Theorem 5.2.3 (Main theorem) *If (5.0.1) is of Briot–Bouquet type and* $\rho(0) \notin \mathbb{N}^*$ *then we have:*

(I) (Holomorphic solutions) *The equation (5.0.1) has a unique solution* $u_0(t,x)$ *holomorphic near the origin of* $\mathbb{C} \times \mathbb{C}^n$ *satisfying* $u_0(0,x) \equiv 0$.

(II) (Singular solutions) *Denote by* S_+ *(resp.* S_{int}*) the set of all* $\widetilde{\mathcal{O}}_+$*-solutions (resp.* $\widetilde{\mathcal{O}}_{\text{int}}$*-solutions) of (5.0.1). Then*

$$S_{\text{int}} = S_+ = \begin{cases} \{u_0(t,x)\} & \textit{when } \operatorname{Re}\rho(0) \leq 0, \\ \{u_0(t,x)\} \cup \{U(\varphi) : 0 \neq \varphi(x) \in \mathbb{C}\{x\}\} & \textit{when } \operatorname{Re}\rho(0) > 0 \end{cases}$$

where $U(\varphi)$ *is an* $\widetilde{\mathcal{O}}_+$*-solution of (5.0.1) having an expansion of the following form:*

$$U(\varphi) = \sum_{i\geq 1} u_i(x)t^i + \sum_{i+2j\geq k+2, j\geq 1} \varphi_{i,j,k}(x)t^{i+j\rho(x)}(\log t)^k, \qquad \varphi_{0,1,0}(x) = \varphi(x).$$

Note that this theorem is consistent with the one for equation (5.0.2). To see this we have only to recall that (5.0.2) is transformed into the equation

$$t\frac{dw}{dt} = \rho w$$

under the transformation

$$u = w + a_{1,0}t + \sum_{i+j\geq 2} a_{i,j}t^i w^j$$

and therefore in (5.0.3) the condition $u(t) \in \widetilde{\mathcal{O}}_+$ is equivalent to the condition $ct^\rho \in \widetilde{\mathcal{O}}_+$ (see chapter 4 and also [28]). Thus, our equation (5.0.1) is quite similar to the Briot-Bouquet equation (5.0.2) not only in the form of the equation but also in the structure of the solutions.

5.3 Holomorphic solutions

In this section, we will study holomorphic solutions of non linear equations of the form

$$\text{(5.3.1)} \qquad \left(t\frac{\partial}{\partial t} - \rho(x)\right)u = ta(x) + G_2(x)\left(t, t\frac{\partial u}{\partial t}, u, \frac{\partial u}{\partial x_1}, \ldots, \frac{\partial u}{\partial x_n}\right)$$

where $\rho(x)$ and $a(x)$ are holomorphic functions defined in a polydisk D centered at the origin of $\mathbb{C}^n$ and

$$G_2(x)(t, z, X_0, X_1, \ldots, X_n) = \sum_{p+q+|\alpha|\geq 2} a_{p,q,\alpha}(x)t^p z^q (X_0)^{\alpha_0}(X_1)^{\alpha_1}\cdots(X_n)^{\alpha_n}$$

in which the coefficients $\{a_{p,q,\alpha}(x)\}_{p+q+|\alpha|\geq 2}$ are holomorphic in D and

$$|a_{p,q,\alpha}(x)| \leq A_{p,q,\alpha}.$$

Moreover, the power series

$$|G_2|(t,z,X) = \sum_{p+q+|\alpha|\geq 2} A_{p,q,\alpha} t^p z^q X^\alpha$$

is convergent near the origin of $\mathbb{C}_t \times \mathbb{C}_z \times (\mathbb{C}_X)^{n+1}$.

Definition 5.3.1 *By a* holomorphic solution *of (5.3.1) we mean a solution holomorphic near the origin of* $\mathbb{C}_t \times (\mathbb{C}_x)^n$ *satisfying* $u(0,x) \equiv 0$.

Definition 5.3.2 *By a* formal solution *of (5.3.1) we mean a formal power series solution of the form*

$$\sum_{m\geq 1} u_m(x) t^m$$

the coefficients $\{u_m(x)\}_{m\geq 1}$ *of which are holomorphic in a common polydisk centered at the origin of* $(\mathbb{C}_x)^n$.

We have

Theorem 5.3.3 *1. Each formal solution of (5.3.1) is convergent.*

2. If $\rho(0) \notin \mathbb{N}^*$ *(5.3.1) has a unique formal solution which gives a unique holomorphic solution.*

Note that the assertion (I) of the main theorem 5.2.3 follows easily from 2. of theorem 5.3.3.

Proof : The assumption $\rho(0) \notin \mathbb{N}^*$ implies easily that (5.3.1) has a unique formal solution of the form

$$\sum_{m\geq 1} u_m(x) t^m.$$

Moreover, $u_m(x)$ is determined by the following recursive formula:

$$u_1(x) = \frac{a(x)}{1-\rho(x)}$$

and for $m \geq 2$

$$(5.3.2)\quad u_m(x) = \frac{1}{m-\rho(x)} f_m\Big(u_1(x), 2u_2(x), \ldots, (m-1)u_{m-1}(x), u_1(x), u_2(x), \ldots, u_{m-1}(x), \frac{\partial u_1}{\partial x_1}, \ldots, \frac{\partial u_1}{\partial x_n}, \ldots, \frac{\partial u_{m-1}}{\partial x_1}, \ldots, \frac{\partial u_{m-1}}{\partial x_n}, \{a_{p,q,\alpha}(x)\}_{p+q+|\alpha|\leq m}\Big).$$

Hence, if $\rho(0) \notin \mathbb{N}^*$ this formula shows the uniqueness of a formal solution and if we have convergence the uniqueness of a holomorphic solution.

To prove the assertion 2. of theorem 5.3.3 we have only to prove the convergence of this formal solution.

From now we write

$$D_a = \{x \in \mathbb{C}^n : |x_i| \leq a, i = 1, \dots, n\}.$$

We will make use of

Lemma 5.3.4 *If a holomorphic function $u(x)$ in D_R satisfies*

$$|u(x)| \leq \frac{C}{(R-r)^p} \quad \text{on } D_r \text{ for } 0 < r < R$$

then we have

$$\left|\frac{\partial u}{\partial x_i}(x)\right| \leq \frac{Ce(p+1)}{(R-r)^{p+1}} \quad \text{on } D_r \text{ for } 0 < r < R, i = 1, \dots, n$$

(where e is the real number such that $\log e = 1$).

For the proof, see Hörmander ([27] lemma 5.1.3).

Proof of the convergence of the formal solution

$$\sum_{m \geq 1} u_m(x) t^m. \tag{5.3.3}$$

First we assume $\rho(0) \notin \mathbb{N}^*$. Then by taking R sufficiently small we may assume

1. $0 < R < 1$;
2. all the $u_m(x)$ are holomorphic in D_R;
3. we have in D_R

$$\begin{aligned}
|u_1(x)| &\leq A, \\
\left|\frac{\partial u_1}{\partial x_i}(x)\right| &\leq A \qquad \text{for } i = 1, \dots, n, \\
|a_{p,q,\alpha}(x)| &\leq A_{p,q,\alpha} \qquad \text{for } p + q + |\alpha| \geq 2, \\
|m - \rho(x)| &\geq \sigma m \qquad \text{for } m = 1, 2, 3, \dots.
\end{aligned}$$

Consider now the following analytic equation:

$$\sigma Y = \sigma A t + \frac{1}{R-r} \sum_{p+q+|\alpha| \geq 2} \frac{A_{p,q,\alpha}}{(R-r)^{p+q+|\alpha|-2}} t^p Y^q Y^{\alpha_0} (eY)^{\alpha_1} \cdots (eY)^{\alpha_n}.$$

By the implicit function theorem this equation has a unique holomorphic solution of the form

(5.3.4)
$$Y = \sum_{m \geq 1} Y_m(r) t^m$$

and $Y_m(r)$ is determined by the following recursive formula:

$$Y_1 = A$$

and for $m \geq 2$

(5.3.5)
$$\sigma Y_m = \frac{1}{R-r} F_m \left(Y_1, \ldots, Y_{m-1}, eY_1, \ldots, eY_{m-1}, \left\{ \frac{A_{p,q,\alpha}}{(R-r)^{p+q+|\alpha|-2}} \right\}_{p+q+|\alpha| \leq m} \right).$$

Moreover, by induction on m we can see that $Y_m(r)$ is expressed in the form

(5.3.6)
$$Y_m(r) = \frac{C_m}{(R-r)^{m-1}}, \quad m = 1, 2, \ldots$$

with constants $C_1 = A$ and $C_m \geq 0$ for $m \geq 2$.

Let us show that this power series (5.3.4) is a majorant power series for the formal solution (5.3.3). To do so it is sufficient to prove the following inequalities for all m:

(5.3.7) $|u_m(x)| \leq |m u_m(x)| \leq Y_m(r)$ on D_r for $0 < r < R$,

(5.3.8) $\left| \frac{\partial u_m}{\partial x_i}(x) \right| \leq eY_m(r)$ on D_r for $0 < r < R, i = 1, \ldots, n$.

Since $Y_1 = A$ the case $m = 1$ is clear from the definition of A. We will prove the general case by induction on m.

Let $m \geq 2$. Assume that (5.3.7) and (5.3.8) with m replaced by p are already known for all $p < m$. Then from (5.3.2) and (5.3.5) we have

(5.3.9)
$$\begin{aligned} |u_m(x)| \leq \; & \frac{1}{|m-\rho(x)|} f_m \Bigg(|u_1|, |2u_2|, \ldots, |(m-1)u_{m-1}|, |u_1|, |u_2|, \ldots, |u_{m-1}|, \\ & \left| \frac{\partial u_1}{\partial x_1} \right|, \ldots, \left| \frac{\partial u_1}{\partial x_n} \right|, \ldots, \left| \frac{\partial u_{m-1}}{\partial x_1} \right|, \ldots, \left| \frac{\partial u_{m-1}}{\partial x_n} \right|, \\ & \{|a_{p,q,\alpha}|\}_{p+q+|\alpha| \leq m} \Bigg) \\ \leq \; & \frac{1}{\sigma m} f_m \Bigg(Y_1, \ldots, Y_{m-1}, eY_1, \ldots, eY_1, \ldots, eY_{m-1}, \ldots, eY_{m-1}, \\ & \left\{ \frac{A_{p,q,\alpha}}{(R-r)^{p+q+|\alpha|-2}} \right\}_{p+q+|\alpha| \leq m} \Bigg) \end{aligned}$$

$$= \frac{1}{\sigma m} F_m \Bigg(Y_1, \dots, Y_{m-1}, eY_1, \dots, eY_{m-1}, \left\{ \frac{A_{p,q,\alpha}}{(R-r)^{p+q+|\alpha|-2}} \right\}_{p+q+|\alpha| \le m} \Bigg)$$

$$= \frac{1}{m}(R-r)Y_m(r)$$

which yields (5.3.7). Since $Y_m(r)$ has the form (5.3.6) the above inequality (5.3.9) is written as

$$|u_m(x)| \le \frac{C_m}{m(R-r)^{m-2}} \quad \text{on } D_r \text{ for } 0 < r < R$$

and, therefore, by lemma 5.3.4 we obtain

$$\left| \frac{\partial u_m}{\partial x_i}(x) \right| \le \frac{(m-1)eC_m}{m(R-r)^{m-1}} \le eY_m(r) \quad \text{on } D_r \text{ for } 0 < r < R, i = 1, \dots, n$$

which implies (5.3.8).

Thus, by summing up we have obtained the convergence of the formal solution (5.3.3) under the assumption $\rho(0) \notin \mathbb{N}^*$.

Now assume that $\rho(0) = k \in \mathbb{N}^*$. Then the equation for $u_k(x)$ takes the form

(5.3.10)

$$(k - \rho(x))u_k(x) = f_k \Bigg(u_1(x), 2u_2(x), \dots, (k-1)u_{k-1}(x), u_1(x), \dots, u_{k-1}(x), \frac{\partial u_1}{\partial x_1}, \dots, \frac{\partial u_1}{\partial x_n}, \dots, \frac{\partial u_{k-1}}{\partial x_1}, \dots, \frac{\partial u_{k-1}}{\partial x_n}, \{a_{p,q,\alpha}\}_{p+q+|\alpha| \le k} \Bigg).$$

When $\rho(x) \not\equiv k$ and if (5.3.1) has a formal solution (5.3.10) implies that $u_k(x)$ is determined uniquely. When $\rho(x) \equiv k$ and if (5.3.1) has a formal solution we must have from (5.3.10)

$$0 \equiv f_k \Bigg(u_1(x), 2u_2(x), \dots, (k-1)u_{k-1}(x), u_1(x), \dots, u_{k-1}(x), \frac{\partial u_1}{\partial x_1}, \dots, \frac{\partial u_1}{\partial x_n}, \dots, \frac{\partial u_{k-1}}{\partial x_1}, \dots, \frac{\partial u_{k-1}}{\partial x_n}, \{a_{p,q,\alpha}\}_{p+q+|\alpha| \le k} \Bigg)$$

on D_R which implies that $u_k(x)$ can be taken arbitrarily. Moreover, if $u_k(x)$ satisfies (5.3.7) and (5.3.8) for $m = k$ the proof given before can be applied and the formal solution is convergent. Hence, to have the convergence of the formal solution we have only to notice the following fact: for a given $u_k(x)$ we can modify $\{A_{p,q,\alpha}\}_{p+q+|\alpha| \le k}$ so that (5.3.7) and (5.3.8) for $m = k$ are satisfied. Thus, the proof of theorem 5.3.3 is completed. ∎

Denote by $\mathbb{C}\{x\}[[t]]_0$ the ring of formal power series of the form

$$\sum_{m\geq 1} u_m(x)t^m$$

the coefficients $u_m(x)$ of which are holomorphic in a common disk centered at the origin of $(\mathbb{C}_x)^n$ and by $\mathbb{C}\{x\}\{t\}_0$ the subring of convergent series.

Definition 5.3.5 *An operator*

$$D : \mathbb{C}\{x\}[[t]]_0 \longrightarrow \mathbb{C}\{x\}[[t]]_0$$

is called regular singular *if and only if $u \in \mathbb{C}\{x\}[[t]]_0$ and $Du \in \mathbb{C}\{x\}\{t\}_0$ imply $u \in \mathbb{C}\{x\}\{t\}_0$.*

Then we have

Corollary 5.3.6 *The operator*

$$D : u \longrightarrow Du = \left(t\frac{\partial}{\partial t} - \rho(x)\right) u - ta(x) - G_2(x)\left(t, t\frac{\partial u}{\partial t}, u, \frac{\partial u}{\partial x_1}, \ldots, \frac{\partial u}{\partial x_n}\right)$$

is regular singular.

Remark 5.3.7 *Theorem 5.3.3 is not true for a partial differential equation containing higher derivatives with respect to x. For example let us consider*

$$\left(t\frac{\partial}{\partial t} - \rho\right) u = \frac{t}{1-x} + t\frac{\partial^2 u}{\partial x^2} \qquad (5.3.11)$$

where ρ is a constant and $\rho \notin \mathbb{N}^*$. The equation (5.3.11) has the following formal solution

$$u(t,x) = \sum_{m\geq 1} \frac{(2m-2)!}{(1-\rho)\cdots(m-\rho)} \frac{t^m}{(1-x)^{2m-1}}$$

which clearly diverges.

Remarks 5.3.8

1. *When $\rho(0) = k \in \mathbb{N}^*$ and $\rho(x) \not\equiv k$ hold and if (5.3.1) has a formal solution then theorem 5.3.3 gives a unique holomorphic solution.*
2. *When $\rho(x) \equiv k \in \mathbb{N}^*$ holds near the origin of $(\mathbb{C}_x)^n$ and if (5.3.1) has a formal solution then theorem 5.3.3 gives a family of holomorphic solutions depending on an arbitrary function $u_k(x)$.*

For some other results on the existence of holomorphic solutions for non linear partial differential equations see chapters 2 and 3 and also [10], [12].

5.4 Singular solutions

In this section, we will construct a family of singular solutions in $\widetilde{\mathcal{O}}_+$ of non linear partial differential equations of the form

$$(5.4.1)\qquad \left(t\frac{\partial}{\partial t}-\rho(x)\right)u = ta(x)+G_2(x)\left(t,u,\frac{\partial u}{\partial x_1},\dots,\frac{\partial u}{\partial x_n}\right)$$

where $\rho(x)$ and $a(x)$ are holomorphic functions defined in a polydisk D centered at the origin of $(\mathbb{C}_x)^n$ and

(5.4.2)

$$G_2(x)\,(t,X_0,X_1,\dots,X_n)=\sum_{p+|\alpha|\geq 2} a_{p,\alpha}(x)t^p\,(X_0)^{\alpha_0}\,(X_1)^{\alpha_1}\dots(X_n)^{\alpha_n}$$

in which the coefficients $\{a_{p,\alpha}(x)\}_{p+|\alpha|\geq 2}$ are holomorphic in D and

$$|a_{p,\alpha}|\leq A_{p,\alpha}.$$

Moreover, the power series

$$|G_2|\,(t,X)=\sum_{p+|\alpha|\geq 2} A_{p,\alpha}t^pX^\alpha$$

is convergent near the origin of $\mathbb{C}_t\times(\mathbb{C}_X)^{n+1}$.

From theorem 5.3.3 we know the following: if $\rho(0)\notin\mathbb{N}^*$ the equation (5.4.1) has a unique holomorphic solution $u(t,x)$ satisfying $u(0,x)\equiv 0$.

Now let us find singular solutions of (5.4.1) of the form

$$(5.4.3)\qquad u(t,x)=w\left(t,t^{\rho(x)},x,\log t\right)$$

with

$$w\,(t_1,t_2,x,y)=\sum_{m_1+m_2\geq 1} w_{m_1,m_2}(x,y)\,(t_1)^{m_1}\,(t_2)^{m_2}\,.$$

To do so we put t_1, t_2, and y as follows:

$$(5.4.4)\qquad t_1=t,\qquad t_2=t^{\rho(x)},\qquad\text{and}\qquad y=\log t.$$

Then under the relations (5.4.3) and (5.4.4) we have

$$\begin{aligned} t\frac{\partial u}{\partial t} &= t_1\frac{\partial w}{\partial t_1}+\rho(x)t_2\frac{\partial w}{\partial t_2}+\frac{\partial w}{\partial y},\\ \frac{\partial u}{\partial x_i} &= \frac{\partial w}{\partial x_i}+\frac{\partial\rho(x)}{\partial x_i}yt_2\frac{\partial w}{\partial t_2}\quad\text{for } i=1,\dots,n. \end{aligned}$$

Therefore, in order that $u(t,x)$ given by (5.4.3) is a solution of (5.4.1) the function $w\,(t_1,t_2,x,y)$ must be a solution of

(5.4.5) $$\left(t_1\frac{\partial}{\partial t_1}+\rho(x)t_2\frac{\partial}{\partial t_2}-\rho(x)+\frac{\partial}{\partial y}\right)w$$
$$=t_1a(x)+G_2(x)\left(t_1,w,\frac{\partial w}{\partial x_1}+\frac{\partial \rho(x)}{\partial x_1}yt_2\frac{\partial w}{\partial t_2},\dots,\frac{\partial w}{\partial x_n}+\frac{\partial \rho(x)}{\partial x_n}yt_2\frac{\partial w}{\partial t_2}\right).$$

Proposition 5.4.1 *1. If the condition*

(5.4.6)
$$m_1+\rho(x)\,(m_2-1)\neq 0 \quad \text{on } D \text{ for any } (m_1,m_2)\in\mathbb{N}^2\setminus\{(0,0),(0,1)\}$$

is satisfied (5.4.5) has a formal solution of the form

$$\sum_{m\geq 1}u_m(x)\,(t_1)^m+\sum_{m_1+2m_2\geq k+2,m_2\geq 1}\varphi_{m_1,m_2,k}(x)\,(t_1)^{m_1}\,(t_2)^{m_2}\,y^k$$

where $\varphi_{0,1,0}(x)$ is arbitrary and all the other coefficients $u_m(x),\varphi_{m_1,m_2,k}(x)$ are uniquely determined by $\varphi_{0,1,0}(x)$. If $\varphi_{0,1,0}(x)$ is holomorphic in D_r then all the coefficients $u_m(x),\varphi_{m_1,m_2,k}(x)$ are also holomorphic in D_r.

2. If $\rho(0)\notin\mathbb{N}^*$ and $\operatorname{Re}\rho(0)>0$ then the formal solution given in 1. is convergent in a region of the form

$$\{(t_1,t_2,x,y)\in\mathbb{C}\times\mathbb{C}\times\mathbb{C}^n\times\mathbb{C}:$$
$$|t_1|<\varepsilon,|t_2|<\varepsilon,|yt_1|<\varepsilon,\left|y^2t_2\right|<\varepsilon,\ \text{and}\ |x_i|<r\quad(i=1,\dots,n)\}$$

for some $\varepsilon>0$ and $r>0$ (depending on $\varphi_{0,1,0}(x)$).

Before the proof let us present some preparatory discussions. First we note that under the relation

$$w\,(t_1,t_2,x,y)=\sum_{m_1+m_2\geq 1}w_{m_1,m_2}(x,y)\,(t_1)^{m_1}\,(t_2)^{m_2}$$

the equation (5.4.5) is equivalent to a recursive family of equations as follows:

(5.4.7) $$\left(1-\rho(x)+\frac{\partial}{\partial y}\right)w_{1,0}\ =\ a(x),$$

(5.4.8) $$\frac{\partial}{\partial y}w_{0,1}\ =\ 0$$

and for $m_1+m_2\geq 2$

(5.4.9) $$\left(m_1+\rho(x)\,(m_2-1)+\frac{\partial}{\partial y}\right)w_{m_1,m_2}$$
$$=f_{m_1,m_2}\Bigg(\{w_{i,j}\}_{i\leq m_1,j\leq m_2,i+j\leq m_1+m_2-1},$$
$$\left\{\frac{\partial w_{i,j}}{\partial x_1}+\frac{\partial\rho}{\partial x_1}yjw_{i,j}\right\}_{i\leq m_1,j\leq m_2,i+j\leq m_1+m_2-1},\dots,$$
$$\left\{\frac{\partial w_{i,j}}{\partial x_n}+\frac{\partial\rho}{\partial x_n}yjw_{i,j}\right\}_{i\leq m_1,j\leq m_2,i+j\leq m_1+m_2-1},\{a_{p,\alpha}\}_{p+\alpha\leq m_1+m_2}\Bigg).$$

Next let us consider the operator

$$L_{m_1,m_2} = m_1 + \rho(x)\,(m_2 - 1) + \frac{\partial}{\partial y}.$$

Denote by $\mathbb{C}_r\{x\}$ the ring of all holomorphic functions in D_r and by $P_{m,r}(x)[y]$ the set of all polynomials $\sum_{0 \le i \le m} \varphi_i(x) y^i$ of degree at most m in y with coefficients belonging to $\mathbb{C}_r\{x\}$. Then if the condition (5.4.6) is satisfied we can easily see the following properties:

1. $L_{0,1}\left(P_{0,r}(x)[y]\right) = 0$;
2. for any $(m_1, m_2) \in \mathbb{N}^2 \setminus \{(0,0), (0,1)\}$ and $m \in \mathbb{N}$ the operator

$$L_{m_1,m_2} : P_{m,r}(x)[y] \longrightarrow P_{m,r}(x)[y]$$

is invertible.

In order to estimate the operator norm of the inverse of L_{m_1,m_2} for $\varphi(x) \in \mathbb{C}_r\{x\}$ we define the norm $\|\varphi\|_r$ by

$$\|\varphi\|_r = \max_{x \in D_r} |\varphi(x)| \tag{5.4.10}$$

and for $w(x,y) = \sum_{0 \le i \le m} \varphi_i(x) y^i \in P_{m,r}(x)[y]$ we define the norm $\|w\|_{r,\lambda}$ by

$$\|w\|_{r,\lambda} = \sum_{0 \le i \le m} \|\varphi_i\|_r \lambda^i.$$

Then we have

Lemma 5.4.2 *If the condition*

$$|m_1 + \rho(x)\,(m_2 - 1)| \ge \sigma\,(m_1 + m_2) \quad \text{on } D_r \text{ for any } (m_1, m_2) \in \mathbb{N}^2 \setminus \{(0,0), (0,1)\}$$

is satisfied for some $\sigma > 0$ we have the following properties for any $(m_1, m_2) \in \mathbb{N}^2 \setminus \{(0,0), (0,1)\}$ and $m \in \mathbb{N}$:

1. *$L_{m_1,m_2} : P_{m,r}(x)[y] \longrightarrow P_{m,r}(x)[y]$ is an invertible operator;*
2. *for any $w(x,y) \in P_{m,r}(x)[y]$ we have*

$$\frac{\sigma}{2}\,(m_1 + m_2)\,\|w\|_{r,\lambda} \le \|L_{m_1,m_2}(w)\|_{r,\lambda} \quad \text{if } \lambda \ge \frac{2m}{\sigma\,(m_1 + m_2)}.$$

Proof : The property 1. is clear. Put $w(x,y) = \sum_{0\le i\le m} \varphi_i(x)y^i \in P_{m,r}(x)[y]$. Then 2. is verified by the following:

$$\begin{aligned}\|L_{m_1,m_2}(w)\|_{r,\lambda} &\geq \sum_{0\le i\le m} \|(m_1+\rho\,(m_2-1))\,\varphi_i\|_r\,\lambda^i - \sum_{0\le i\le m} i\,\|\varphi_i\|_r\,\lambda^{i-1} \\ &\geq \sigma\,(m_1+m_2)\sum_{0\le i\le m}\|\varphi_i\|_r\,\lambda^i - \frac{m}{\lambda}\sum_{0\le i\le m}\|\varphi_i\|_r\,\lambda^i \\ &= \left(\sigma\,(m_1+m_2)-\frac{m}{\lambda}\right)\|w\|_{r,\lambda}\,.\end{aligned}$$

■

This lemma guarantees that the same argument as in section 5.3 can also be applied to (5.4.5).

Proof of proposition 5.4.1: The existence of a formal solution of the form

$$\sum_{m_1+m_2\ge 1} w_{m_1,m_2}(x,y)\,(t_1)^{m_1}\,(t_2)^{m_2} \tag{5.4.11}$$

with

$$\begin{cases} w_{m,0}\in P_{0,r}(x)[y]=\mathbb{C}_r\{x\} & \text{for } m\ge 1,\\ w_{m_1,m_2}\in P_{m_1+2m_2-2,r}(x)[y] & \text{for } m_1+m_2\ge 1 \text{ and } m_2\ge 1\end{cases} \tag{5.4.12}$$

is a consequence of (5.4.7), (5.4.8), (5.4.9), and 1. of lemma 5.4.2.

Let us show the convergence of the formal solution (5.4.11) with (5.4.12). The assumptions $\rho(0)\notin\mathbb{N}^*$ and $\operatorname{Re}\rho(0)>0$ imply that

$$|m_1+\rho(x)\,(m_2-1)|\geq\sigma\,(m_1+m_2) \quad \text{on } D_R \text{ for any } (m_1,m_2)\in\mathbb{N}^2\backslash\{(0,0),(0,1)\}$$

is satisfied for some $\sigma>0$ and $R>0$. Choosing $R>0$ sufficiently small we may assume

1. $0<R<1$;
2. $w_{m,0}\in\mathbb{C}_R\{x\}$ for $m\ge 1$;
3. $w_{m_1,m_2}(x,y)\in P_{m_1+2m_2-2,R}(x)[y]$ for $m_1+m_2\ge 1$ and $m_2\ge 1$;
4. we have the following estimates:

$$\begin{aligned} &\|w_{1,0}\|_R\leq A_1, && \left\|\frac{\partial w_{1,0}}{\partial x_i}\right\|_R\leq A_1 && \text{for } i=1,\dots,n,\\ &\|w_{0,1}\|_R\leq A_2, && \left\|\frac{\partial w_{0,1}}{\partial x_i}\right\|_R\leq A_2 && \text{for } i=1,\dots,n,\\ & && \left\|\frac{\partial \rho}{\partial x_i}\right\|_R\leq c && \text{for } i=1,\dots,n,\\ & && \|a_{p,\alpha}\|_R\leq A_{p,\alpha} && \text{for } p+|\alpha|\ge 2.\end{aligned}$$

Consider now the following analytic equation:

$$\frac{\sigma}{2}Y = \frac{\sigma}{2}(A_1 t_1 + A_2 t_2) + \\ + \frac{1}{R-r}\sum_{p+|\alpha|\geq 2}\frac{A_{p,\alpha}}{(R-r)^{p+|\alpha|-2}}(t_1)^p\, Y^{\alpha_0}(eY + c\lambda Y)^{\alpha_1}\cdots(eY+c\lambda Y)^{\alpha_n}.$$

The implicit function theorem tells us that this equation has a unique holomorphic solution of the form

$$Y = \sum_{m_1+m_2\geq 1} Y_{m_1,m_2}(r,\lambda)\,(t_1)^{m_1}\,(t_2)^{m_2}. \tag{5.4.13}$$

Moreover, it is easy to see that the coefficients $Y_{m_1,m_2}(r,\lambda)$ has the form

$$Y_{m_1,m_2}(r,\lambda) = \frac{C_{m_1,m_2}(\lambda)}{(R-r)^{m_1+m_2-1}}$$

with $C_{1,0}(\lambda) = A_1, C_{0,1}(\lambda) = A_2$ and $C_{m_1,m_2}(\lambda) \geq 0$ (for $m_1 + m_2 \geq 2$).

Hence, by induction on (m_1, m_2) we can see in the same way as in section 5.3 that the following inequalities hold for any $(m_1, m_2) \in \mathbb{N}^2 \setminus \{(0,0)\}, 0 < r < R$, and $\lambda > \frac{4}{\sigma}$:

$$\begin{aligned} \|w_{m_1,m_2}\|_{r,\lambda} &\leq Y_{m_1,m_2}(r,\lambda), \\ \left\|\frac{\partial w_{m_1,m_2}}{\partial x_i}\right\|_{r,\lambda} &\leq eY_{m_1,m_2}(r,\lambda) \qquad \text{for } i = 1,\ldots,n, \\ \left\|\frac{\partial \rho}{\partial x_i} y m_2 w_{m_1,m_2}\right\|_{r,\lambda} &\leq c\lambda Y_{m_1,m_2}(r,\lambda) \quad \text{for } i = 1,\ldots,n. \end{aligned}$$

This implies that the power series Y in (5.4.13) is a majorant series of the formal solution (5.4.11).

If we fix r and λ as above by the convergence of Y we have

$$\|w_{m_1,m_2}\|_{r,\lambda} \leq \frac{M}{\varepsilon^{m_1+m_2}} \quad \text{for } (m_1,m_2) \in \mathbb{N}^2 \setminus \{(0,0)\}$$

for some $M > 0$ and $\varepsilon > 0$. Therefore, if $|x| \leq r$ and $\lambda \geq 1$ by denoting $\eta(y) = \max\{1, |y|\}$ we have

$$\begin{aligned} &\sum_{m_1+m_2\geq 1} |w_{m_1,m_2}(x,y)|\,|t_1|^{m_1}\,|t_2|^{m_2} \\ &\qquad \leq \sum_{m_1+m_2\geq 1} \|w_{m_1,m_2}\|_{r,\lambda}\,(\eta(y))^{m_1+2m_2}\,|t_1|^{m_1}\,|t_2|^{m_2} \\ &\qquad \leq \sum_{m_1+m_2\geq 1} \frac{M}{\varepsilon^{m_1+m_2}}\,(\eta(y)\,|t_1|)^{m_1}\,\left(\eta(y)^2\,|t_2|\right)^{m_2} \end{aligned}$$

which converges for any (t_1, t_2, y) satisfying $\eta(y)\,|t_1| < \varepsilon$ and $\eta(y)^2\,|t_2| < \varepsilon$. This proves 2. of proposition 5.4.1. ∎

Summing up we have

Theorem 5.4.3 *If $\rho(0) \notin \mathbb{N}^*$ and $\operatorname{Re}\rho(0) > 0$ the equation (5.4.1) has a family of $\widetilde{\mathcal{O}}_+$-solutions of the form*

$$\sum_{i\geq 1} u_i(x)t^i + \sum_{i+2j\geq k+2,\, j\geq 1} \varphi_{i,j,k}(x)t^{i+j\rho(x)}(\log t)^k \tag{5.4.14}$$

where $\varphi_{0,1,0}(x) \in \mathbb{C}\{x\}$ can be taken arbitrarily and all the coefficients $u_i(x), \varphi_{i,j,k}(x)$ are holomorphic in a common disk centered at the origin of $(\mathbb{C}_x)^n$. Moreover, the solution (5.4.14) is uniquely determined by $\varphi_{0,1,0}(x) \in \mathbb{C}\{x\}$. If we take $\varphi_{0,1,0}(x) = 0$ the above $\widetilde{\mathcal{O}}_+$-solution is reduced to the unique holomorphic solution of (5.4.1).

Proof : Let

$$\sum_{i\geq 1} u_i(x)t^i + \sum_{i+2j\geq k+2,\, j\geq 1} \varphi_{i,j,k}(x)\,(t_1)^i\,(t_2)^j\,y^k$$

be a holomorphic solution of (5.4.5) obtained in proposition 5.4.1. Then by the assumption $\operatorname{Re}\rho(0) > 0$ and 2. of proposition 5.4.1 we can easily see that the series

$$\sum_{i\geq 1} u_i(x)t^i + \sum_{i+2j\geq k+2,\, j\geq 1} \varphi_{i,j,k}(x)t^{i+j\rho(x)}(\log t)^k$$

is convergent in $\widetilde{\mathcal{O}}_+$ and therefore it gives an $\widetilde{\mathcal{O}}_+$-solution of (5.4.1).

Since the existence of the unique holomorphic solution of (5.4.1) is already known to complete the proof of theorem 5.4.3 it is sufficient to prove that the formal solution of (5.4.1) of the form (5.4.14) is uniquely determined by $\varphi_{0,1,0} \in \mathbb{C}\{x\}$.

Let $\hat{u}(t,x)$ be a formal solution of (5.4.1) of the form

$$\hat{u}(t,x) = \sum_{i\geq 1} u_i(x)t^i + \sum_{i+2j\geq k+2,\, j\geq 1} \varphi_{i,j,k}(x)t^{i+j\rho(x)}(\log t)^k \tag{5.4.15}$$

where $u_i(x), \varphi_{i,j,k}(x) \in \mathbb{C}_r\{x\}$.

When $\rho(x)$ is not a constant function in $\mathbb{C}\{x\}$ or when $\rho(x) \equiv$ constant $\notin \mathbb{Q}$ then any finite subset of

$$\left\{t^{i+j\rho(x)}(\log t)^k : i+j+k \geq 1\right\}$$

is linearly independent over the ring $\mathbb{C}_r\{x\}$ as a family of functions and therefore all the coefficients $u_i(x), \varphi_{i,j,k}(x) \in \mathbb{C}_r\{x\}$ of the formal solution $\hat{u}(t,x)$ are uniquely determined by $\varphi_{0,1,0}(x)$.

When $\rho(x) \equiv$ constant $\in \mathbb{Q}$ $\rho(x)$ is expressed as $\rho(x) = \frac{p}{q}$ for some $p, q \in \mathbb{N}^*$ relatively prime. Since $\rho(0) \notin \mathbb{N}^*$ is assumed we have $q \geq 2$. Then $\hat{u}(t,x)$ in (5.4.15) can be rewritten in the form

$$(5.4.16)\qquad \hat{u}(t,x) = \sum_{l\geq 1, 0\leq k\leq 2(l-1)} \Psi_{l,k}(x) t^{\frac{l}{q}} (\log t)^k$$

where $\Psi_{p,0}(x) = \varphi_{0,1,0}(x)$. Since $\hat{u}(t,x)$ is a formal solution of (5.4.1) with $\rho(x) \equiv \frac{p}{q}$ and since any finite subset of $\left\{t^{\frac{l}{q}}(\log t)^k : l+k \geq 1\right\}$ is linearly independent over the ring $\mathbb{C}_r\{x\}$ an easy calculation shows that logarithmic terms do not appear in the formal solution of the form (5.4.16) and therefore $\hat{u}(t,x)$ is expressed as

$$(5.4.17)\qquad \hat{u}(t,x) = \sum_{l\geq 1} \Psi_{l,0}(x) t^{\frac{l}{q}}.$$

Thus if we rewrite $\hat{u}(t,x)$ into the form (5.4.17) as above we can conclude that all the coefficients $\Psi_{l,0}(x)$ are uniquely determined by $\Psi_{p,0}(x) = \varphi_{0,1,0}(x)$.

Thus, the proof of theorem 5.4.3 is completed. ■

5.5 Uniqueness of the solution

In this section, we will discuss the uniqueness of the $\widetilde{\mathcal{O}}_{\mathrm{int}}$-solution of the equation (5.4.1).

As is proved in theorem 5.4.3 we already know the following uniqueness result: if $\rho(0) \notin \mathbb{N}^*$ and $\mathrm{Re}\,\rho(0) > 0$, if $u_1(t,x) \in \widetilde{\mathcal{O}}_+$ and $u_2(t,x) \in \widetilde{\mathcal{O}}_+$ are solutions of (5.4.1) having the expansion of the form

(5.5.1)
$$u_p(t,x) = \sum_{i\geq 1} u_i^{(p)}(x) t^i + \sum_{i+2j\geq k+2, j\geq 1} \varphi_{i,j,k}^{(p)}(x) t^{i+j\rho(x)} (\log t)^k \quad \text{for } p = 1,2$$

with $u_i^{(p)}(x), \varphi_{i,j,k}^{(p)}(x) \in \mathbb{C}_r\{x\}$ for some $r > 0$, and if $\varphi_{0,1,0}^{(1)}(x) = \varphi_{0,1,0}^{(2)}(x)$ holds in $\mathbb{C}_r\{x\}$ then we have $u_1(t,x) = u_2(t,x)$ in $\widetilde{\mathcal{O}}_+$.

The purpose of this section is to remove the assumption that $u_p(t,x)$ has an expansion of the form (5.5.1).

Theorem 5.5.1 *Let $u_1(t,x) \in \widetilde{\mathcal{O}}_{\mathrm{int}}$ and $u_2(t,x) \in \widetilde{\mathcal{O}}_{\mathrm{int}}$ be solutions of (5.4.1). Then we have:*

1. *For any $a < \mathrm{Re}\,\rho(0)$ we have $t^{-a}(u_1 - u_2) \in \widetilde{\mathcal{O}}_+$.*

2. *If $t^{-b}(u_1 - u_2) \in \widetilde{\mathcal{O}}_{\mathrm{int}}$ for some $b \geq \mathrm{Re}\,\rho(0)$ we have $u_1(t,x) = u_2(t,x)$ in $\widetilde{\mathcal{O}}_{\mathrm{int}}$.*

Proof : Let $u_1(t,x) \in \widetilde{\mathcal{O}}_{\mathrm{int}}$ and $u_2(t,x) \in \widetilde{\mathcal{O}}_{\mathrm{int}}$ be two solutions of (5.4.1). Set $w = (u_1 - u_2) \in \widetilde{\mathcal{O}}_{\mathrm{int}}$. Then $w(t,x)$ is a solution of

$$\left(t\frac{\partial}{\partial t}-\rho(x)\right)w=A_0(t,x)w+\sum_{1\le i\le n}A_i(t,x)\frac{\partial w}{\partial x_i} \tag{5.5.2}$$

with

$$A_i(t,x)=\int_0^1\frac{\partial G_2}{\partial X_i}(x)\left(t,sw+u_2,s\frac{\partial w}{\partial x_1}+\frac{\partial u_2}{\partial x_1},\dots,s\frac{\partial w}{\partial x_n}+\frac{\partial u_2}{\partial x_n}\right)ds$$

for $i=0,1,\dots,n$

where $G_2(x)\,(t,X_0,X_1,\dots,X_n)$ is as in (5.4.2). Since w and u_2 belong to $\widetilde{\mathcal{O}}_{\text{int}}$ from (5.4.2) we have $A_i(t,x)\in\widetilde{\mathcal{O}}_{\text{int}}$ for $i=0,1,\dots,n$. Therefore, by the definition of $\widetilde{\mathcal{O}}_{\text{int}}$ there are $\varepsilon(s),R>0$, and $\mu(s)$ satisfying ($\mu1$), ($\mu2$), and ($\mu3$) such that for $i=0,1,\dots,n$ we have

1. $A_i(t,x)$ is holomorphic in $S(\varepsilon(s))\times D_R$;
2. for any $\theta>0$ there are $\delta>0$ and $M>0$ such that
$$\max_{x\in D_R}|A_i(t,x)|\le M\mu(|t|)$$
holds for any $t\in S_\theta(\delta)=\{t\in S_\theta:0<|t|\le\delta\}$; in addition, by ($\mu3$) we may assume $\int_0^\delta\frac{\mu(s)}{s}ds<\frac{R}{3M}$.

Take any $\theta>0$ and fix it. Let $R>0,\mu(s),M>0$, and $\delta>0$ be as above. For $t_0\in S_\theta(\delta)$ we write

$$L(t_0,\delta)=\left\{t:t=st_0\text{ for }0<s\le\frac{\delta}{|t_0|}\right\}.$$

Lemma 5.5.2 *Let $0<R_2<R_1<R$ be such that $R_2+\frac{R}{3}<R_1$ and $R_1+\frac{R}{3}<R$. Then we have:*

1. *For any $(t_0,x^0)\in S_\theta(\delta)\times D_{R_1}$ the equation*
$$\begin{cases}\frac{dx(t)}{dt}=-\frac{1}{t}\left(A_1(t,x(t)),\dots,A_n(t,x(t))\right),\\ x(t_0)=x^0\end{cases} \tag{5.5.3}$$
has a unique holomorphic solution $x(t)$ defined near $L(t_0,\delta)$ such that $x(t)\in D_R$ holds for any $t\in L(t_0,\delta)$ and that $x(t)$ converges to some point in D_R as t tends to zero in $L(t_0,\delta)$.
2. *Denote by $x(t,t_0,x^0)$ the unique solution of (5.5.3) and put*
$$\Gamma(t_0,x^0)=\left\{\left(t,x(t,t_0,x^0)\right):t\in L(t_0,\delta)\right\}.$$
Then we have for any $t_0\in S_\theta(\delta)$
$$\bigcup_{x^0\in D_{R_1}}\Gamma(t_0,x^0)\supset L(t_0,\delta)\times D_{R_2}.$$

To see this we have only to notice the conditions $R_1 + \frac{R}{3} < R$, $R_2 + \frac{R}{3} < R_1$ and the following fact: for any $t \in L(t_0, \delta)$ we have

$$|x_i(t) - x_i^0| \leq \left| \int_C \frac{A_i(\tau, x(\tau))}{\tau} d\tau \right| \leq M \int_0^\delta \frac{\mu(s)}{s} ds \leq \frac{R}{3}$$

where C is the segment in the complex plane joining the two points t and t_0.

By using this lemma we can investigate the behavior of $w(t, x(t))$ as t tends to zero in $L(t_0, \delta)$. In fact, if we put

$$W(t) = w\left(t, x\left(t, t_0, x^0\right)\right) \quad \text{for } t \in L(t_0, \delta) \text{ and } \left(t_0, x^0\right) \in S_\theta(\delta) \times D_{R_1}$$

by (5.5.2) and (5.5.3) we have

$$t\frac{dW}{dt} - (\rho(x(t)) + A_0(t, x(t)))\, W = 0 \tag{5.5.4}$$

with $x(t) = x\left(t, t_0, x^0\right)$ and, therefore, by integrating (5.5.4) we obtain

$$W(t) = W(t_0) \exp\left[- \int_C \frac{\rho(x(\tau)) + A_0(\tau, x(\tau))}{\tau} d\tau \right]. \tag{5.5.5}$$

Now let us prove 1. of theorem 5.5.1. Take any $a < \operatorname{Re}\rho(0)$. We may assume that

$$a + \varepsilon < \operatorname{Re}\rho(x) \quad \text{on } D_R$$

holds for some $\varepsilon > 0$. Then by (5.5.5) we have for any $t \in L(t_0, \delta)$ satisfying $|t| \leq |t_0|$

$$\begin{aligned} \left|t^{-a-\varepsilon} W(t)\right| &\leq |t_0|^{-a-\varepsilon} |W(t_0)| \exp\left[\left| \int_C \frac{A_0(\tau, x(\tau))}{\tau} d\tau \right|\right] \times \\ &\times \qquad \exp\left[- \int_{\frac{|t|}{|t_0|}}^1 \frac{\operatorname{Re}\rho(x(st_0)) - (a+\varepsilon)}{s} ds \right] \\ &\leq |t_0|^{-a-\varepsilon} |W(t_0)| \exp\left[M \int_0^\delta \frac{\mu(s)}{s} ds \right]. \end{aligned}$$

Hence, by putting $t_0 = \delta e^{i\eta}$ for $|\eta| \leq \frac{\theta}{2}$ and, by using 2. of lemma 5.5.2, we obtain for any $(t, x) \in S_{\frac{\theta}{2}}(\delta) \times D_{R_2}$

$$\left|t^{-a-\varepsilon} w(t, x)\right| \leq \delta^{-a-\varepsilon} \exp\left[M \int_0^\delta \frac{\mu(s)}{s} ds \right] \times \max_{|\eta| \leq \frac{\theta}{2}, x \in D_{R_1}} \left|w\left(\delta e^{i\eta}, x\right)\right|.$$

This implies that $t^{-a}w \in \widetilde{\mathcal{O}}_+$ holds since $\theta > 0$ is taken arbitrarily. Thus, 1. of theorem 5.5.1 is proved.

Next let us prove 2. of theorem 5.5.1. Assume that $t^{-b}w = t^{-b}(u_1 - u_2) \in \widetilde{\mathcal{O}}_{\text{int}}$ holds for some $b \geq \operatorname{Re}\rho(0)$. Then by the definition of $\widetilde{\mathcal{O}}_{\text{int}}$ we see the following: if $R > 0$ is sufficiently small we have

$$\max_{x \in D_R} \left| t^{-b} w(t,x) \right| = o(1) \quad \text{as } t \text{ tends to zero in } S_\theta \tag{5.5.6}$$

for any $\theta > 0$.

When $b > \operatorname{Re}\rho(0)$ or when $b \equiv \operatorname{Re}\rho(x)$ we may assume that

$$b \geq \operatorname{Re}\rho(x) \quad \text{on } D_R.$$

Then by (5.5.5) and (5.5.6) we have

$$\begin{aligned} \left| w\left(t_0, x^0\right) \right| &\leq |t_0|^b \left| t^{-b} w\left(t, x\left(t, t_0, x^0\right)\right) \right| \exp\left[M \int_0^\delta \frac{\mu(s)}{s} ds \right] \times \\ &\times \quad \exp\left[\int_{\frac{|t|}{|t_0|}}^1 \frac{\operatorname{Re}\rho\left(x\left(st_0\right)\right) - b}{s} ds \right] \\ &= o(1) \quad \text{as } t \text{ tends to zero in } L\left(t_0, \delta\right) \end{aligned}$$

and, therefore, we obtain $w\left(t_0, x^0\right) = 0$. This implies that $w \equiv 0$ in $\widetilde{\mathcal{O}}_{\text{int}}$ since $\left(t_0, x^0\right) \in S_\theta(\delta) \times D_{R_1}$ is taken arbitrarily.

When $b = \operatorname{Re}\rho(0)$ and $b \not\equiv \operatorname{Re}\rho(x)$ hold we can find an $x^* \in \mathbb{C}^n$ sufficiently close to the origin such that $b > \operatorname{Re}\rho\left(x^*\right)$ holds. Then by the above considerations we have $w(t,x) = 0$ on $S_\theta(\delta) \times \{x : |x - x^*| \leq R\}$ for sufficiently small $R > 0$. Hence, we can obtain $w \equiv 0$ in $\widetilde{\mathcal{O}}_{\text{int}}$ since w is a holomorphic function. Thus, 2. of theorem 5.5.1 is proved. ∎

5.6 Proof of the main theorem 5.2.3

Assume that (5.0.1) is of Briot–Bouquet type with respect to t and let $\rho(x)$ be the characteristic exponent of (5.0.1). Then (5.0.1) can be rewritten into an equation of the form (5.4.1). Therefore, we already know the following results:

(C1) (by theorem 5.3.3) If $\rho(0) \notin \mathbb{N}^*$ the equation (5.0.1) has a unique holomorphic solution $u_0(t,x)$ satisfying $u_0(0,x) \equiv 0$.

(C2) (by theorem 5.4.3) If $\rho(0) \notin \mathbb{N}^*$ and $\operatorname{Re}\rho(0) > 0$, for any $\varphi(x) \in \mathbb{C}\{x\}$ the equation (5.0.1) has a unique $\widetilde{\mathcal{O}}_+$-solution $U(\varphi)(t,x)$ having an expansion of the form

$$U(\varphi) = \sum_{i\geq 1} u_i(x)t^i + \sum_{i+2j\geq k+2, j\geq 1} \varphi_{i,j,k}(x)t^{i+j\rho(x)}(\log t)^k \tag{5.6.1}$$

with $\varphi_{0,1,0}(x) = \varphi(x)$ where all the coefficients $u_i(x), \varphi_{i,j,k}(x)$ are holomorphic in a common disk centered at the origin of $(\mathbb{C}_x)^n$. If we take $\varphi = 0 \in \mathbb{C}\{x\}$ the solution $U(\varphi)(t,x)$ is reduced to the unique holomorphic solution u_0 in (C1).

(C3) (by calculation) If $\rho(0) \notin \mathbb{N}^*$ and $\operatorname{Re}\rho(0) > 0$, if $b > 0$ satisfies $\operatorname{Re}\rho(0) < b < \min\{2\operatorname{Re}\rho(0), \operatorname{Re}\rho(0)+1\}$ and if a solution $u(t,x) \in \widetilde{\mathcal{O}}_{\text{int}}$ of (5.0.1) is expressed in the form

$$t^{-b}\left(u(t,x) - \sum_{1\leq i\leq b} u_i(x)t^i - \varphi(x)t^{\rho(x)}\right) \in \widetilde{\mathcal{O}}_+$$

then the coefficients $\{u_i(x) : 1 \leq i \leq b\}$ are uniquely determined by the equation (5.0.1) and they are independent of $\varphi(x)$.

Now denote by S_+ (resp. S_{int}) the set of all $\widetilde{\mathcal{O}}_+$-solutions (resp. $\widetilde{\mathcal{O}}_{\text{int}}$-solutions) of (5.0.1). If $\rho(0) \notin \mathbb{N}^*$ by (C1) and (C2) we have

(5.6.2)

$$S_{\text{int}} \supset S_+ \supset \begin{cases} \{u_0\} & \text{when } \operatorname{Re}\rho(0) \leq 0\,, \\ \{u_0\} \cup \{U(\varphi) : 0 \neq \varphi(x) \in \mathbb{C}\{x\}\} & \text{when } \operatorname{Re}\rho(0) > 0. \end{cases}$$

Hence, to complete the proof of the main theorem it is sufficient to prove the following proposition:

Proposition 5.6.1 *Assume (A1), (A2), and (A3). Let $u_0, U(\varphi)$, and S_{int} be as above. Then we have:*

1. *If $\operatorname{Re}\rho(0) \leq 0$ and $u \in S_{\text{int}}$ we have $u \equiv u_0$ in $\widetilde{\mathcal{O}}_{\text{int}}$.*
2. *If $\rho(0) \notin \mathbb{N}^*, \operatorname{Re}\rho(0) > 0$, and $u \in S_{\text{int}}$ we can find a $\varphi(x) \in \mathbb{C}\{x\}$ such that $u \equiv U(\varphi)$ holds in $\widetilde{\mathcal{O}}_{\text{int}}$.*

Proof : First let us show 1.: Assume $\operatorname{Re}\rho(0) \leq 0$ and $u \in S_{\text{int}}$. Then by putting $b = 0$ we have $t^{-b}(u - u_0) = (u - u_0) \in \widetilde{\mathcal{O}}_{\text{int}}$ and $\operatorname{Re}\rho(0) \leq b$. Hence, by 2. of theorem 5.5.1 we obtain $u \equiv u_0$ in $\widetilde{\mathcal{O}}_{\text{int}}$.

Next let us show 2.: Assume $\rho(0) \notin \mathbb{N}^*, \operatorname{Re}\rho(0) > 0$, and $u \in S_{\text{int}}$. Set $w = (u - u_0) \in \widetilde{\mathcal{O}}_{\text{int}}$. Since u and u_0 are solutions of (5.0.1) w satisfies the following equation:

$$\left(t\frac{\partial}{\partial t} - \rho(x)\right) w = f(t,x)$$

with

(5.6.3)

$$f(t,x) = -\rho(x)w + F\left(t, x, w + u_0, \frac{\partial w}{\partial x} + \frac{\partial u_0}{\partial x}\right) - F\left(t, x, u_0, \frac{\partial u_0}{\partial x}\right).$$

Take $0 < a < b$ such that

(5.6.4) $$a < \operatorname{Re}\rho(0) < b < \min\{2a, a+1\}.$$

Then by 1. of theorem 5.5.1 we have $t^{-a}w = t^{-a}(u - u_0) \in \widetilde{\mathcal{O}}_+$ and therefore by (5.6.3) we obtain $t^{-b}f \in \widetilde{\mathcal{O}}_+$. It is easy to see that $W(t,x)$ defined by

$$W(t,x) = \int_0^1 s^{-\rho(x)-1} f(st,x)ds$$

satisfies $t^{-b}W \in \widetilde{\mathcal{O}}_+$ and $\left(t\frac{\partial}{\partial t} - \rho(x)\right) W = f(t,x)$. Hence, by solving the equation $\left(t\frac{\partial}{\partial t} - \rho(x)\right)(w - W) = 0$ we can see that $w - W$ is expressed in the form

$$(w - W)(t,x) = \varphi(x)t^{\rho(x)}$$

for some $\varphi(x) \in \mathbb{C}\{x\}$. Thus, by summing up we obtain

$$t^{-b}\left(u(t,x) - u_0(t,x) - \varphi(x)t^{\rho(x)}\right) = t^{-b}W \in \widetilde{\mathcal{O}}_+.$$

On the other hand, if we use the same $\varphi(x)$ as above by (5.6.1) and (5.6.4) we have

$$t^{-b}\left(U(\varphi)(t,x) - \sum_{1\le i\le b} u_i(x)t^i - \varphi(x)t^{\rho(x)}\right) \in \widetilde{\mathcal{O}}_+.$$

Hence, by (C3) we have $t^{-b}(u - U(\varphi)) \in \widetilde{\mathcal{O}}_+$ and, therefore, by 2. of theorem 5.5.1 we obtain $u \equiv U(\varphi)$ in $\widetilde{\mathcal{O}}_+$. ∎

Thus, by (5.6.2) and proposition 5.6.1 we obtain easily the main theorem. See also Tahara [46], [47]. ∎

5.7 Remarks

In this chapter, we restricted ourselves to the study of singular solutions in $\widetilde{\mathcal{O}}_+$ or $\widetilde{\mathcal{O}}_{\text{int}}$. But there seems to be a possibility that (5.0.1) has singular solutions which do not belong to the class $\widetilde{\mathcal{O}}_{\text{int}}$ as is seen in the following example.

Let us consider

$$t\frac{\partial u}{\partial t} = \rho u + u\frac{\partial u}{\partial x} \tag{5.7.1}$$

where $(t, x) \in \mathbb{C}^2$ and $\rho \in \mathbb{C}$ is constant. By the separation of variables we can see that (5.7.1) has a family of solutions of "meromorphic type" as follows:

$$u_{a,b,c}(t,x) = \begin{cases} \frac{\rho t^\rho (ax+b)}{\rho c - a t^\rho} & \text{when } \rho \neq 0, \\ \frac{ax+b}{c - a\log t} & \text{when } \rho = 0, \end{cases}$$

where $a, b, c \in \mathbb{C}$ are arbitrary. Note that in the case $\operatorname{Re}\rho \leq 0$ and $a \neq 0$ the solution $u_{a,b,c}(t,x)$ does not belong to the class $\widetilde{\mathcal{O}}_{\text{int}}$.

5.8 Supplementary result

In chapter 4, we have studied differential equations of the form

$$xy' = f(x,y)$$

and in particular we have proved that if $f(0,0) \neq 0$ this equation is analytically equivalent to

$$xz' = f(0,0).$$

Using the same idea as in the proof of theorem 5.3.3 Gérard and Sibuya [17] have proved the following result:

Let

$$\tau = \sum_{1\leq j\leq n} a_j(x)\frac{\partial}{\partial x_j} \quad \text{with } x = (x_1, \dots, x_n) \in \mathbb{C}^n$$

be a vector field with coefficients $a_j(x)$ holomorphic in a neighborhood of the origin of $\mathbb{C}^n$ and consider a system of differential equations

$$\tau y = f(x,y) \quad \text{with } y = (y_1, \dots, y_m) \in \mathbb{C}^m \text{ and } x = (x_1, \dots, x_n) \in \mathbb{C}^n \tag{5.8.1}$$

where the entries of the m–dimensional vector $f(x,y)$ are holomorphic in a polydisk centered at the origin of $\mathbb{C}^{n+m}$. Then we have

Theorem 5.8.1 *If the vector $f(0,0)$ is non zero then system (5.8.1) is analytically equivalent to the system*

$$\tau z = f(0,0) \quad \textit{with } z = (z_1, \ldots, z_m) \in \mathbb{C}^m . \tag{5.8.2}$$

Consider a system of non linear partial differential equations of the form

$$\tau y + \frac{\partial y}{\partial z_1} = f(x, y) \tag{5.8.3}$$

where $x = (x_1, \ldots, x_n)\,, y = (y_1, \ldots, y_m)\,, \tau = \sum_{1 \leq j \leq n} a_j(x) \frac{\partial}{\partial x_j}$ with coefficients $a_j(x)$ holomorphic in a neighborhood of the origin of $\mathbb{C}^n$, and

$$f(x,y) = (f_1(x,y), \ldots, f_m(x,y))$$

where the entries of the m–dimensional vector $f(x,y)$ are holomorphic in a polydisk centered at the origin of $\mathbb{C}^{n+m}$. Assume, moreover, that $f(0,0) \neq 0$.

By using a linear transformation of coordinates we can assume that $f(0,0) = (1, 0, \ldots, 0)$. By the transformation $y = z + z_1 u$ we get

$$z_1 \tau u + f(0,0) + u + z_1 \frac{\partial u}{\partial z_1} = f\left(x, z + z_1 u\right). \tag{5.8.4}$$

Then following the proof of theorem 5.3.3 we get

Theorem 5.8.2 *System (5.8.4) admits a unique convergent power series solution*

$$u = \sum_{p \geq 0} u_p\left(x, z'\right)\left(z_1\right)^p$$

where $z' = (z_2, \ldots, z_m)$ and the coefficients $u_p\left(x, z'\right)$ are holomorphic in a common polydisk centered at the origin of $\mathbb{C}^{n+m-1}$.

Now let us prove theorem 5.8.1.

Proof of theorem 5.8.1: Assume that $f(0,0) = (1, 0, \ldots, 0)$ and let us prove that there exists a holomorphic transformation $y = \varphi(x, z)$ which reduces the system (5.8.1) to the system (5.8.2). The transformation φ must be a solution of the system

$$\sum_{1 \leq j \leq n} a_j(x) \frac{\partial \varphi}{\partial x_j} + \sum_{1 \leq j \leq m} \frac{\partial \varphi}{\partial z_j} \tau z_j = f(x, \varphi(x, z)).$$

The assumption $f(0,0) = (1, 0, \ldots, 0)$ implies that $\tau z_j = 0$ for $2 \leq j \leq m$ and $\tau z_1 = 1$. It follows that φ must be a solution of

$$\sum_{1 \leq j \leq n} a_j(x) \frac{\partial \varphi}{\partial x_j} + \frac{\partial \varphi}{\partial z_1} = f(x, \varphi(x, z)).$$

Theorem 5.8.2 tells us that this equation has a unique holomorphic solution of the form $\varphi = z + z_1 u$ and it is easy to see looking at the formal solution that this solution is a transformation reducing (5.8.1) to (5.8.2). ∎

6 Maillet's type theorems for non linear singular partial differential equations

Introduction

Near an irregular singular point of an ordinary linear differential equation there may exist formal power series solutions $\sum_{k\geq 0} u_k t^k$ that diverge everywhere, but it is relatively easy to show that such solutions are in some formal Gevrey class i.e. the coefficients u_k grow at most like a fixed power of $k!$ as k tends to infinity.

In the non linear case of algebraic ordinary differential equations, Maillet [36] showed that all formal solutions must still be in some formal Gevrey class (we will call such a result one of *Maillet's type*). Mahler [35] proved also this result in a somewhat more general algebraic setting. The proofs of the above results follow by direct estimation of u_k from recurrence relations. Using a majorant argument we have proved in chapter 1 (and [14]) an other result of Maillet's type for certain more general "algebraic" operator equations and as an application we got the classical Maillet's theorem but the equation must satisfy a certain type of non degeneracy condition which is satisfied for an equation having a linear part.

With a different method Malgrange [37] has proved a generalization of Maillet's theorem for "analytic" ordinary differential equations and he gives a better estimation for the formal Gevrey index of a formal solution.

In this chapter, we are studying formal solutions of non linear singular partial differential equations of the type

(6.0.1)
$$\left(\tau^l + c_{l-1}(x)\tau^{l-1} + \cdots + c_0(x)\right) u = \sum_{1\leq i\leq d} t_i B_i(x) + G_2(x)\,(t_1, t_2, \ldots, t_d, D^m u)$$

where $l \in \mathrm{N}$ and τ is the singular triangular vector field

$$\tau = \sum_{1\leq i\leq d} \left(\sum_{1\leq j\leq i} a_{i,j}(x) t_j \right) \frac{\partial}{\partial t_i}$$

(for the precise meaning of the notations used in (6.0.1) see section 6.2).

In this chapter, by a *formal series* we mean a formal power series of the form

$$\sum_{k_1+\cdots+k_d\geq 1} u_{(k_1,\ldots,k_d)}(x)\,(t_1)^{k_1} \cdots (t_d)^{k_d} \qquad (6.0.2)$$

the coefficients $\{u_{(k_1,\ldots,k_d)}(x)\}_{k_1+\cdots+k_d\geq 1}$ of which are holomorphic in a common disk centered at the origin of $(\mathbb{C}_x)^n$. A *formal solution* of (6.0.1) is a formal series of the form (6.0.2) which satisfies formally the equation (6.0.1). If a formal series $\hat{u}(t,x)$ of the form (6.0.2) satisfies the condition that for some $s \in \mathbb{R}, s \geq 1$

$$\sum_{k_1+\cdots+k_d\geq 1} \frac{u_{(k_1,\ldots,k_d)}(x)}{((k_1+\cdots+k_d)!)^{s-1}} (t_1)^{k_1} \cdots (t_d)^{k_d}$$

is convergent near the origin of $(\mathbb{C}_x)^n \times (\mathbb{C}_t)^d$ we say that $\hat{u}(t,x)$ belongs to the *formal Gevrey class* $\mathcal{E}^{\{s\}}$. It is clear that if $\hat{u}(t,x) \in \mathcal{E}^{\{s\}}$ then $\hat{u}(t,x) \in \mathcal{E}^{\{s'\}}$ for all $s' \geq s$. Let us define the *formal Gevrey index* of $\hat{u}(t,x)$ as the number

$$s_0 = \inf\left\{s \in \mathbb{R} : s \geq 1, \hat{u}(t,x) \in \mathcal{E}^{\{s\}}\right\}.$$

We define (see section 6.3)

$$s_l = 1 + \max\left\{0, \sup_{(\mu,\nu)\in J}\left(\frac{m_\nu - l}{|\mu| + |\nu| - 1}\right)\right\}.$$

The main result for (6.0.1) is as follows:

if the vector field τ satisfies the Poincaré condition [41] then each formal solution of (6.0.1) is in the formal Gevrey class $\mathcal{E}^{\{s\}}$ for any $s \geq s_l$.

Let $\hat{u}(t,x)$ be a formal solution of (6.0.1) written in the form $\sum_{p\geq 1} u_p(t,x)$ where for all p the term $u_p(t,x)$ is a homogeneous polynomial in t of degree p with holomorphic coefficients in x and let q be the valuation of $\hat{u}(t,x)$ (*i.e.* $u_p \equiv 0$ for $p < q$ and $u_q \not\equiv 0$). We obtain the following result:

the formal solution $\hat{u}(t,x)$ is in the formal Gevrey class $\mathcal{E}^{\{s\}}$ for any

$$s \geq s_l(\hat{u}(t,x)) = 1 + \max\left\{0, \sup_{(\mu,\nu)\in J}\left(\frac{m_\nu - l}{|\mu| + q(|\nu| - 1)}\right)\right\}.$$

If we are restricting ourselves to the case of ordinary differential equation (*i.e.* t is one variable and x does not occur) then we have

$$s_l(\hat{u}(t)) = 1 + \frac{1}{r}$$

where r is the number defined by Malgrange in [37].

This chapter is organized in the following way. In section 6.2 we are studying the case $l = 1$, in section 6.3 the general case, and in section 6.4 we are showing that for some particular equations we have the exactness of the formal Gevrey index and give examples to illustrate our theory.

We have to mention that in the linear case Miyake [38] has obtained interesting results concerning the formal Gevrey indices for a certain class of integro–differential equations.

As a conclusion we formulate the following conjecture:

The number $s_l(\hat{u}(t,x))$ is the exact formal Gevrey index of each formal power series solution $\hat{u}(t,x)$ of (6.0.1).

In the next chapter, we are developping the same theory for partial differential equations without linear part.

6.1 Implicit function theorem

In this chapter, we will study the convergence of various types of formal power series solutions of partial differential equations.

The basic plan of our proofs of the convergence is as follows: first we seek an analytic equation which has as a solution a majorant series of our formal solution and then by using an implicit function theorem we will get the convergence of the majorant series.

For the convenience of readers let us recall here the implicit function theorem utilized in this chapter.

Consider the following analytic equation:

$$y = ta(x) + G_2(t,x,y) \tag{6.1.1}$$

where $y \in \mathbb{C}$, $t \in \mathbb{C}$, $x = (x_1, x_2, \ldots, x_n) \in \mathbb{C}^n$, $a(x)$ is a holomorphic function defined near the origin of $(\mathbb{C}_x)^n$, and $G_2(t,x,y)$ is a holomorphic function defined near the origin of $\mathbb{C}_t \times (\mathbb{C}_x)^n \times \mathbb{C}_y$ of the form

$$G_2(t,x,y) = \sum_{k+l \geq 2} a_{k,l}(x) t^k y^l.$$

Then we have the classical

Theorem 6.1.1 (Implicit function theorem) *The equation (6.1.1) has a unique formal solution of the form*

$$y^*(t,x) = \sum_{1 \leq k < \infty} y_k(x) t^k \tag{6.1.2}$$

where the coefficients $\{y_k(x)\}_{k \geq 1}$ are holomorphic in a common disk centered at the origin of $\mathbb{C}^n$ and this solution is holomorphic at the origin of $\mathbb{C}_t \times (\mathbb{C}_x)^n$.

Let us consider now the following equation:

$$c_p(x)y^p + c_{p-1}(x)ty^{p-1} + \cdots + c_1(x)t^{p-1}y + c_0(x)t^p = G_{p+1}(t,x,y) \tag{6.1.3}$$

where $p \in \mathbb{N}^* = \mathbb{N} \setminus \{0\}$, $c_i(x)$ for $i = 0, 1, \ldots, p$ are holomorphic functions near the origin of $(\mathbb{C}_x)^n$, and $G_{p+1}(t,x,y)$ is a holomorphic function defined near the origin of $\mathbb{C}_t \times (\mathbb{C}_x)^n \times \mathbb{C}_y$ of the form

$$G_{p+1}(t,x,y) = \sum_{k+l \geq p+1} a_{k,l}(x) t^k y^l.$$

Denote by $H(x,X)$ and $C(x,X)$ the following polynomials:

$$\begin{aligned} H(x,X) &= c_p(x)X^p + c_{p-1}(x)X^{p-1} + \cdots + c_1(x)X + c_0(x), \\ C(x,X) &= \frac{\partial}{\partial X}H(x,X) = pc_p(x)X^{p-1} + (p-1)c_{p-1}(x)X^{p-2} + \cdots + c_1(x). \end{aligned}$$

Then we have

Theorem 6.1.2 (Modified implicit function theorem) *Assume that the two equations $H(0,X)=0$ and $C(0,X)=0$ have no common roots. Then if (6.1.3) has a formal solution $y = y^*(t,x)$ of the form (6.1.2) it must be convergent near the origin of $\mathbb{C}_t \times (\mathbb{C}_x)^n$.*

Proof : Let $y^*(t,x)$ be a formal solution of (6.1.3) of the form (6.1.2) and put

$$Z(t,x) = \sum_{k\geq 2} y_k(x)t^{k-1}.$$

Then $y^* = t\,(y_1(x) + Z(t,x))$ and the equation (6.1.3) is decomposed into the following two equations:

$$H\,(x,y_1) = 0, \tag{6.1.4}$$

$$\begin{aligned} C\,(x,y_1)\,Z = &-\sum_{2\leq i\leq p}\ \sum_{2\leq j\leq i} \frac{i!}{j!(i-j)!}c_i(x)\,(y_1(x))^{i-j}\,Z^j \\ &+ \sum_{k+l\geq p+1} a_{k,l}(x)t^{k+l-p}\,(y_1(x)+Z)^l\,. \end{aligned} \tag{6.1.5}$$

Since $y^*(t,x)$ is a formal solution of (6.1.3) $y_1(x)$ satisfies the equation (6.1.4) and $Z(t,x)$ is a formal solution of (6.1.5). Since $H(0,X)$ and $C(0,X)$ have no common roots by (6.1.4) we have $C\,(0,y_1(0)) \neq 0$ and, therefore, we see that the equation (6.1.5) in Z is of type (6.1.1). Thus, by applying the result for (6.1.1) to (6.1.5) we obtain the convergence of $Z(t,x)$ which proves the modified implicit function theorem. ■

6.2 Non linear equations with first order linear part

In this section, we will study the formal solutions of non linear singular partial differential equations having a first order linear part and show that these formal solutions are in some formal Gevrey class.

Let $d,n \in \mathbb{N}^*, t=(t_1,\ldots,t_d)\in\mathbb{C}^d, x=(x_1,\ldots,x_n)\in\mathbb{C}^n, \Delta$ a polydisk centered at the origin of $\mathbb{C}^n$, and τ a vector field of the form

$$\tau = \sum_{1\leq i\leq d}\left(\sum_{1\leq j\leq i} a_{i,j}(x)t_j\right)\frac{\partial}{\partial t_i}$$

where the coefficients $a_{i,j}(x)$ are holomorphic in Δ.

For all $m \in \mathbb{N}^*$ we define

$$I_m = \{(a,b,c) : |a| = |b| \leq m, |b| + |c| \leq m\}$$

where $a = (a_1, \dots, a_d) \in \mathbb{N}^d, b = (b_1, \dots, b_d) \in \mathbb{N}^d, c = (c_1, \dots, c_n) \in \mathbb{N}^n, |a| = a_1 + \cdots + a_d, |b| = b_1 + \cdots + b_d, |c| = c_1 + \cdots + c_n$. Let δ be the cardinal of I_m and denote $t^a = (t_1)^{a_1} \cdots (t_d)^{a_d}, (D_t)^b = \left(\frac{\partial}{\partial t_1}\right)^{b_1} \cdots \left(\frac{\partial}{\partial t_d}\right)^{b_d}, (D_x)^c = \left(\frac{\partial}{\partial x_1}\right)^{c_1} \cdots \left(\frac{\partial}{\partial x_n}\right)^{c_n}$, and

$$D^m u = \left\{ \left(t^a (D_t)^b (D_x)^c \right) u : (a,b,c) \in I_m \right\}.$$

For $1 \leq i \leq d, \mu = (\mu_1, \dots, \mu_d) \in \mathbb{N}^d, \nu = \{\nu_{a,b,c} \in \mathbb{N} : (a,b,c) \in I_m\}$ let $B_i(x)$ and $g_{\mu,\nu}(x)$ be holomorphic functions in Δ. Denote

$$Z = \{Z_{a,b,c} \in \mathbb{C} : (a,b,c) \in I_m\}, \qquad Z^\nu = \prod_{(a,b,c)\in I_m} (Z_{a,b,c})^{\nu_{a,b,c}},$$

$t^\mu = (t_1)^{\mu_1} \cdots (t_d)^{\mu_d}$, and

$$\|g_{\mu,\nu}\| = \sup_{x\in\Delta} |g_{\mu,\nu}(x)|.$$

Let

$$G_2(x)(t,Z) = \sum_{|\mu|+|\nu|\geq 2} g_{\mu,\nu}(x) t^\mu Z^\nu$$

be such that $\sum_{|\mu|+|\nu|\geq 2} \|g_{\mu,\nu}\| \, t^\mu Z^\nu$ is convergent in a neighborhood of the origin of $\mathbb{C}^d \times \mathbb{C}^\delta$.

In this section, we are considering partial differential equations of the form

$$\text{(6.2.1)} \quad \left(\sum_{1\leq i\leq d} \left(\sum_{1\leq j\leq i} a_{i,j}(x) t_j \right) \frac{\partial}{\partial t_i} - \rho(x) \right) u = \sum_{1\leq i\leq d} t_i B_i(x) + G_2(x)\left(t_1, t_2, \dots, t_d, D^m u\right)$$

where $\rho(x)$ is a holomorphic function in Δ.

Definition 6.2.1 *By a* formal series *we mean a formal power series of the form*

$$\text{(6.2.2)} \quad \sum_{|k|\geq 1} u_k(x) t^k$$

where the coefficients $\{u_k(x)\}_{|k|\geq 1}$ *are holomorphic in a common disk centered at the origin of* $\mathbb{C}^n$.

Definition 6.2.2 *For all $s \in \mathbb{R}, s \geq 1$ we define $\mathcal{E}^{\{s\}}$ as the set of formal series of the form (6.2.2) such that the series*

$$\sum_{|k|\geq 1} \frac{u_k(x)}{(|k|!)^{s-1}} t^k$$

is convergent.

In this case, we will say that the formal series is in the *formal Gevrey class* $\mathcal{E}^{\{s\}}$.

Let us denote by $\widehat{G}_2(x)(t, Z)$ the formal power series in t, Z associated with $G_2(x)(t, Z)$ and introduce the formal equation

$$\text{(6.2.3)} \quad \left(\sum_{1\leq i\leq d} \left(\sum_{1\leq j\leq i} a_{i,j}(x) t_j \right) \frac{\partial}{\partial t_i} - \rho(x) \right) u = \sum_{1\leq i\leq d} t_i B_i(x) + \widehat{G}_2(x)\left(t_1, t_2, \ldots, t_d, D^m u\right).$$

Definition 6.2.3 *By a* formal solution *of (6.2.1) we mean a formal power series of the form (6.2.2) which satisfies the equation (6.2.3).*

Introduce

$$J = \{(\mu, \nu) : |\nu| \geq 1 \text{ and } g_{\mu,\nu}(x) \not\equiv 0\}$$

and for $\nu = \{\nu_{a,b,c} : (a, b, c) \in I_m\}$

$$m_\nu = \max\{|b| + |c| : \nu_{a,b,c} \neq 0\};$$

then define

$$s_1 = 1 + \max\left\{0, \sup_{(\mu,\nu)\in J} \left(\frac{m_\nu - 1}{|\mu| + |\nu| - 1} \right) \right\}.$$

The main result for (6.2.1) is stated as follows:

Theorem 6.2.4 *If the Poincaré condition ([41])*

$$\text{(6.2.4)} \quad \left| \sum_{1\leq i\leq d} k_i a_{i,i}(0) - \rho(0) \right| \geq \sigma (k_1 + \cdots + k_d)$$

$$\text{for any } (k_1, \ldots, k_d) \in \mathbb{N}^d \setminus \{(0, 0, \ldots, 0)\}$$

is satisfied for some $\sigma > 0$ then we have

1. *(6.2.1) has a unique formal solution $\hat{u}(t, x)$ of the form (6.2.2);*
2. *moreover, the formal solution $\hat{u}(t, x)$ is in the formal Gevrey class $\mathcal{E}^{\{s\}}$ for any $s \geq s_1$.*

Before the proof let us make some preparatory discussions and remarks.

Remark 6.2.5 *If $s_1 = 1$ then $m_\nu \le 1$ for all ν and we are reduced to the case that was studied in chapter 5 (see also [18] and [20]) and all formal solutions of the equation (6.2.1) are convergent.*

First we note that the above Poincaré condition (6.2.4) is equivalent to the uniform condition as follows:

Lemma 6.2.6 *If the condition (6.2.4) is satisfied we can find $R > 0$ and $\sigma_0 > 0$ such that for all $x \in D_R = \{x \in \mathbb{C}^n : |x_i| \le R \text{ for } i = 1, \ldots, n\}$ we have*

$$(6.2.5) \qquad \left| \sum_{1\le i\le d} k_i a_{i,i}(x) - \rho(x) \right| \ge \sigma_0 \, (k_1 + \cdots + k_d)$$

$$\textit{for any } (k_1, \ldots, k_d) \in \mathbb{N}^d \setminus \{(0,0,\ldots,0)\}.$$

Proof : Assume (6.2.4). Denote by $C(x)$ the convex hull of the set $\{a_{1,1}(x), \ldots, a_{d,d}(x)\}$ in $\mathbb{C}$. Then we have

$$(6.2.6) \qquad C(0) \not\ni 0.$$

In fact, this is verified as follows. Suppose $C(0) \ni 0$. Then for $\varepsilon = \frac{\sigma}{2}$ we can take $(p_1, \ldots, p_d) \in \mathbb{Q}^d$ such that $p_1 + \cdots + p_d = 1, p_i \ge 0$ for $i = 1, \ldots, d$, and

$$\left| \sum_{1\le i\le d} p_i a_{i,i}(0) \right| < \varepsilon = \frac{\sigma}{2}.$$

Therefore, if we choose $L \in \mathbb{N}^*$ so that $p_1, \ldots, p_d \in \mathbb{Z}/L$ and $|\rho(0)| < L\frac{\sigma}{2}$ then by putting $k_i = p_i L \in \mathbb{N}$ for $i = 1, \ldots, d$ we have

$$\frac{\left| \sum_{1\le i\le d} k_i a_{i,i}(0) - \rho(0) \right|}{k_1 + \cdots + k_d} = \left| \sum_{1\le i\le d} p_i a_{i,i}(0) - \frac{\rho(0)}{L} \right| < \varepsilon + \frac{|\rho(0)|}{L} < \sigma$$

which contradicts the condition (6.2.4).

Since $C(0)$ is compact in $\mathbb{C}$ the condition (6.2.6) implies

$$h = \operatorname{dist}(\{0\}, C(0)) > 0.$$

Choose $r > 0$ sufficiently small and $K > 0$ sufficiently large. Then we have $\operatorname{dist}(\{0\}, C(x)) > \frac{2h}{3}$, and $\frac{|\rho(x)|}{K} < \frac{h}{3}$ for any $x \in D_r$ and, therefore, for any $(k_1, \ldots, k_d) \in \mathbb{N}^d \setminus \{(0, \ldots, 0)\}$ satisfying $k_1 + \cdots + k_d > K$ we have

$$\frac{\left| \sum_{1\le i\le d} k_i a_{i,i}(x) - \rho(x) \right|}{k_1 + \cdots + k_d} \ge \left| \frac{\sum_{1\le i\le d} k_i a_{i,i}(x)}{k_1 + \cdots + k_d} \right| - \frac{|\rho(x)|}{k_1 + \cdots + k_d} > \frac{2h}{3} - \frac{h}{3} = \frac{h}{3} \quad \text{on } D_r.$$

On the other hand, since the number of elements of the set

$$\{(k_1,\dots,k_d)\in\mathbb{N}^d\setminus\{(0,\dots,0)\} : k_1+\cdots+k_d\le K\}$$

is finite we can find an $s>0$ such that

$$\left|\sum_{1\le i\le d} k_i a_{i,i}(x)-\rho(x)\right| \ge \frac{\sigma}{2}(k_1+\cdots+k_d)$$

holds on D_s for any $(k_1,\dots,k_d)\in\mathbb{N}^d\setminus\{(0,\dots,0)\}$ satisfying $k_1+\cdots+k_d\le K$.

Thus, by taking $R=\min\{r,s\}$ and $\sigma_0=\min\left\{\frac{h}{3},\frac{\sigma}{2}\right\}$ we obtain the conclusion of lemma 6.2.6. ■

Remark 6.2.7 *The lemma 6.2.6 is also true if $\rho(x)$ is only a continuous function of x.*

Remark 6.2.8 *We have assumed that in the equation (6.2.1) the vector field τ is in triangular form. If τ is a linear vector field of the form*

$$\tau=\sum_{1\le i\le d}\left(\sum_{1\le j\le d} a_{i,j}(x)t_j\right)\frac{\partial}{\partial t_i}$$

satisfying the Poincaré condition (6.2.4) we can see using the same method as in [15] that by a linear transformation with respect to the variables $t_1,t_2,\dots,t_d$ the vector field τ can be transformed into a vector field in triangular form.

In the proof of theorem 6.2.4, we may assume that the condition (6.2.5) is satisfied for some $R>0$ and $\sigma_0>0$. Moreover, without loss of generality we may assume that

$$\sum_{1\le i\le d}\left(\sum_{j<i}\max_{x\in D_R}|a_{i,j}(x)|\right)<\frac{\sigma_0}{2}. \tag{6.2.7}$$

In fact, if we apply the change of variables

$$(t_1,\dots,t_d)\in\mathbb{C}^d\longrightarrow(c_1t_1,\dots,c_dt_d)\in\mathbb{C}^d \tag{6.2.8}$$

the equation (6.2.1) is transformed into

$$\left(\sum_{1\le i\le d}\left(\sum_{1\le j\le i} a^*_{i,j}(x)t_j\right)\frac{\partial}{\partial t_i}-\rho(x)\right)u^* = \sum_{1\le i\le d} t_iB^*_i(x)+G^*_2(x)(t_1,t_2,\dots,t_d,D^m u^*) \tag{6.2.9}$$

in which

$$a^*_{i,j}(x) = \frac{c_j}{c_i} a_{i,j}(x), \qquad B^*_i(x) = c_i B_i(x),$$

$$G^*_2(x)(t_1, \ldots, t_d, Z) = G_2(x)\left(c_1 t_1, \ldots, c_d t_d, \left\{\frac{c^a}{c^b} Z_{a,b,c} : (a,b,c) \in I_m\right\}\right).$$

Therefore, by putting $c_1 = c, \ldots, c_d = c^d$ for sufficiently large $c > 0$ we can see that (6.2.9) satisfies the condition (6.2.7) with $a_{i,j}(x)$ replaced by $a^*_{i,j}(x)$. Note that the condition (6.2.5) is invariant under the change of coordinates (6.2.8).

Next let us recall some notations. Denote by

- $\mathbb{C}\{x\}$ the ring of germs of holomorphic functions at the origin of $(\mathbb{C}_x)^n$,
- $\mathbb{C}_r\{x\}$ the subring of holomorphic functions on $D_r = \{x \in \mathbb{C}^n : |x_i| \le r$ for $i = 1, \ldots, n\}$,
- $H_{k,r}\{x\}[t_1, \ldots, t_d]$ the set of all homogeneous polynomials of degree k in $t_1, \ldots, t_d$ with coefficients in $\mathbb{C}_r\{x\}$,
- $\mathbb{C}_r\{x\}[[t_1, \ldots, t_d]]$ the ring of formal power series in $t_1, \ldots, t_d$ with coefficients in $\mathbb{C}_r\{x\}$, and
- $\mathfrak{M}\mathbb{C}_r\{x\}[[t_1, \ldots, t_d]]$ the subring of all $Y \in \mathbb{C}_r\{x\}[[t_1, \ldots, t_d]]$ of the form
$$Y = \sum_{|k|\ge 1} y_k(x) t^k.$$

Then each element $Y \in \mathfrak{M}\mathbb{C}_r\{x\}[[t_1, \ldots, t_d]]$ can be uniquely expressed in the form

$$Y = \sum_{k\ge 1} Y_k \quad \text{where } Y_k \in H_{k,r}\{x\}[t_1, \ldots, t_d] \text{ for } k = 1, 2, \ldots. \tag{6.2.10}$$

For $\varphi(x) \in \mathbb{C}_r\{x\}$ we define $\|\varphi\|_r = \max_{x \in D_r} |\varphi(x)|$ and for

$$u_k = \sum_{k_1+\cdots+k_d=k} \varphi_{(k_1,\ldots,k_d)}(x)\,(t_1)^{k_1} \cdots (t_d)^{k_d} \in H_{k,r}\{x\}[t_1, \ldots, t_d] \tag{6.2.11}$$

we define the norm

$$\|u_k\|_{r,k} = \sum_{k_1+\cdots+k_d=k} \|\varphi_{(k_1,\ldots,k_d)}\|_r.$$

With the norm $\|\cdot\|_{r,k}$ the space $H_{k,r}\{x\}[t_1, \ldots, t_d]$ becomes a Banach space. Moreover, if $f \in H_{k,r}\{x\}[t_1, \ldots, t_d]$ and $g \in H_{l,r}\{x\}[t_1, \ldots, t_d]$ we have

$$\|f \cdot g\|_{r,k+l} \le \|f\|_{r,k} \|g\|_{r,l}. \tag{6.2.12}$$

For simplicity if $u_k \in H_{k,r}\{x\}[t_1, \ldots, t_d]$ we write $\|u_k\|_r$ instead of $\|u_k\|_{r,k}$.

The following lemma may be proved in the same way as in the book of Hörmander [27] lemma 5.1.3.

Lemma 6.2.9 *If $u_k \in H_{k,R}\{x\}[t_1,\ldots,t_d]$ satisfies*

$$\|u_k\|_r \leq \frac{C}{(R-r)^p} \quad \text{for } 0<r<R$$

for some $p \geq 0$ and $C>0$ then we have

$$\left\|\frac{\partial}{\partial x_i}u_k\right\|_r \leq \frac{(p+1)eC}{(R-r)^{p+1}} \quad \text{for } 0<r<R \text{ and } i=1,2,\ldots,n.$$

Now let us consider the following linear partial differential operator

$$L = \sum_{1\leq i\leq d}\left(\sum_{1\leq j\leq i} a_{i,j}(x)t_j\right)\frac{\partial}{\partial t_i} - \rho(x)$$

where $a_{i,j}(x), \rho(x) \in \mathbb{C}_R\{x\}$. Then for any $k \in \mathbb{N}^*$ we have

$$L\left(H_{k,R}\{x\}[t_1,\ldots,t_d]\right) \subset H_{k,R}\{x\}[t_1,\ldots,t_d]$$

and, therefore,

$$L\left(\mathfrak{M}\mathbb{C}_r\{x\}[[t_1,\ldots,t_d]]\right) \subset \mathfrak{M}\mathbb{C}_r\{x\}[[t_1,\ldots,t_d]].$$

Denote by L_k the restriction of L to $H_{k,R}\{x\}[t_1,\ldots,t_d]$.

If $Y \in \mathfrak{M}\mathbb{C}_r\{x\}[[t_1,\ldots,t_d]]$ is expressed in the form (6.2.10) then $L(Y)$ is given by $L(Y) = \sum_{k\geq 1} L_k(Y_k)$.

It is easy to see that $H_{k,R}\{x\}[t_1,\ldots,t_d]$ is a free $\mathbb{C}_R\{x\}$–module of finite rank and that L_k is a $\mathbb{C}_R\{x\}$–linear operator in $H_{k,R}\{x\}[t_1,\ldots,t_d]$.

Lemma 6.2.10 *If (6.2.5) and (6.2.7) are satisfied for some $R>0$ and $\sigma_0>0$ we have the following properties for all $k \in \mathbb{N}^*$:*

1. *$L_k : H_{k,R}\{x\}[t_1,\ldots,t_d] \longrightarrow H_{k,R}\{x\}[t_1,\ldots,t_d]$ is an invertible operator,*
2. *for any $u_k \in H_{k,R}\{x\}[t_1,\ldots,t_d]$ we have*

$$\|L_k u_k\|_r \geq \frac{\sigma_0}{2}k\,\|u_k\|_r \quad \text{for } 0<r<R.$$

Proof : For one variable t_1 we put

$$e_1(t_1) = t_1 \quad \text{and} \quad e_p(t_1) = t_1 e_{p-1}(t_1) = (t_1)^p \quad \text{for } p \geq 2.$$

For two variables t_1, t_2 we put

$$e_1(t_1,t_2) = (t_1,t_2) \quad \text{and} \quad e_p(t_1,t_2) = \left(t_1 e_{p-1}(t_1,t_2), t_2 e_{p-1}(t_2)\right) \quad \text{for } p \geq 2.$$

In general, for d variables $t_1,\ldots,t_d$ we put

$$\begin{aligned} e_1(t_1,\dots,t_d) &= (t_1,\dots,t_d), \\ e_p(t_1,\dots,t_d) &= (t_1 e_{p-1}(t_1,\dots,t_d), t_2 e_{p-1}(t_2,\dots,t_d),\dots,t_d e_{p-1}(t_d)) \\ &\qquad \text{for } p \geq 2. \end{aligned}$$

Then $e_k(t_1,\dots,t_d)$ is a basis of the free $\mathbb{C}_R\{x\}$–module $H_{k,R}\{x\}[t_1,\dots,t_d]$. It is easy to see that the matrix $A_k(x)$ of L_k in this basis is upper triangular and the diagonal elements are

$$\sum_{1\leq i\leq d} k_i a_{i,i}(x) - \rho(x) \quad \text{with } k_1+\dots+k_d = k.$$

Hence, 1. of lemma 6.2.10 is a consequence of (6.2.5).

Next let us show 2. Assume that u_k is expressed in the form (6.2.11). Then by (6.2.5) and (6.2.7) we have

$$\begin{aligned} \|L_k u_k\|_r &\geq \sum_{k_1+\dots+k_d=k} \left\| \left(\sum_{1\leq i\leq d} k_i a_{i,i} - \rho \right) \varphi_{(k_1,\dots,k_d)} \right\|_r \\ &\quad - \sum_{k_1+\dots+k_d=k} \left(\sum_{1\leq i\leq d} \sum_{j<i} k_i \|a_{i,j}\|_r \right) \|\varphi_{(k_1,\dots,k_d)}\|_r \\ &\geq \sigma_0 k \|u_k\|_r - \frac{\sigma_0}{2} k \|u_k\|_r \\ &= \frac{\sigma_0}{2} k \|u_k\|_r \end{aligned}$$

which proves the condition 2. of lemma 6.2.10. ∎

Proof of theorem 6.2.4: As it was seen above we may assume that (6.2.5) and (6.2.7) are satisfied for some $0 < R < 1$ and $\sigma_0 > 0$ and, therefore, the assertion of lemma 6.2.10 is valid. The case $s_1 = 1$ being known (see remark 6.2.5) we assume that $s_1 > 1$.

Put $A_{\mu,\nu} = \|g_{\mu,\nu}\|_R$. By the assumption we see that $\sum_{|\mu|+|\nu|\geq 2} A_{\mu,\nu} t^\mu Z^\nu$ is convergent near the origin of $\mathbb{C}^d \times \mathbb{C}^\delta$.

Since L_k is an invertible operator in $H_{k,R}\{x\}[t_1,\dots,t_d]$ we can easily see that (6.2.1) has a unique formal solution of the form

$$\sum_{k\geq 1} u_k(t_1,\dots,t_d,x)$$

with $u_k(t_1,\dots,t_d,x) \in H_{k,R}\{x\}[t_1,\dots,t_d]$ for all k.

Moreover, $u_k(t_1,\dots,t_d,x)$ is determined by the following recursive formula:

$$L_1 u_1 = \sum_{1\leq i\leq d} t_i B_i(x) \tag{6.2.13}$$

and for $k \geq 2$

(6.2.14)

$$L_k u_k = \sum_{2\leq|\mu|+|\nu|\leq k} g_{\mu,\nu}(x)t^{\mu}\left[\sum_{k^*+|\mu|=k} 1\times \prod_{(a,b,c),|b|+|c|\leq m_\nu} \left(D_{a,b,c}u_{k_{a,b,c}(1)}(t,x)\right)\times\right.$$
$$\left.\times\left(D_{a,b,c}u_{k_{a,b,c}(2)}(t,x)\right)\times\cdots\times\left(D_{a,b,c}u_{k_{a,b,c}(\nu_{a,b,c})}(t,x)\right)\right]$$

where

$$\begin{aligned}
k_{a,b,c} &= \left(k_{a,b,c}(1), k_{a,b,c}(2), \ldots, k_{a,b,c}\left(\nu_{a,b,c}\right)\right),\\
k^* &= \sum_{(a,b,c),|b|+|c|\leq m_\nu} \left|k_{a,b,c}\right|,\\
\left|k_{a,b,c}\right| &= k_{a,b,c}(1)+k_{a,b,c}(2)+\cdots+k_{a,b,c}\left(\nu_{a,b,c}\right),
\end{aligned}$$

and $D_{a,b,c} = t^a\left(D_t\right)^b\left(D_x\right)^c$.

Take any $s \geq s_1 > 1, 0 < r < R$ and introduce

$$v_k(t,x) = \frac{u_k(t,x)}{((k-1)!)^{s-1}}.$$

Using the properties of the norm $\|\cdot\|_{r,k}$ in particular

$$\left\|g_{\mu,\nu}(x)t^{\mu}\right\|_r = \left\|g_{\mu,\nu}\right\|_r \leq A_{\mu,\nu}$$

and (6.2.12) we have

$$\begin{aligned}
\frac{\sigma_0}{2}\left\|v_k(t,x)\right\|_r &\leq \frac{1}{k((k-1)!)^{s-1}}\left\|L_k u_k(t,x)\right\|_r\\
&\leq \frac{1}{k((k-1)!)^{s-1}}\sum_{2\leq|\mu|+|\nu|\leq k} A_{\mu,\nu}\left[\sum_{k*+|\mu|=k} 1\times\right.\\
&\quad\times\prod_{(a,b,c),|b|+|c|\leq m_\nu}\left(\left(k_{a,b,c}(1)-1\right)!\right)^{s-1}\times\\
&\quad\times\left(\left(k_{a,b,c}(2)-1\right)!\right)^{s-1}\times\cdots\times\left(\left(k_{a,b,c}\left(\nu_{a,b,c}\right)-1\right)!\right)^{s-1}\times\\
&\quad\times\prod_{(a,b,c),|b|+|c|\leq m_\nu}\left(\left\|D_{a,b,c}v_{k_{a,b,c}(1)}(t,x)\right\|_r\right)\times\\
&\quad\times\left(\left\|D_{a,b,c}v_{k_{a,b,c}(2)}(t,x)\right\|_r\right)\times\cdots\times\\
&\quad\left.\times\left(\left\|D_{a,b,c}v_{k_{a,b,c}(\nu_{a,b,c})}(t,x)\right\|_r\right)\right].
\end{aligned}$$

Using the inequality $(p_1!)(p_2!)\cdots(p_m!) \leq (p_1+p_2+\cdots+p_m)!$ for any finite sequence of positive integers $p_1, p_2, \ldots, p_m$ we get

$$\frac{\sigma_0}{2}\left\|v_k(t,x)\right\|_r \leq \sum_{2\leq|\mu|+|\nu|\leq k} A_{\mu,\nu}\Big[\sum_{k*+|\mu|=k} \frac{((k-|\mu|-|\nu|)!)^{s-1}}{k\,((k-1)!)^{s-1}} \times$$

$$\times \prod_{(a,b,c),|b|+|c|\leq m_\nu} \left(\left\|D_{a,b,c}v_{k_{a,b,c}(1)}(t,x)\right\|_r\right) \times$$

$$\times \left(\left\|D_{a,b,c}v_{k_{a,b,c}(2)}(t,x)\right\|_r\right) \times \cdots \times$$

$$\times \left(\left\|D_{a,b,c}v_{k_{a,b,c}(\nu_{a,b,c})}(t,x)\right\|_r\right)\Big].$$

Now introduce

$$L_{|\mu|,|\nu|} = \min\left\{|\mu|+|\nu|\,,2+\left[\frac{m-1}{s_1-1}\right]\right\}$$

where $\left[\frac{m-1}{s_1-1}\right]$ is the integral part of $\frac{m-1}{s_1-1}$.

As $|\mu|+|\nu|\geq 2$ we have clearly

$$2\leq L_{|\mu|,|\nu|}\leq 2+\left[\frac{m-1}{s_1-1}\right],$$

$$((k-|\mu|-|\nu|)!)^{s-1}\leq \left((k-L_{|\mu|,|\nu|})!\right)^{s-1},$$

and

$$\begin{aligned}\frac{((k-|\mu|-|\nu|)!)^{s-1}}{k\,((k-1)!)^{s-1}} &\leq \frac{1}{k\left[(k-1)\cdots\left(k-L_{|\mu|,|\nu|}+1\right)\right]^{s-1}}\\ &\leq \frac{C}{k^{1+\left(L_{|\mu|,|\nu|}-1\right)(s-1)}}\end{aligned}$$

where C is a constant which does not depend on μ and ν.

We have

$$\frac{1}{k^{1+\left(L_{|\mu|,|\nu|}-1\right)(s-1)}}\leq\frac{1}{k^{m_\nu}}.$$

To prove this it is enough to prove that

$$1+\left(L_{|\mu|,|\nu|}-1\right)(s-1)\geq m_\nu$$

which is equivalent to prove that

$$|\mu|+|\nu|\geq\frac{m_\nu-1}{s-1}+1$$

and

$$2+\left[\frac{m-1}{s_1-1}\right]\geq\frac{m_\nu-1}{s-1}+1.$$

The last two inequalities follow from $s\geq s_1$ and a simple calculation.

Finally we get

$$\frac{\sigma_0}{2}\left\|v_k(t,x)\right\|_r \leq \sum_{2\leq|\mu|+|\nu|\leq k} A_{\mu,\nu}\Bigg[\sum_{k*+|\mu|=k}\frac{C}{k^{m_\nu}}\times \times \prod_{(a,b,c),|b|+|c|\leq m_\nu}\left(\left\|D_{a,b,c}v_{k_{a,b,c}(1)}(t,x)\right\|_r\right)\times \times\left(\left\|D_{a,b,c}v_{k_{a,b,c}(2)}(t,x)\right\|_r\right)\times\cdots\times \times\left(\left\|D_{a,b,c}v_{k_{a,b,c}(\nu_{a,b,c})}(t,x)\right\|_r\right)\Bigg].$$

Let us remember that our aim is to prove that the series $\sum_{k\geq1} v_k(t,x)$ is convergent near the origin of $(\mathbb{C}_t)^d \times (\mathbb{C}_x)^n$. For doing this let us introduce the majorant equation

(6.2.15)

$$\frac{\sigma_0}{2}Y = \frac{\sigma_0}{2}AX + \frac{C}{(R-r)^m}\sum_{|\mu|+|\nu|\geq2}\frac{A_{\mu,\nu}\left(|\mu|+|\nu|\right)^m}{(R-r)^{m(|\mu|+|\nu|-2)}}X^{|\mu|}(BY)^{|\nu|}$$

where $A>0$ and $B\geq1$ are constants chosen so that

$$\left\|D_{a,b,c}v_1\right\|_R \leq A \quad \text{for } (a,b,c)\in I_m \quad \text{and} \quad B = e^m m^m.$$

Since $\sum_{|\mu|+|\nu|\geq2} A_{\mu,\nu}t^\mu Z^\nu$ is a convergent series it is easy to see that (6.2.15) is an analytic equation with respect to X and Y.

The implicit function theorem tells us that for any $0<r<R$ this equation (6.2.15) has a unique holomorphic solution Y of the form

$$Y = \sum_{k\geq1} Y_k(r)X^k.$$

Moreover, $Y_k(r)$ is determined by the following recursive formula:

$$Y_1(r) = A$$

and for $k\geq2$

$$\text{(6.2.16)}\quad \frac{\sigma_0}{2}Y_k(r) = \frac{C}{(R-r)^m}\sum_{2\leq|\mu|+|\nu|\leq k}\frac{A_{\mu,\nu}\left(|\mu|+|\nu|\right)^m}{(R-r)^{m(|\mu|+|\nu|-2)}}\times \times\sum_{k_1+\cdots+k_{|\nu|}=k-|\mu|}\left(BY_{k_1}\right)\times\cdots\times\left(BY_{k_{|\nu|}}\right).$$

By induction we can see that $Y_k(r)$ has the form

$$\text{(6.2.17)}\qquad Y_k(r) = \frac{C_k}{(R-r)^{m(k-1)}}$$

for some constants $C_1 = A$ and $C_k \geq 0$ for $k \geq 2$. In fact, if (6.2.17) with k replaced by p is true for all $p < k$ from (6.2.16) we have

$$\begin{aligned}\frac{\sigma_0}{2} Y_k(r) &= \frac{C}{(R-r)^m} \sum_{2 \leq |\mu|+|\nu| \leq k} \frac{A_{\mu,\nu}\left(|\mu|+|\nu|\right)^m}{(R-r)^{m(|\mu|+|\nu|-2)}} \times \\ &\times \sum_{k_1+\cdots+k_{|\nu|}=k-|\mu|} \frac{(BC_{k_1})}{(R-r)^{m(k_1-1)}} \times \cdots \times \frac{(BC_{k_{|\nu|}})}{(R-r)^{m\left(k_{|\nu|}-1\right)}} \\ &= C \frac{\sum_{|\mu|+|\nu| \geq 2} \sum_{k_1+\cdots+k_{|\nu|}=k-|\mu|} A_{\mu,\nu}\left(|\mu|+|\nu|\right)^m (BC_{k_1}) \times \cdots \times (BC_{k_{|\nu|}})}{(R-r)^{m(k-1)}}\end{aligned}$$

which implies that (6.2.17) is also true for k.

Now let us show that

$$(6.2.18) \qquad \left\| t^a (D_t)^b (D_x)^c v_k \right\|_r = \left\| D_{a,b,c} v_k \right\|_r \leq \frac{B}{k^{m-|b|-|c|}} Y_k(r) \quad \text{for } 0 < r < R \text{ and } (a,b,c) \in I_m$$

holds for any $k \in \mathbb{N}^*$. Since $Y_1(r) = A$ and $B > 1$ the case $k = 1$ is clear from the definition of A. The general case is proved by induction as follows.

Suppose that $k \geq 2$ and that (6.2.18) with k replaced by p is true for all $p < k$. Then

$$\begin{aligned}\frac{\sigma_0}{2} \left\| v_k(t,x) \right\|_r \leq C &\sum_{2 \leq |\mu|+|\nu| \leq k} A_{\mu,\nu} \left[\sum_{k^*+|\mu|=k} \frac{1}{k^{m_\nu}} \times \right. \\ &\times \prod_{(a,b,c),|b|+|c| \leq m_\nu} \frac{\left(BY_{k_{a,b,c}(1)}(r)\right)}{\left(k_{a,b,c}(1)\right)^{m-|b|-|c|}} \times \\ &\left. \times \frac{\left(BY_{k_{a,b,c}(2)}(r)\right)}{\left(k_{a,b,c}(2)\right)^{m-|b|-|c|}} \times \cdots \times \frac{\left(BY_{k_{a,b,c}(\nu_{a,b,c})}(r)\right)}{\left(k_{a,b,c}\left(\nu_{a,b,c}\right)\right)^{m-|b|-|c|}} \right].\end{aligned}$$

As $m - |b| - |c| \geq m - m_\nu$ we have

$$\begin{aligned}\frac{\sigma_0}{2} \left\| v_k(t,x) \right\|_r \leq \frac{C}{k^m} &\sum_{2 \leq |\mu|+|\nu| \leq k} A_{\mu,\nu} \left[\sum_{k^*+|\mu|=k} 1 \times \right. \\ &\times \left(\frac{k}{\prod_{(a,b,c),|b|+|c| \leq m_\nu} k_{a,b,c}(1) k_{a,b,c}(2) \cdots k_{a,b,c}\left(\nu_{a,b,c}\right)} \right)^{m-m_\nu} \times \\ &\left. \times \prod_{(a,b,c),|b|+|c| \leq m_\nu} \left(BY_{k_{a,b,c}(1)}(r)\right) \left(BY_{k_{a,b,c}(2)}(r)\right) \cdots \left(BY_{k_{a,b,c}(\nu_{a,b,c})}(r)\right) \right].\end{aligned}$$

Consider an expression of the form $\frac{k}{i_1 i_2 \cdots i_{|\nu|}}$ where all the entries are integers, $i_l \geq 1$ for all $l = 1, 2, \ldots, |\nu|$ and $i_1 + i_2 + \cdots + i_{|\nu|} + |\mu| = k$. Then $i_l \leq i_1 i_2 \cdots i_{|\nu|}$ for all $l = 1, 2, \ldots, |\nu|$ implies $i_1 + i_2 + \cdots + i_{|\nu|} \leq |\nu| \, i_1 i_2 \cdots i_{|\nu|}$ and

$$k \leq |\nu| \, i_1 i_2 \cdots i_{|\nu|} + |\mu| \leq (|\mu| + |\nu|) \, i_1 i_2 \cdots i_{|\nu|}$$

and finally

$$\frac{k}{i_1 i_2 \cdots i_{|\nu|}} \leq (|\mu| + |\nu|) \, .$$

Applying this inequality and $m - m_\nu \leq m$ we get

$$\|v_k(t,x)\|_r \leq \frac{1}{k^m} \left[\frac{2C}{\sigma_0} \sum_{2 \leq |\mu| + |\nu| \leq k} A_{\mu,\nu} \left(\sum_{k^* + |\mu| = k} (|\mu| + |\nu|)^m \times \right. \right.$$
$$\left. \left. \times \prod_{(a,b,c), |b| + |c| \leq m_\nu} \left(BY_{k_{a,b,c}(1)}(r)\right) \left(BY_{k_{a,b,c}(2)}(r)\right) \cdots \left(BY_{k_{a,b,c}(\nu_{a,b,c})}(r)\right) \right) \right] .$$

As $0 < R < 1$ we have $1 \leq \frac{1}{(R-r)^{m(|\mu|+|\nu|-2)}}$ and so

$$\|v_k(t,x)\|_r \leq \frac{1}{k^m} \left[\frac{2C}{\sigma_0} \sum_{2 \leq |\mu| + |\nu| \leq k} \frac{A_{\mu,\nu} \, (|\mu| + |\nu|)^m}{(R-r)^{m(|\mu|+|\nu|-2)}} \times \right.$$
$$\times \sum_{k^* + |\mu| = k} \prod_{(a,b,c), |b| + |c| \leq m_\nu} \left(BY_{k_{a,b,c}(1)}(r)\right) \times$$
$$\left. \times \left(BY_{k_{a,b,c}(2)}(r)\right) \cdots \left(BY_{k_{a,b,c}(\nu_{a,b,c})}(r)\right) \right] .$$

This means that

$$\|v_k(t,x)\|_r \leq \frac{(R-r)^m}{k^m} Y_k(r) = \frac{1}{k^m} \frac{C_k}{(R-r)^{m(k-2)}} .$$

As $(R - r) < 1 < B$ we get that (6.2.18) is true for $|b| = 0$ and $|c| = 0$.

Let us prove now that (6.2.18) is true for all $(a, b, c) \in I_m$. It is easy to see that

$$\|D_{a,b,c} v_k(t,x)\|_r \leq k^{|b|} \, \|(D_x)^c \, v_k(t,x)\|_r \, .$$

By repeated application of lemma 6.2.9 we have

$$\|D_{a,b,c} v_k(t,x)\|_r \leq k^{|b|} \frac{1}{k^m} C_k e^{|c|} \frac{(m(k-2)+1) \cdots (m(k-2)+|c|)}{(R-r)^{m(k-2)+|c|}} .$$

As $|c| \leq m$ and $m(k-2) + l \leq mk$ for all $l = 1, 2, \ldots, |c|$ we get

$$\begin{aligned}\|D_{a,b,c}v_k(t,x)\|_r &\le k^{|b|}\frac{1}{k^m}C_k\frac{e^{|c|}m^{|c|}k^{|c|}}{(R-r)^{m(k-1)}}\\ &\le C_k\frac{1}{k^{m-|b|-|c|}}\frac{e^m m^m}{(R-r)^{m(k-1)}}\\ &= \frac{B}{k^{m-|b|-|c|}}Y_k(r).\end{aligned}$$

Thus, (6.2.18) is proved for all $k \in \mathbb{N}^*$.

By (6.2.18) we see that the series Y is a majorant series of

$$\sum_{k\ge1}\|v_k\|_r X^k \tag{6.2.19}$$

and, therefore, the convergence of (6.2.19). Thus, we have obtained the convergence of $\sum_{k\ge1} v_k(t,x)$ near the origin of $(\mathbb{C}_t)^d \times (\mathbb{C}_x)^n$. ■

The proof of theorem 6.2.4 is completed. But (6.2.18) implies also the convergence of the series

$$\sum_{k\ge1}\|D_{a,b,c}v_k(t,x)\|_r X^k.$$

Let $\hat{u}(t,x) = \sum_{|k|\ge1} u_k(x)t^k$ be a formal power series which can be written in the form $\sum_{k\ge1} u_k(t,x)$ where for all k

$$u_k(t,x) \in H_{k,r}\{x\}[t_1,\dots,t_d].$$

Definition 6.2.11 *The* valuation $\nu(\hat{u})$ *of a formal power series $\hat{u}(t,x)$ is defined as the integer q such that $u_k(t,x)=0$ for all $k<q$ and $u_q(t,x)\ne 0$.*

Lastly let us give a modified and more precise version of theorem 6.2.4.

Theorem 6.2.12 *Assume that 0 does not belong to the convex hull of the set $\{a_{1,1}(0), a_{2,2}(0), \dots, a_{d,d}(0)\}$ in $\mathbb{C}$. Then if (6.2.1) has a formal solution $\hat{u}(t,x)$ of valuation q it is in the formal Gevrey class $\mathcal{E}^{\{s\}}$ for any*

$$s \ge s_1(\hat{u}(t,x)) = 1+\max\left\{0,\ \sup_{(\mu,\nu)\in J}\left(\frac{m_\nu - 1}{|\mu|+q(|\nu|-1)}\right)\right\}.$$

Proof : Assume that 0 does not belong to the convex hull of the set $\{a_{1,1}(0), a_{2,2}(0),\dots,a_{d,d}(0)\}$ in $\mathbb{C}$. Then as is seen in the proof of lemma 6.2.6 there exist $R>0, \sigma_0>0$, and $K\in\mathbb{N}^*$ such that

$$\left|\sum_{1\le i\le d} k_i a_{i,i}(x) - \rho(x)\right| \ge \sigma_0(k_1+\dots+k_d) \quad \text{on } D_R$$

for $(k_1,\ldots,k_d) \in \mathbb{N}^d$ satisfying $k_1+\cdots+k_d \geq K$.

Let $\sum_{k\geq q} u_k$ where $u_k \in H_{k,R}\{x\}[t_1,t_2,\ldots,t_d]$ for $k \geq q$ be a formal solution of valuation q of (6.2.1). Let $s \geq s_1(\hat{u})$ and define

$$v_k(t,x) = \frac{u_k(t,x)}{((k-q)!)^{s-1}}.$$

Introduce

$$L_{|\mu|,|\nu|}(q) = \min\left\{|\mu| + q\,|\nu|\,, q+1+\left[\frac{m-1}{s_1(\hat{u})-1}\right]\right\}$$

where $\left[\frac{m-1}{s_1(\hat{u})-1}\right]$ is the integral part of $\frac{m-1}{s_1(\hat{u})-1}$.

As $|\mu|+|\nu| \geq 2$ and $|\nu| \geq 1$ hold for $(\mu,\nu) \in J$ we have

$$q+1 \leq L_{|\mu|,|\nu|}(q) \leq q+1+\left[\frac{m-1}{s_1(\hat{u})-1}\right],$$

$$((k-|\mu|-q\,|\nu|)!)^{s-1} \leq \left((k-L_{|\mu|,|\nu|}(q))!\right)^{s-1}.$$

Finally we get

$$\frac{1}{k^{1+\left(L_{|\mu|,|\nu|}(q)-q\right)(s-1)}} \leq \frac{1}{k^{m_\nu}}.$$

Then if we set $v_k = 0$ for $k < q$ and if (6.2.18) is satisfied for $k = 1,2,\ldots,K$ we can prove theorem 6.2.12 by the same argument as theorem 6.2.4. Thus, to have theorem 6.2.12 it is sufficient to notice the following: we can modify $\{A_{\mu,\nu}\}_{|\mu|+|\nu|\leq K}$ so that (6.2.18) is satisfied for $k = 1,2,\ldots,K$. ∎

Denote by $\mathfrak{M}\mathbb{C}\{x\}[[t_1,t_2,\ldots,t_d]]$ the ring of formal power series in $t_1,\ldots,t_d$ without constant term the coefficients of which are germs of holomorphic functions in x and by $\mathfrak{M}\mathbb{C}\{x\}\{t_1,t_2,\ldots,t_d\}$ the subring of convergent series.

Definition 6.2.13 *An operator*

$$D : \mathfrak{M}\mathbb{C}\{x\}[[t_1,t_2,\ldots,t_d]] \longrightarrow \mathfrak{M}\mathbb{C}\{x\}[[t_1,t_2,\ldots,t_d]]$$

is called regular singular *if* $u \in \mathfrak{M}\mathbb{C}\{x\}[[t_1,t_2,\ldots,t_d]]$ *and* $Du \in \mathfrak{M}\mathbb{C}\{x\}\{t_1,t_2,\ldots,t_d\}$ *imply* $u \in \mathfrak{M}\mathbb{C}\{x\}\{t_1,t_2,\ldots,t_d\}$.

Corollary 6.2.14

1. *Assume $s_1 = 1$ (this is the case when $m = 1$) and the same condition as in theorem 6.2.4. Then (6.2.1) has a unique solution $u(t,x)$ holomorphic near the origin of $(\mathbb{C}_t)^d \times (\mathbb{C}_x)^n$ satisfying $u(0,x) \equiv 0$.*

2. *Assume $s_1 = 1$ and the same condition as in theorem 6.2.12. Then every formal solution of (6.2.1) is convergent and the operator*

$$u \longmapsto \left(\sum_{1\leq i\leq d}\left(\sum_{1\leq j\leq i} a_{i,j}(x)t_j\right)\frac{\partial}{\partial t_i} - \rho(x)\right)u - G_2(x)\left(t_1,\ldots,t_d,D^1u\right)$$

is regular singular.

6.3 Non linear equations with higher order linear part

The notations are the same as in section 6.2. Let $l \in \mathbb{N}$ and $m \in \mathbb{N}$.

In this section, we will study formal solutions of non linear singular partial differential equations of the form

(6.3.1)
$$\left(\tau^l + c_{l-1}(x)\tau^{l-1} + \cdots + c_0(x)\right) u = \sum_{1 \le i \le d} t_i B_i(x) + G_2(x)\left(t_1, t_2, \ldots, t_d, D^m u\right)$$

where τ is the singular triangular vector field

$$\tau = \sum_{1 \le i \le d} \left(\sum_{1 \le j \le i} a_{i,j}(x) t_j \right) \frac{\partial}{\partial t_i},$$

$a_{i,j}(x), B_i(x)$, and $c_i(x)$ are holomorphic functions defined in a polydisk Δ centered at the origin of $(\mathbb{C}_x)^n$. Define for all $l \ge 0$

$$s_l = 1 + \max\left\{0, \sup_{(\mu,\nu)\in J} \left(\frac{m_\nu - l}{|\mu| + |\nu| - 1} \right)\right\}.$$

Denote by $\rho_1(x), \rho_2(x), \ldots, \rho_l(x)$ the roots of

$$\rho^l + c_{l-1}(x)\rho^{l-1} + \cdots + c_0(x) = 0.$$

These roots are not necessarily distinct or holomorphic but they are continuous functions of x and lemma 6.2.6 is also true in this case.

Then the main result for (6.3.1) is stated as follows:

Theorem 6.3.1 *If there exists $\sigma > 0$ such that for $p = 1, 2, \ldots, l$ we have*

(6.3.2)
$$\left| \sum_{1 \le i \le d} k_i a_{i,i}(0) - \rho_p(0) \right| \ge \sigma\left(k_1 + k_2 + \cdots + k_d\right)$$

for any $(k_1, k_2, \ldots, k_d) \in \mathbb{N}^d \setminus \{(0, 0, \ldots, 0)\}$

then

1. *(6.3.1) has a unique formal solution $\hat{u}(t, x)$ of the form (6.2.2);*
2. *moreover, the formal solution $\hat{u}(t, x)$ is in the formal Gevrey class $\mathcal{E}^{\{s\}}$ for any $s \ge s_l$.*

Proof : As is seen in lemma 6.2.6 the condition (6.3.2) is equivalent to the condition that there are $R > 0$ and $\sigma_0 > 0$ such that for $p = 1, 2, \ldots, l$ and for any $(k_1, k_2, \ldots, k_d) \in \mathbb{N}^d \setminus \{(0, 0, \ldots, 0)\}$

$$\left|\sum_{1\leq i\leq d} k_i a_{i,i}(x) - \rho_p(x)\right| \geq \sigma_0 \left(k_1 + k_2 + \cdots + k_d\right) \quad \text{on } D_R.$$

Moreover, we may assume that (6.2.7) is valid.

Put

$$L = \tau^l + c_{l-1}(x)\tau^{l-1} + \cdots + c_0(x).$$

Then by the definition of $\rho_1(x), \rho_2(x), \ldots, \rho_l(x)$ we have

$$L = (\tau - \rho_1(x))(\tau - \rho_2(x)) \cdots (\tau - \rho_l(x)).$$

Therefore, if we denote by L_k the restriction of L to $H_{k,R}\{x\}[t_1, t_2, \ldots, t_d]$ we have for $k \geq 1$

1. $L_k : H_{k,R}\{x\}[t_1, t_2, \ldots, t_d] \longrightarrow H_{k,R}\{x\}[t_1, t_2, \ldots, t_d]$ is an invertible operator,
2. for any $u_k \in H_{k,R}\{x\}[t_1, t_2, \ldots, t_d]$ we have

$$\|L_k u_k\|_r \geq \left(\frac{\sigma_0}{2} k\right)^l \|u_k\|_r \quad \text{for } 0 < r < R.$$

Hence, by the same argument as in the proof of theorem 6.2.4 we obtain theorem 6.3.1. ∎

Similarly we have

Theorem 6.3.2 *Assume that 0 does not belong to the convex hull of the set $\{a_{1,1}(0), a_{2,2}(0), \ldots, a_{d,d}(0)\}$ in $\mathbb{C}$. Then if (6.3.1) has a formal solution $\hat{u}(t,x)$ of valuation q it is in the formal Gevrey class $\mathcal{E}^{\{s\}}$ for any*

$$s \geq s_l(\hat{u}(t,x)) = 1 + \max\left\{0, \sup_{(\mu,\nu)\in J}\left(\frac{m_\nu - l}{|\mu| + q(|\nu| - 1)}\right)\right\}.$$

The proof of theorem 6.3.2 is easily obtained by following the proof of theorem 6.2.12. In the particular case when $l = 0$ or $m = 0$, we have the following

Corollary 6.3.3 *1. If $c_0(0) \neq 0$ and $m \geq 0$ the equation*

$$c_0(x)u = \sum_{1\leq i\leq d} t_i B_i(x) + G_2(x)\left(t_1, \ldots, t_d, D^m u\right)$$

has a unique formal solution $\hat{u}(t,x)$ of the form (6.2.2). Moreover, the solution $\hat{u}(t,x)$ is in the formal Gevrey class $\mathcal{E}^{\{s\}}$ for any $s \geq s_0$.

2. When $m = 0$ the result of 1. is nothing more than the implicit function theorem.

Corollary 6.3.4 *1. Assume $s_l = 1$ (this is the case when $l \geq m$) and the same condition as in theorem 6.3.1. Then (6.3.1) has a unique solution $u(t,x)$ holomorphic near the origin of $(\mathbb{C}_t)^d \times (\mathbb{C}_x)^n$ satisfying $u(0,x) \equiv 0$.*

2. Assume $s_l = 1$ and the same condition as in theorem 6.3.2. Then every formal solution of (6.3.1) is convergent and the operator

$$u \longmapsto \left(\tau^l + c_{l-1}(x)\tau^{l-1} + \cdots + c_0(x)\right) u - G_2(x)\left(t_1, \ldots, t_d, D^m u\right)$$

is regular singular.

Corollary 6.3.5 *Assume that $d = 1$ and that x does not occur in (6.3.1). Then for every formal solution of $\hat{u}(t)$ we have*

$$s_l\left(\hat{u}(t)\right) = 1 + \frac{1}{r}$$

where r is the number introduced by Malgrange in [37].

6.4 Formal Gevrey index for a particular type of equations — Examples

Consider now the following equation in the two variables (t,x):

$$\left(t\frac{\partial}{\partial t} - \rho\right) u = \frac{t}{1-x} + R_2\left(t, D^{*m}u\right) \tag{6.4.1}$$

where

$$\begin{aligned} D^{*m}u &= \left\{(tD_t)^j D_x^a u : j + a \leq m\right\}, \\ R_2(t,Z) &= \sum_{p+|\nu|\geq 2} a_{p,\nu} t^p Z^\nu \end{aligned}$$

is a convergent power series in the product of the two disks

$$|t| \leq r \qquad \text{and} \qquad |Z| \leq R.$$

We are assuming that ρ is a real number satisfying $\rho < 1$ and, moreover, that $a_{p,\nu}$ are all real and non negative. Now we have

$$\begin{aligned} m_\nu &= \max\left\{j + a : \nu_{j,a} \neq 0\right\} \quad \text{where } \nu = \left\{\nu_{j,a} : j + a \leq m\right\}, \\ J &= \left\{(p,\nu) : |\nu| \geq 1 \text{ and } a_{p,\nu} \neq 0\right\}, \\ s_1 &= 1 + \max\left\{0, \sup_{(p,\nu)\in J}\left(\frac{m_\nu - 1}{p + |\nu| - 1}\right)\right\}. \end{aligned}$$

We know that the equation (6.4.1) has a unique formal solution $\hat{u}(x,t)$ of valuation 1 belonging to $\mathcal{E}^{\{s\}}$ for any $s \geq s_1$.

Proposition 6.4.1 *If the formal solution $\hat{u}(t,x)$ of (6.4.1) is in $\mathcal{E}^{\{s\}}$ for some $s \geq 1$ then $s \geq s_1$.*

This proposition asserts that the formal Gevrey index of $\hat{u}(t,x)$ is equal to s_1.

Proof : The formal solution of (6.4.1) is of the form

$$\hat{u}(t,x) = tu_1(x) + t^2 [\ldots]$$

where

$$u_1(x) = \frac{1}{(1-\rho)(1-x)}.$$

Take $(p,\nu) \in J$ and fix it. Take also (j,a) such that $j+a = m_\nu$ and fix it. Now consider the equation

$$\left(t\frac{\partial}{\partial t} - \rho\right) w = \frac{t}{1-x} + a_{p,\nu} t^p \left[\prod_{(i,b)\neq(j,a)} \left((tD_t)^i D_x^b u_1(x) t \right)^{\nu_{i,b}} \right] \times \times \left((tD_t)^j D_x^a u_1(x) t \right)^{\nu_{j,a}-1} \times \left((tD_t)^j D_x^a w \right). \tag{6.4.2}$$

It is easy to see that the equation (6.4.2) has a unique formal power series w as a solution and that this solution is dominated by the formal power series solution of (6.4.1). The equation (6.4.2) can be written in the form

$$\left(t\frac{\partial}{\partial t} - \rho\right) w = \frac{t}{1-x} + a_{p,\nu} t^{p+|\nu|-1} \left[\prod_{(i,b)\neq(j,a)} \left(D_x^b u_1(x) \right)^{\nu_{i,b}} \right] \times \times \left(D_x^a u_1(x) \right)^{\nu_{j,a}-1} \times \left((tD_t)^j D_x^a w \right) \tag{6.4.3}$$

where $u_1(x) = \frac{1}{(1-\rho)(1-x)}$ and $D_x^a(u_1(x)) = \frac{A_a}{(1-x)^{a+1}}$ where A_a is a positive constant. Now the equation (6.4.3) can be written in the form

$$\left(t\frac{\partial}{\partial t} - \rho\right) w = \frac{t}{1-x} + t^{p+|\nu|-1} \frac{A}{(1-x)^d} \left((tD_t)^j D_x^a w \right) \tag{6.4.4}$$

where $A > 0$ and $d \in \mathrm{N}$ are constants depending upon p, ν, j, a.

Lemma 6.4.2 *Consider an equation of the form*

$$\left(t\frac{\partial}{\partial t} - \rho\right) w = \frac{t}{1-x} + t^l \frac{A}{(1-x)^d} \left((tD_t)^j D_x^a w \right).$$

This equation

1. *has a unique formal solution $w(t,x)$ and*

2. *$w(t,x) \in \mathcal{E}^{\{s\}}$ implies that $s - 1 \geq \frac{j+a-1}{l}$.*

Proof : By an explicit calculation we get the formal solution

$$w(t,x) = \sum_{1 \leq k < \infty} w_k(x) t^k$$

where

1. if k is of the form $lq+1$ then $w_k(x) = \frac{C_q}{(1-x)^{q(d+a)+1}}$ with

$$C_0 = \frac{1}{1-\rho},$$
$$C_q = C_{q-1}\left[\frac{A}{ql+1-\rho}\right]((q-1)l+1)^j \frac{((q-1)(d+a)+a)!}{((q-1)(d+a))!},$$

2. if k is not of the form $lq+1$ then $w_k(x) = 0$.

Now it is easy to see that as q tends to infinity the part C_q^* of C_q that affects the convergence is as follows:

$$C_q^* \sim (q!)^{j+a-1} \qquad \text{or} \qquad C_q^* \sim (k!)^{\frac{j+a-1}{l}}.$$

This implies that $w(t,x) \in \mathcal{E}^{\{s\}}$ yields $s \geq 1 + \frac{j+a-1}{l}$ and the lemma is proved. ■

Applying this result to the equation (6.4.4) we get that (6.4.4) has a unique formal solution $w(t,x)$ and that $w(t,x) \in \mathcal{E}^{\{s\}}$ implies

$$s \geq 1 + \frac{j+a-1}{p+|\nu|-1}.$$

This proves proposition 6.4.1. ■

Examples 6.4.3 *1. Consider the equation*

$$(6.4.5) \qquad (tD_t+1)\,u = \frac{t}{1-x} + t^p\,(D_x)^m\,u\,(D_x)^n\,u$$

where p, m, n are integers, $p \geq 0, m \geq n \geq 1$.

Then this equation has a unique formal solution $\hat{u}$ of valuation 1 and $\hat{u}(t,x) \in \mathcal{E}^{\{s\}}$ is equivalent to $s \geq 1 + \frac{m-1}{p+1}$.

2. Consider the equation

$$(6.4.6) \qquad (tD_t+1)\,u = \frac{t^\mu}{1-x} + t^p\,(D_x)^m\,u\,(D_x)^n\,u$$

where μ, p, m, n are integers, $\mu \geq 1, p \geq 0, m \geq n \geq 1$.

Then this equation has a unique formal solution $\hat{u}(t,x)$ of valuation μ and with formal Gevrey index $1 + \frac{m-1}{p+\mu}$.

To see this last result set $u(t,x) = t^{\mu-1}v(t,x)$ and apply the proposition 6.4.1. This example leads us to the following conjecture:

> *The number $s_l\,(\hat{u}(t,x))$ is the exact formal Gevrey index of each formal power series solution $\hat{u}(t,x)$ of (6.3.1).*

6.5 Supplementary results

Let us consider again the equation

$$\left(\sum_{1\le i\le d}\left(\sum_{1\le j\le i} a_{i,j}(x)t_j\right)\frac{\partial}{\partial t_i} - \rho(x)\right)u = \sum_{1\le i\le d} t_iB_i(x) + G_2(x)\left(t_1,t_2,\ldots,t_d,D^m u\right). \tag{6.5.1}$$

We have assumed in section 6.2 that the power series

$$|G_2|\,(t,Z) = \sum_{|\mu|+|\nu|\ge 2} \|g_{\mu,\nu}\|\, t^\mu Z^\nu$$

is convergent near the origin of $(\mathbb{C}_t)^d \times (\mathbb{C}_Z)^\delta$.

In this section, we consider the equation (6.5.1) as a formal equation and assume only that there exists a real number $s' \ge 1$ such that

$$|G_2|^*\,(t,Z) = \sum_{|\mu|+|\nu|\ge 2} \frac{\|g_{\mu,\nu}\|}{((|\mu|+|\nu|)!)^{s'-1}} t^\mu Z^\nu$$

is convergent. This means that $G_2(x)(t,Z)$ is in $\mathcal{E}^{\{s'\}}$ with respect to the variables (t,Z).

The definition of the number s_1 in section 6.2 is formal and does not depend upon the convergence of the series $|G_2|\,(t,Z)$.

Then define

$$s_1' = \max\{s', s_1\}.$$

We have

Theorem 6.5.1 *If the Poincaré condition*

$$\left|\sum_{1\le i\le d} k_i a_{i,i}(0) - \rho(0)\right| \ge \sigma\,(k_1+\cdots+k_d)$$

$$\textit{for any}\ (k_1,\ldots,k_d)\in\mathbb{N}^d\setminus\{(0,0,\ldots,0)\}$$

is satisfied for some $\sigma > 0$ then we have

1. *(6.5.1) has a unique formal solution* $\hat{u}(t,x)$*;*
2. *moreover, the formal solution* $\hat{u}(t,x)$ *is in the formal Gevrey class* $\mathcal{E}^{\{s\}}$ *for any* $s \geq s_1'$.

Proof : In the proof of theorem 6.2.4, choose $s \geq s_1'$. Then we have

$$\frac{\sigma_0}{2}\left\|v_k(t,x)\right\|_r \leq \sum_{2\leq|\mu|+|\nu|\leq k} A_{\mu,\nu}\left[\sum_{k^*+|\mu|=k} \frac{((k-|\mu|-|\nu|)!)^{s-1}}{k\,((k-1)!)^{s-1}} \times H\right]$$

where

$$H = \prod_{(a,b,c),|b|+|c|\leq m_\nu} \left(\left\|D_{a,b,c}v_{k_{a,b,c}(1)}(t,x)\right\|_r\right) \times$$
$$\times \left(\left\|D_{a,b,c}v_{k_{a,b,c}(2)}(t,x)\right\|_r\right) \times \cdots \times \left(\left\|D_{a,b,c}v_{k_{a,b,c}(\nu_{a,b,c})}(t,x)\right\|_r\right).$$

Now we have easily

$$\begin{aligned}
&\frac{\sigma_0}{2}\left\|v_k(t,x)\right\|_r \\
&\leq \sum_{2\leq|\mu|+|\nu|\leq k} A_{\mu,\nu}\frac{((|\mu|+|\nu|)!)^{s-1}}{((|\mu|+|\nu|)!)^{s'-1}}\frac{1}{((|\mu|+|\nu|-L_{|\mu|,|\nu|})!)^{s-1}} \times \\
&\quad\times\left[\sum_{k^*+|\mu|=k}((|\mu|+|\nu|-L_{|\mu|,|\nu|})!)^{s-1}\frac{((k-|\mu|-|\nu|)!)^{s-1}}{k\,((k-1)!)^{s-1}} \times H\right] \\
&\leq \sum_{2\leq|\mu|+|\nu|\leq k}\frac{A^*_{\mu,\nu}}{((|\mu|+|\nu|)!)^{s'-1}} \times \left[\sum_{k^*+|\mu|=k}\frac{((k-L_{|\mu|,|\nu|})!)^{s-1}}{k\,((k-1)!)^{s-1}} \times H\right]
\end{aligned}$$

in which

$$A^*_{\mu,\nu} = \begin{cases} A_{\mu,\nu}\,((|\mu|+|\nu|)!)^{s-1} & \text{if } L_{|\mu|,|\nu|} = |\mu|+|\nu|, \\ \frac{A_{\mu,\nu}((|\mu|+|\nu|)!)^{s-1}}{((|\mu|+|\nu|-L_{|\mu|,|\nu|})!)^{s-1}} & \text{if } L_{|\mu|,|\nu|} < |\mu|+|\nu| \end{cases}$$

and

$$L_{|\mu|,|\nu|} = \min\left\{|\mu|+|\nu|\,,2+\left[\frac{m-1}{s_1-1}\right]\right\}.$$

We know that

$$2 \leq L_{|\mu|,|\nu|} \leq 2+\left[\frac{m-1}{s_1-1}\right].$$

There are only a finite number of $A^*_{\mu,\nu}$ of the form $A_{\mu,\nu}\,((|\mu|+|\nu|)!)^{s-1}$. Moreover, $L_{|\mu|,|\nu|}$ is bounded as $|\mu|+|\nu|$ tends to infinity. This remark implies the convergence of the power series

$$\sum_{|\mu|+|\nu|\geq 2} \frac{A^*_{\mu,\nu}}{((|\mu|+|\nu|)!)^{s'-1}} t^\mu Z^\nu .$$

Now the end of the proof of theorem 6.5.1 is the same as the end of the proof of theorem 6.2.4 with $A_{\mu,\nu}$ replaced by $\frac{A^*_{\mu,\nu}}{((|\mu|+|\nu|)!)^{s'-1}}$. ■

With the same argument we have

Theorem 6.5.2 *Assume that* 0 *does not belong to the convex hull of the set* $\{a_{1,1}(0), a_{2,2}(0), \ldots, a_{d,d}(0)\}$ *in* $\mathbb{C}$. *Then if (6.5.1) has a formal solution* $\hat{u}(t,x)$ *of valuation* q *it is in the formal Gevrey class* $\mathcal{E}^{\{s\}}$ *for any* $s \geq s_1'(\hat{u}) = \max\{s', s_1(\hat{u})\}$.

Remark 6.5.3 *We have the same type of results for the equation (6.3.1) with higher order linear part.*

7 Maillet's type theorems for non linear singular partial differential equations without linear part

Introduction

In chapter 6, we have studied formal solutions of a class (6.3.1) of non linear singular partial differential equations of order l and we have proved that each formal power series solution is in some formal Gevrey class $\mathcal{E}^{\{s\}}$ for $s = s_l$. The number s_l can be computed explicitly by looking at the coefficients of (6.3.1). Moreover, each formal solution $\hat{u}$ of the equation (6.3.1) is in some formal Gevrey class $\mathcal{E}^{\{s\}}$ for $s = s_l(\hat{u}) \leq s_l$ and $s_l(\hat{u})$ can be calculated explicitly. We showed also in [19] (see also section 6.4 of chapter 6) that for some particular equations the number $s_l(\hat{u})$ is the exact formal Gevrey index of $\hat{u}$. The main assumption in chapter 6 was that the equation (6.3.1) has a linear part.

In the present chapter, we are studying the same problem for a class of singular partial differential equations of order l without linear part.

7.1 Notations and definitions

In this chapter, we will consider non linear singular partial differential equations of the form

(7.1.1) $$F_p(x)\left(t_1,\dots,t_d,u,\tau u,\dots\tau^l u\right) = G_{p+1}(x)\left(t_1,\dots,t_d,D^m u\right)$$

where $p, l, d, m \in \mathbb{N}^* = \{1, 2, \dots\}$,

$$\tau = \sum_{1\leq i\leq d}\left(\sum_{1\leq j\leq i} a_{i,j}(x)t_j\right)\frac{\partial}{\partial t_i},$$

$t = (t_1,\dots,t_d) \in \mathbb{C}^d, x = (x_1,\dots,x_n) \in \mathbb{C}^n, a_{i,j}(x)$ are holomorphic functions defined in a polydisk Δ centered at the origin of $(\mathbb{C}_x)^n$, and

$$D^m u = \left(\left(t^a\left(D_t\right)^b\left(D_x\right)^c\right)u : |a| = |b| \leq m \text{ and } |b| + |c| \leq m\right)$$

in which

$$\begin{aligned} a &= (a_1, a_2, \dots, a_d) \in \mathbb{N}^d, & |a| &= a_1 + a_2 + \dots + a_d, \\ b &= (b_1, b_2, \dots, b_d) \in \mathbb{N}^d, & |b| &= b_1 + b_2 + \dots + b_d, \\ c &= (c_1, c_2, \dots, c_n) \in \mathbb{N}^n, & |c| &= c_1 + c_2 + \dots + c_n, \end{aligned}$$

and

$$\begin{aligned} t^a &= (t_1)^{a_1}(t_2)^{a_2}\cdots(t_d)^{a_d}, \\ (D_t)^b &= \left(\frac{\partial}{\partial t_1}\right)^{b_1}\left(\frac{\partial}{\partial t_2}\right)^{b_2}\cdots\left(\frac{\partial}{\partial t_d}\right)^{b_d}, \\ (D_x)^c &= \left(\frac{\partial}{\partial x_1}\right)^{c_1}\left(\frac{\partial}{\partial x_2}\right)^{c_2}\cdots\left(\frac{\partial}{\partial x_n}\right)^{c_n}. \end{aligned}$$

Set

$$I(D^m) = \left\{(a,b,c) \in \mathbb{N}^d \times \mathbb{N}^d \times \mathbb{N}^n : |a| = |b| \leq m \text{ and } |b| + |c| \leq m\right\}$$

and denote by δ the cardinal of $I(D^m)$.

$$F_p(x)(t_1,\ldots,t_d,X_0,X_1,\ldots,X_l) = \sum_{|k|+|L|=p} c_{k,L}(x)(t_1)^{k_1}\cdots(t_d)^{k_d}(X_0)^{L_0}\cdots(X_l)^{L_l}$$

and

$$G_{p+1}(x)(t,Z) = \sum_{|\mu|+|\nu|\geq p+1} g_{\mu,\nu}(x)t^\mu Z^\nu$$

in which

$$\begin{aligned} Z &= (Z_{a,b,c} : (a,b,c) \in I(D^m)) \in \mathbb{C}^\delta, \\ \nu &= (\nu_{a,b,c} : (a,b,c) \in I(D^m)) \in \mathbb{N}^\delta, \end{aligned}$$

the coefficients $\{c_{k,L}(x)\}_{|k|+|L|=p}$ and $\{g_{\mu,\nu}(x)\}_{|\mu|+|\nu|\geq p+1}$ are holomorphic in Δ and

$$|g_{\mu,\nu}(x)| \leq A_{\mu,\nu} \quad \text{in } \Delta.$$

Moreover, the power series

$$|G_{p+1}|(t,Z) = \sum_{|\mu|+|\nu|\geq p+1} A_{\mu,\nu}t^\mu Z^\nu$$

is convergent near the origin of $(\mathbb{C}_t)^d \times (\mathbb{C}_Z)^\delta$.

By a *formal solution* of (7.1.1) we mean a formal power series solution of the form

$$\sum_{k_1+\cdots+k_d\geq 1} u_{(k_1,\ldots,k_d)}(x)(t_1)^{k_1}\cdots(t_d)^{k_d} \tag{7.1.2}$$

the coefficients $\left\{u_{(k_1,\ldots,k_d)}(x)\right\}_{k_1+\cdots+k_d\geq 1}$ of which are holomorphic in a common disk centered at the origin of $(\mathbb{C}_x)^n$. If a formal solution $\hat{u}(t,x)$ of the form (7.1.2) satisfies the condition that

$$\sum_{k_1+\cdots+k_d\geq 1} \frac{u_{(k_1,\ldots,k_d)}(x)}{((k_1+\cdots+k_d)!)^{s-1}} (t_1)^{k_1} \cdots (t_d)^{k_d}$$

is convergent near the origin of $(\mathbb{C}_x)^n \times (\mathbb{C}_t)^d$ for some $s \geq 1$ we say that $\hat{u}(t,x)$ is a formal solution in the *formal Gevrey class* $\mathcal{E}^{\{s\}}$.

Next, let us introduce some notations. Denote by $\mathbb{C}\{x\}$ the ring of germs of holomorphic functions at the origin of $(\mathbb{C}_x)^n$, by $\mathbb{C}_r\{x\}$ the subring of holomorphic functions on $D_r = \{x \in \mathbb{C}^n : |x_i| \leq r \text{ for } i = 1,\ldots,n\}$, by $H_{k,r}\{x\}[t_1,\ldots,t_d]$ the set of all homogeneous polynomials of degree k in $t_1,\ldots,t_d$ with coefficients in $\mathbb{C}_r\{x\}$, by $\mathbb{C}_r\{x\}[[t_1,\ldots,t_d]]$ the ring of formal power series in $t_1,\ldots,t_d$ with coefficients in $\mathbb{C}_r\{x\}$, and by $\mathfrak{M}_r\{x\}[[t_1,\ldots,t_d]]$ the ring of all $Y \in \mathbb{C}_r\{x\}[[t_1,\ldots,t_d]]$ of the form

$$Y = \sum_{k_1+\cdots+k_d\geq 1} y_{(k_1,\ldots,k_d)}(x)\,(t_1)^{k_1} \cdots (t_d)^{k_d}.$$

Then each element $Y \in \mathfrak{M}_r\{x\}[[t_1,\ldots,t_d]]$ can be uniquely expressed in the form

(7.1.3) $\quad Y = \sum_{k\geq 1} Y_k \qquad \text{where } Y_k \in H_{k,r}\{x\}[t_1,\ldots,t_d] \quad \text{for } k = 1,2,\ldots.$

We say that Y has *valuation* q if $Y_k = 0$ for all $k < q$ and $Y_q \neq 0$.

7.2 Assumptions and results

In this section, we state the results concerning formal solutions of the equation (7.1.1).

Remark 7.2.1 *In general, the equation (7.1.1) has no formal power series solutions of the form* $\sum_{|k|\geq 1} u_k(x)t^k$ *except when it has a particular form as is seen in the examples given later. This is the reason why later on we will restrict ourselves to some more particular equations with some additional conditions.*

Remark 7.2.2 *When* $p = 1$ *the equation (7.1.1) is reduced to an equation (6.3.1) studied in chapter 6 and we can apply the theorems of section 6.3.*

Hence, from now we assume

(C1) $p \geq 2$.

And also assume that the equation (7.1.1) has the following form:

(7.2.1) $\quad F_p(x)\,(t_1, u, \tau u, \ldots, \tau^l u) = G_{p+1}(x)\,(t_1,\ldots,t_d, D^m u)$

(there is only t_1 in the terms of lowest degree).

Note the following:

Example 7.2.3 *Consider the following equations:*

$$(7.2.2)\qquad \left(t_1\frac{\partial u}{\partial t_1}+t_2\frac{\partial u}{\partial t_2}\right)^2 = (t_1)^2+\left(t_1\frac{\partial u}{\partial t_2}\right)^3,$$

$$(7.2.3)\qquad \left(t_1\frac{\partial u}{\partial t_1}+t_2\frac{\partial u}{\partial t_2}\right)^2 = (t_1)^2+\left(t_2\frac{\partial u}{\partial t_1}\right)^3,$$

$$(7.2.4)\qquad \left(t_1\frac{\partial u}{\partial t_1}+t_2\frac{\partial u}{\partial t_2}\right)^2 = x\,(t_1)^2+\left(t_1\frac{\partial u}{\partial t_1}\right)^3.$$

By simple calculation we can see

1. *(7.2.2) has two holomorphic solutions* $u=t_1$ *and* $u=-t_1$*;*
2. *(7.2.3) and (7.2.4) have no formal solutions in* $\mathfrak{M}_r\{x\}\,[[t_1,t_2]]$ *for* $r>0$.

This example tells us that to have formal solutions we need some additional conditions on (7.2.1).

For $\lambda\in\mathbb{C}$ we write

$$\begin{aligned} t(\lambda) &= (\lambda t_1,t_2,\ldots,t_d)\,,\\ Z(\lambda) &= \left(Z_{a,b,c}(\lambda):(a,b,c)\in I\left(D^m\right)\right)\end{aligned}$$

in which

$$Z_{a,b,c}(\lambda)=\begin{cases}\lambda^{a_1+1}Z_{a,b,c} & \text{when } b_1=0,\\ \lambda^{a_1}Z_{a,b,c} & \text{when } b_1>0.\end{cases}$$

Then the following condition (C2) excludes such equations as (7.2.3):

(C2) $G_{p+1}(x)\,(t(\lambda),Z(\lambda))=O\left(\lambda^p\right)$ as $\lambda\longrightarrow 0$ in $\mathbb{C}$ uniformly in (x,t,Z) near the origin of $\left(\mathbb{C}_x\right)^n\times\left(\mathbb{C}_t\right)^d\times\left(\mathbb{C}_Z\right)^\delta$.

Put

$$\begin{aligned} H(x,X) &= F_p(x)\left(1,X,a_{1,1}(x)X,\ldots,\left(a_{1,1}(x)\right)^l X\right),\\ h_1(x) &= F_p(x)\left(0,1,a_{1,1}(x),\ldots,\left(a_{1,1}(x)\right)^l\right).\end{aligned}$$

It is easy to see that $H(x,X)$ is a polynomial of degree p in X with coefficients in $\mathbb{C}\{x\}$ and that the coefficient of X^p is $h_1(x)$. Then the following condition (C3) excludes such equations as (7.2.4):

(C3) (a) $h_1(0)\neq 0$;

(b) $H(0,X)$ has p distinct roots.

Remark 7.2.4 *The condition (b) implies that the equation* $H(x,X)=0$ *has* p *distinct holomorphic solutions. But it can be seen in the proof given below that it is enough to assume that the equation* $H(x,X)=0$ *has holomorphic solutions which are not necessarily distinct.*

Now let us formulate our "*Poincaré condition*" for (7.2.1) under the conditions (C1), (C2), and (C3). Note that by (C3) the equation $H(x, X) = 0$ in X has p holomorphic solutions $X = \varphi^{(1)}(x), \ldots, \varphi^{(p)}(x)$ belonging to $\mathbb{C}\{x\}$. For each $\varphi^{(j)}(x)$ we define

$$C_i^{(j)}(x) = \frac{\partial F_p}{\partial X_i}(x)\left(1, \varphi^{(j)}(x), a_{1,1}(x)\varphi^{(j)}(x), \ldots, (a_{1,1}(x))^l \varphi^{(j)}(x)\right)$$

$$\text{for } i = 0, 1, 2, \ldots, l$$

and consider the following equation:

$$C_l^{(j)}(x)\rho^l + C_{l-1}^{(j)}(x)\rho^{l-1} + \cdots + C_0^{(j)}(x) = 0.$$

If $C_l^{(j)}(0) \neq 0$ this equation in ρ has l solutions $\rho_1^{(j)}(x), \rho_2^{(j)}(x), \ldots, \rho_l^{(j)}(x)$ continuous near the origin of $(\mathbb{C}_x)^n$. Then the "*Poincaré condition*" for (7.2.1) with respect to $\varphi^{(j)}(x)$ is formulated as follows:

(C4) (a) $C_l^{(j)}(0) \neq 0$;

(b) there is a $\sigma > 0$ such that

$$\left|\sum_{1 \leq i \leq d} k_i a_{i,i}(0) - \rho_h^{(j)}(0)\right| \geq \sigma\,(k_1 + \cdots + k_d)$$

for any $(k_1, \ldots, k_d) \in \mathbb{N}^* \times \mathbb{N}^{d-1} \setminus \{(1, 0, \ldots, 0)\}, h = 1, \ldots, l.$

If we have this condition for all $j = 1, \ldots, p$ we will say that (7.2.1) satisfies the *Poincaré condition.* Then the main result for (7.2.1) is stated as follows:

Theorem 7.2.5 *Assume the conditions (C1), (C2), (C3), and (C4) for all $j = 1, \ldots, p$. Then we have*

1. *(7.2.1) has p formal solutions $\hat{u}_1(t, x), \ldots, \hat{u}_p(t, x)$ of the form*

$$\sum_{k_1 \geq 1, k_2, \ldots, k_d \geq 0} \varphi_{(k_1, \ldots, k_d)}(x)\,(t_1)^{k_1} \cdots (t_d)^{k_d}\,; \tag{7.2.5}$$

2. *moreover, the formal solutions $\hat{u}_1(t, x), \ldots, \hat{u}_p(t, x)$ are in the formal Gevrey class $\mathcal{E}^{\{s\}}$ for any $s \geq s'_l$ where the number s'_l can be computed explicitly from the first term of the formal solutions and the coefficients of the equation.*

The following is a typical example which illustrates our assumption:

Example 7.2.6 *Consider*

$$\left(t_1\frac{\partial u}{\partial t_1} + t_2\frac{\partial u}{\partial t_2}\right)^2 - (t_1)^2 = (t_1)^{k_1}(t_2)^{k_2}\left(t_1\frac{\partial u}{\partial t_1}\right)^{n_1}\left(t_2\frac{\partial u}{\partial t_1}\right)^{n_2}\left(t_1\frac{\partial u}{\partial t_2}\right)^{m_1}\left(t_2\frac{\partial u}{\partial t_2}\right)^{m_2}. \tag{7.2.6}$$

If the following two conditions:

1. $k_1 + k_2 + n_1 + n_2 + m_1 + m_2 \geq 3$,
2. $k_1 + n_1 + 2m_1 + m_2 \geq 2$

are satisfied we can apply theorem 7.2.5 to (7.2.6) under $p = 2, m = 1$, *and* $l = 1$. *Hence, we can know that (7.2.6) has two holomorphic solutions in* $t_1 \cdot \mathbb{C}\{t_1, t_2\}$. *Note that the above condition 2. corresponds to the condition (C2).*

By a calculation we can see

Lemma 7.2.7 *Let* $G_{p+1}(x)(t, Z)$ *be as in (7.2.1) with* $p \geq 2$. *Then*

1. *if* $u \in \mathfrak{M}_r\{x\}[[t_1, \ldots, t_d]]$ *we have*
$$G_{p+1}(x)(t_1, \ldots, t_d, D^m u) \in \bigoplus_{i \geq p+1} H_{i,r}\{x\}[t_1, \ldots, t_d];$$
2. *if* $u \in t_1 \cdot \mathbb{C}_r\{x\}[[t_1, \ldots, t_d]]$ *and if the condition (C2) is satisfied we have*
$$G_{p+1}(x)(t_1, \ldots, t_d, D^m u) \in \bigoplus_{k \geq 1} (t_1)^p H_{k,r}\{x\}[t_1, \ldots, t_d].$$

This plays an essential role in the proof of theorem 7.2.5 given later.

7.3 A basic lemma

In chapter 6, section 6.3, we have considered equations of the form

(7.3.1)
$$\left(\tau^l + c_{l-1}(x)\tau^{l-1} + \cdots + c_0(x)\right) u = \sum_{1 \leq i \leq d} t_i B_i(x) + G_2(x)(t_1, \ldots, t_d, D^m u)$$

with

$$G_2(x)(t, Z) = \sum_{|\mu|+|\nu| \geq 2} g_{\mu,\nu}(x) t^\mu Z^\nu$$

where $l, m \in \mathbb{N}^*$, $c_i(x), \tau$, and $G_2(x)(t, Z)$ are the same as in (7.2.1) for $p = 1$.

We have introduced

$$\begin{aligned} \Delta &= \{(\mu, \nu) : |\nu| \geq 1 \text{ and } g_{\mu,\nu}(x) \not\equiv 0\}, \\ m_\nu &= \max\{|b| + |c| : \nu_{a,b,c} \neq 0\} \quad \text{for } \nu = \{\nu_{a,b,c} : (a, b, c) \in I(D^m)\} \end{aligned}$$

and have defined for all $l \geq 0$

$$s_l = s_l(7.3.1) = 1 + \max\left\{0, \sup_{(\mu,\nu) \in \Delta} \left(\frac{m_\nu - l}{|\mu| + |\nu| - 1}\right)\right\}.$$

Let $\rho_1(x), \rho_2(x), \ldots, \rho_l(x)$ be the roots of

$$\rho^l + c_{l-1}(x)\rho^{l-1} + \cdots + c_0(x) = 0.$$

Then the result for (7.3.1) was stated as follows:

Theorem 7.3.1 *If the condition (Poincaré condition)*

$$\left|\sum_{1\le i\le d} k_i a_{i,i}(0) - \rho_p(0)\right| \ge \sigma\,(k_1 + k_2 + \cdots + k_d)$$

$$\text{for } p = 1, 2, \ldots, l \text{ and } (k_1, k_2, \ldots, k_d) \in \mathbb{N}^d \setminus \{(0, 0, \ldots, 0)\}$$

is satisfied for some $\sigma > 0$ then we have

1. *(7.3.1) has a unique formal solution $\hat{u}(t, x)$ of the form (7.1.2);*
2. *moreover, the formal solution $\hat{u}(t, x)$ is in the formal Gevrey class $\mathcal{E}^{\{s\}}$ for any $s \ge s_l$.*

Consider now an equation of the form

$$\text{(7.3.2)}\quad \left(\tau^l + c_{l-1}(x)\tau^{l-1} + \cdots + c_0(x)\right) u = G_2(x)\,(t_1, \ldots, t_d, D^m u) + R(x)\,(t_1, \ldots, t_d, {}^*D^m(\varphi + u))$$

where $l, m \in \mathbb{N}^*$, $c_i(x), \tau$, and $G_2(x)(t, Z)$ are the same as in (7.2.1) for $p = 1, \varphi(x) \in \mathbb{C}_r\{x\}$ for some fixed $r > 0$, and

$${}^*D^m w = \left((D_t)^b\,(D_x)^c\,w : |b| + |c| \le m\right).$$

Put

$$J\,({}^*D^m) = \left\{(b, c) \in \mathbb{N}^d \times \mathbb{N}^n : |b| + |c| \le m\right\}.$$

This is a finite set. Denote by ξ the cardinal of $J\,({}^*D^m)$. We write

$$\begin{aligned} y &= (y_{b,c} : (b, c) \in J\,({}^*D^m)) \in \mathbb{C}^\xi, \\ \pi &= (\pi_{b,c} : (b, c) \in J\,({}^*D^m)) \in \mathbb{N}^\xi, \\ L(\pi) &= \sum_{(b,c)\in J({}^*D^m)} |b|\,\pi_{b,c}, \end{aligned}$$

and for $p \in \mathbb{N}$ we denote by Δ_p the set of all $(\mu, \pi) \in \mathbb{N}^d \times \mathbb{N}^\xi$ satisfying the following:

i) $|\mu| \ge L(\pi) + 1$ and

ii) $|\mu| + p \ge L(\pi) + |\pi|$.

Note that by ii) we have $(|\mu| - L(\pi)) + p \ge |\pi|$ and therefore for any $k \in \mathbb{N}^*$ the set $\{(\mu, \pi) \in \Delta_p : |\mu| - L(\pi) = k\}$ is a finite set .

For $R(x)(t, y)$ we assume the following condition:

$$R(x)(t, y) = \sum_{(\mu,\pi)\in\Delta_p} b_{\mu,\pi}(x) t^\mu y^\pi$$

for some $p \in \mathbb{N}$, where $b_{\mu,\pi}(x) \in \mathbb{C}_r\{x\}$ and

$$|b_{\mu,\pi}(x)| \le B_{\mu,\pi} \quad \text{in } D_r \text{ for } (\mu,\pi) \in \Delta_p.$$

Moreover, the series

$$\sum_{(\mu,\pi)\in\Delta_p} B_{\mu,\pi}\lambda^{|\mu|-L(\pi)}$$

is a convergent power series in λ.

Then we have

Lemma 7.3.2 *Theorem 7.3.1 is valid for (7.3.2). This means that if the Poincaré condition is satisfied then each formal solution of (7.3.2) is in the formal Gevrey class $\mathcal{E}^{\{s\}}$ for any $s \ge s_l(7.3.2)$.*

Proof : Since $|\mu| \ge L(\pi)+1$ is valid for $(\mu,\pi) \in \Delta_p$ we can rewrite $R(x)(t, {}^*D^m(\varphi+u))$ into the form

$$R(x)\,(t_1,\ldots,t_d, {}^*D^m(\varphi+u)) = \sum_{1\le i\le d} t_i G^{(i)}(x)\,(t_1,\ldots,t_d, D^m u)$$

with

$$G^{(i)}(x)(t,Z) = \sum_{|\alpha|+|\nu|\ge 0} a^{(i)}_{\alpha,\nu}(x) t^\alpha Z^\nu$$

for some $a^{(i)}_{\alpha,\nu} \in \mathbb{C}_r\{x\}$. Of course, this expression is not unique. But in any expression the inequality

$$(7.3.3)\quad \sum_{1\le i\le d}\ \sum_{|\alpha|+|\nu|\ge 0} \left\|a^{(i)}_{\alpha,\nu}\right\|_r \lambda^{|\alpha|+|\nu|+1} \le \sum_{(\mu,\pi)\in\Delta_p} B_{\mu,\pi}\lambda^{|\mu|-L(\pi)}(C+\lambda)^{|\pi|}$$

is valid for any $\lambda > 0$ and $C > 0$ satisfying

$$C \ge \|(D_x)^c\,\varphi\|_r \quad \text{for } |c| \le m,$$

where for $\varphi(x) \in \mathbb{C}_r\{x\}$

$$\|\varphi\|_r = \max_{x\in D_r} |\varphi(x)|\,.$$

In order to see that we can apply theorem 7.3.1 to the equation

$$\begin{aligned}&\left(\tau^l + c_{l-1}(x)\tau^{l-1} + \cdots + c_0(x)\right) u \\ &\qquad = G_2(x)\,(t_1,\ldots,t_d, D^m u) + \sum_{1\le i\le d} t_i G^{(i)}(x)\,(t_1,\ldots,t_d, D^m u)\end{aligned}$$

it is sufficient to note that the series

$$(7.3.4)\qquad \sum_{1\le i\le d}\ \sum_{|\alpha|+|\nu|\ge 0} \left\|a^{(i)}_{\alpha,\nu}\right\|_r \lambda^{|\alpha|+|\nu|+1}$$

is a convergent power series in λ. Let us show this now.

Put $h = C + 1 \geq 1$. Using $|\mu| + p \geq L(\pi) + |\pi|$ we have for any $0 < \lambda < 1$

$$\begin{aligned}(7.3.5)\qquad \lambda^{|\mu|-L(\pi)}(C+\lambda)^{|\pi|} &= \frac{1}{\lambda^p}\lambda^{(|\mu|+p)-(L(\pi)+|\pi|)}\left(C\lambda+\lambda^2\right)^{|\pi|} \\ &\leq \frac{1}{\lambda^p}(h\lambda)^{(|\mu|+p)-(L(\pi)+|\pi|)}(h\lambda)^{|\pi|} \\ &= h^p(h\lambda)^{|\mu|-L(\pi)}.\end{aligned}$$

Hence, by (7.3.3) and (7.3.5) we obtain

$$\sum_{1\leq i\leq d}\ \sum_{|\alpha|+|\nu|\geq 0} \left\| a^{(i)}_{\alpha,\nu}\right\|_r \lambda^{|\alpha|+|\nu|+1} \leq h^p \sum_{(\mu,\pi)\in\Delta_p} B_{\mu,\pi}(h\lambda)^{|\mu|-L(\pi)} < +\infty$$

for sufficiently small $\lambda > 0$, which proves that the series (7.3.4) is a convergent power series in λ. The proof of lemma 7.3.2 is completed. ■

7.4 Proof of theorem 7.2.5

Our plan of the proof of theorem 7.2.5 is as follows: first we transform the equation (7.2.1) into an equation with respect to $\varphi \in \mathbb{C}_r\{x\}$ and $w \in \mathfrak{M}_r\{x\}\,[[t_1,\ldots,t_d]]$ under the relation

$$(7.4.1)\qquad u = t_1(\varphi + w)$$

and then we will apply lemma 7.3.2 to the reduced equation.

First we substitute (7.4.1) into both sides of (7.2.1). The left hand side of (7.2.1) is expressed as follows:

$$\begin{aligned}(7.4.2)\quad & F_p(x)\left(t_1, u, \tau u, \ldots, \tau^l u\right) \\ &= (t_1)^p \cdot F_p(x)\left(1, \varphi + w, (a_{1,1})\,\varphi + (\tau + a_{1,1})\,w, \ldots, (a_{1,1})^l\,\varphi + (\tau + a_{1,1})^l\,w\right) \\ &= (t_1)^p \cdot \Bigg\{ F_p(x)\left(1, \varphi, (a_{1,1})\,\varphi, \ldots, (a_{1,1})^l\,\varphi\right) + \\ &\quad + \sum_{0\leq i\leq l} \frac{\partial F_p}{\partial X_i}(x)\left(1, \varphi, (a_{1,1})\,\varphi, \ldots, (a_{1,1})^l\,\varphi\right)(\tau + a_{1,1})^i\,w + K_2\left(\varphi, D^l w\right)\Bigg\}\end{aligned}$$

with

$$\begin{aligned} & K_2\left(\varphi, D^l w\right) \\ &= \sum_{i_0+\cdots+i_l\geq 2} \frac{1}{i_0!\cdots i_l!}\frac{\partial^{i_0+\cdots+i_l}F_p}{(\partial X_0)^{i_0}\cdots(\partial X_l)^{i_l}}(x)\left(1, \varphi, (a_{1,1})\,\varphi, \ldots, (a_{1,1})^l\,\varphi\right) \times \\ &\qquad\qquad \times (w)^{i_0}\left((\tau + a_{1,1})\,w\right)^{i_1}\cdots\left((\tau + a_{1,1})^l\,w\right)^{i_l}.\end{aligned}$$

Since (C2) is assumed the assertion 2. of lemma 7.2.7 is valid. Therefore, the right hand side of (7.2.1) is expressed as follows:

$$G_{p+1}(x)\,(t_1,\ldots,t_d,D^m u) = (t_1)^p \cdot R(x)\,(t_1,\ldots,t_d,{}^*D^m(\varphi+w)) \tag{7.4.3}$$

with

$$\begin{aligned}
&R(x)\,(t_1,\ldots,t_d,{}^*D^m(\varphi+w))\\
&= \sum_{|\mu|+|\nu|\geq p+1} g_{\mu,\nu}(x) t^{\mu+s(\nu)+(|\nu|_0-p)e_1} \times \prod_{(a,b,c)\in I_0} \left[(D_t)^b (D_x)^c (\varphi+w)\right]^{\nu_{a,b,c}} \times \\
&\quad \times \prod_{(a,b,c)\in I_+} \left[t_1 (D_t)^b (D_x)^c (\varphi+w) + b_1 (D_t)^{b-e_1} (D_x)^c (\varphi+w)\right]^{\nu_{a,b,c}}
\end{aligned}$$

where

$$\begin{aligned}
e_1 &= (1,0,\ldots,0) \in \mathbb{N}^d,\\
s_i(\nu) &= \sum_{(a,b,c)\in I(D^m)} a_i \nu_{a,b,c}, \quad \text{for } i=1,2,\ldots,d,\\
s(\nu) &= (s_1(\nu),\ldots,s_d(\nu)),\\
I_0 &= \{(a,b,c)\in I(D^m) : b_1 = 0\},\\
I_+ &= \{(a,b,c)\in I(D^m) : b_1 > 0\},\\
|\nu|_0 &= \sum_{(a,b,c)\in I_0} \nu_{a,b,c}.
\end{aligned}$$

Note that the condition (C2) is equivalent to the following: if $g_{\mu,\nu}(x) \not\equiv 0$ then $\mu_1 + s_1(\nu) + |\nu|_0 \geq p$ holds. If $\varphi \in \mathbb{C}_r\{x\}$ and $w \in \mathfrak{M}_r\{x\}\,[[t_1,\ldots,t_d]]$ it is easy to see

$$R(x)\,(t_1,\ldots,t_d,{}^*D^m(\varphi+w)) \in \bigoplus_{k\geq 1} H_{k,r}\{x\}\,[t_1,\ldots,t_d]\,.$$

Hence, by (7.4.2) and (7.4.3) we see that (7.2.1) is equivalent to the following equations with unknown $\varphi \in \mathbb{C}_r\{x\}$ and $w \in \mathfrak{M}_r\{x\}\,[[t_1,\ldots,t_d]]$:

$$F_p(x)\left(1,\varphi,(a_{1,1})\varphi,\ldots,(a_{1,1})^l\varphi\right) = 0, \tag{7.4.4}$$

$$\begin{aligned}
&\sum_{0\leq i\leq l} \frac{\partial F_p}{\partial X_i}(x)\left(1,\varphi,(a_{1,1})\varphi,\ldots,(a_{1,1})^l\varphi\right)(\tau+a_{1,1})^i w\\
&\qquad = -K_2\left(\varphi,D^l w\right) + R(x)\,(t_1,\ldots,t_d,{}^*D^m(\varphi+w))\,.
\end{aligned} \tag{7.4.5}$$

Now let us solve (7.4.4) and (7.4.5).

By the definition of $H(x,X)$ the equation (7.4.4) is nothing but the equation $H(x,\varphi) = 0$. Since (C3) is assumed $H(x,X) = 0$ has p holomorphic solutions $X = \varphi^{(1)}(x),\ldots,\varphi^{(p)}(x) \in \mathbb{C}\{x\}$.

Thus, (7.4.4) is equivalent to the condition that φ is equal to one of $\varphi^{(1)}(x), \dots, \varphi^{(p)}(x)$.

Put $\varphi = \varphi^{(j)}(x)$ and let us solve (7.4.5). By the definition of $C_i^{(j)}(x)$ (7.4.5) is nothing but the equation

$$(7.4.6) \quad \left(C_l^{(j)}(x)\,(\tau + a_{1,1})^l + C_{l-1}^{(j)}(x)\,(\tau + a_{1,1})^{l-1} + \cdots + C_0^{(j)}(x)\right) w \\ = -K_2\left(\varphi, D^l w\right) + R(x)\,(t_1, \dots, t_d, {}^*D^m(\varphi + w)).$$

It is easy to see that $K_2\left(\varphi, D^l w\right)$ is a particular form of $G_2(x)\,(t_1, \dots, t_d, D^l w)$ in (7.3.2) and that $R(x)\,(t_1, \dots, t_d, {}^*D^m(\varphi + w))$ satisfies the same condition as in (7.3.2). Since for this j (C4) is assumed $C_l^{(j)}(x) \neq 0$ holds near the origin of $(C_x)^n$ and the condition

$$\left| \sum_{1 \le i \le d} k_i a_{i,i}(0) + a_{1,1}(0) - \rho_h^{(j)}(0) \right| \ge \sigma\,(k_1 + \cdots + k_d)$$

$$\text{for any } (k_1, \dots, k_d) \in \mathbb{N}^d \setminus \{(0, \dots, 0)\} \text{ and } h = 1, 2, \dots, l$$

is satisfied for some $\sigma > 0$. Note that the following relation holds:

$$C_l^{(j)}(x)\,(\tau + a_{1,1}(x))^l + C_{l-1}^{(j)}(x)\,(\tau + a_{1,1}(x))^{l-1} + \cdots + C_0^{(j)}(x) \\ = C_l^{(j)}(x)\left(\tau + a_{1,1}(x) - \rho_1^{(j)}(x)\right) \cdots \left(\tau + a_{1,1}(x) - \rho_l^{(j)}(x)\right).$$

Hence, by lemma 7.3.2 we obtain

1. (7.4.6) has a unique formal solution $w_j^*(t,x) \in \mathfrak{M}_r\{x\}\,[[t_1, \dots, t_d]]$;
2. moreover, the $w_j^*(t,x)$ is in the formal Gevrey class $\mathcal{E}^{\{s\}}$ for any $s \ge s_l(7.4.6)$.

Above we put

$$\hat{u} = t_1\,(\varphi + w^*)$$

and have seen that our equation (7.2.1) has p formal solutions $\hat{u}_j = t_1\left(\varphi^{(j)} + w_j^*\right)$. The formal power series w_j^* is a formal solution of the equation (7.4.6) which is of the form (7.3.2) and satisfies the Poincaré condition (C4) for this j. As a consequence w_j^* is in $\mathcal{E}^{\{s\}}$ with $s \ge s_l(7.4.6)$ and the same is true for the formal solution $\hat{u}_j$ of (7.2.1). Then the number s_l' is given by

$$s_l' = s_l(7.2.1) = \max_{j=1,\dots,p} \{s_l(7.4.6)\}.$$

And the proof of theorem 7.2.5 is completed. ■

7.5 Complementary results

In chapter 6, we proved also for the equation (7.3.1) the following:

Theorem 7.5.1 *Assume that* 0 *does not belong to the convex hull of the set* $\{a_{1,1}(0), a_{2,2}(0), \dots, a_{d,d}(0)\}$ *in* $\mathbb{C}$. *Then if (7.3.1) has a formal solution* $\hat{u}(t,x)$ *of valuation* q *it is in the formal Gevrey class* $\mathcal{E}^{\{s\}}$ *for any* $s \geq s_l(\hat{u})$ *where*

$$s_l(\hat{u}) = s_l(7.3.1)(\hat{u}) = 1 + \max\left\{0, \sup_{(\mu,\nu)\in\Delta}\left(\frac{m_\nu - l}{|\mu| + q(|\nu| - 1)}\right)\right\}.$$

Put

$$C_l(x,X) = \frac{\partial F_p}{\partial X_l}(x)\left(1, X, (a_{1,1}(x))X, \dots, (a_{1,1}(x))^l X\right)$$

which is a polynomial of degree $p-1$ with coefficients in $\mathbb{C}\{x\}$.

Then we also have

Theorem 7.5.2 *Assume (C1), (C2), that the two equations* $H(0,X) = 0$ *and* $C_l(0,X) = 0$ *have no common roots, and that* 0 *does not belong to the convex hull of the set* $\{a_{1,1}(0), \dots, a_{d,d}(0)\}$ *in* $\mathbb{C}$. *Then if (7.2.1) has a formal solution of the form (7.2.5) it must be in the formal Gevrey class* $\mathcal{E}^{\{s\}}$ *for any* $s \geq s_l(7.5.2)(w^*)$ *where* w^* *and (7.5.2) are defined later.*

Proof : Let $\hat{u}(t,x)$ be a formal solution of (7.2.1) of the form (7.2.5). Then $\hat{u}$ is expressed in the form

$$\hat{u} = t_1(\varphi + w^*)$$

for some $\varphi \in \mathbb{C}_r\{x\}$ and $w^* \in \mathfrak{M}_r\{x\}[[t_1, \dots, t_d]]$. Since (7.2.1) is equivalent to the equations (7.4.4) and (7.4.5) we can see that $\varphi(x)$ satisfies the equation

$$H(x, \varphi(x)) = 0 \tag{7.5.1}$$

near the origin of $(\mathbb{C}_x)^n$ and that w^* is a formal solution of the equation

$$\begin{aligned}(7.5.2)\quad & C_l(x,\varphi(x))(\tau + a_{1,1})^l w \\ & + \sum_{0\leq i\leq l-1} \frac{\partial F_p}{\partial X_i}(x)\left(1, \varphi, (a_{1,1})\varphi, \dots, (a_{1,1})^l\varphi\right)(\tau + a_{1,1})^i w \\ & = -K_2\left(\varphi, D^l w\right) + R(x)\left(t_1, \dots, t_d, {}^*D^m(\varphi + w)\right).\end{aligned}$$

Since $H(0,X) = 0$ and $C_l(0,X) = 0$ have no common roots (7.5.1) implies $C_l(0,\varphi(0)) \neq 0$ and, therefore, we see that (7.5.2) is an equation of type (7.3.2). Thus by lemma 7.3.2 and theorem 7.5.1 we can see that the formal solution w^* of (7.5.2) must be in the formal Gevrey class $\mathcal{E}^{\{s\}}$ for any $s \geq s_l(7.5.2)(w^*)$ which proves theorem 7.5.2. ∎

From now we write

$$s'_l(\hat{u}) = s_l(7.2.1)(\hat{u}) = s_l(7.5.2)(w^*).$$

In this chapter, we always assumed that $l \geq 1$ and $m \geq 1$. But our discussion is valid also in the case $l = 0$ or $m = 0$ as is stated in the following.

Let $m \geq 0$, let us consider

$$(7.5.3) \quad c_p(x)u^p + c_{p-1}(x)t_1u^{p-1} + \cdots + c_1(x)(t_1)^{p-1}u + c_0(x)(t_1)^p = G_{p+1}(x)(t_1, \ldots, t_d, D^m u)$$

and put

$$\begin{aligned} H(x, X) &= c_p(x)X^p + c_{p-1}(x)X^{p-1} + \cdots + c_1(x)X + c_0(x), \\ C(x, X) &= \frac{\partial H}{\partial X}(x, X) = pc_p(x)X^{p-1} + (p-1)c_{p-1}(x)X^{p-2} + \cdots + c_1(x). \end{aligned}$$

Corollary 7.5.3 *Assume that (C1) and (C2) hold. Then we have:*

(1) If $c_p(0) \neq 0$, if $H(0, X) = 0$ has p distinct roots $\lambda_1, \ldots, \lambda_p$, and if $C(0, \lambda_i) \neq 0$ for $i = 1, \ldots, p$ then (7.5.3) has p formal solutions $\hat{u}_1(t, x), \ldots, \hat{u}_p(t, x)$ of the form (7.2.5). Moreover, these formal solutions are in the formal Gevrey class $\mathcal{E}^{\{s\}}$ for any $s \geq s_0(7.5.3)$.

(1) If the two equations $H(0, X) = 0$ and $C(0, X) = 0$ have no common roots then every formal solution $\hat{u}(t, x)$ of (7.5.3) of the form (7.2.5) is in the formal Gevrey class $\mathcal{E}^{\{s\}}$ for any $s \geq s_0(7.5.3)(\hat{u})$.*

(2) When $m = 0$ the above result (1) (resp. (1)) is nothing but the implicit function theorem (resp. the modified implicit function theorem) (see chapter 6, section 6.1).*

Corollary 7.5.4 *1. Assume $s_l(7.2.1) = 1$ (this is the case when $l \geq m$) and the same condition as in theorem 7.2.5. Then (7.2.1) has p solutions $u_1(t, x), \ldots, u_p(t, x)$ holomorphic near the origin of $(\mathbb{C}_t)^d \times (\mathbb{C}_x)^n$ satisfying $u_i(0, x) = 0$.*

2. Assume $s_l(7.2.1) = 1$ and the same condition as in theorem 7.5.2. Then every formal solution of (7.2.1) is convergent and the operator

$$u \longmapsto F_p(x)(t_1, u, \tau u, \ldots, \tau^l u) - G_{p+1}(x)(t_1, \ldots, t_d, D^m u)$$

is regular singular.

By *regular singular* we mean the following: if $\hat{u} \in \mathfrak{M}\{x\}[[t_1, \ldots, t_d]]$ is such that $F_p(x)(t_1, \hat{u}, \tau\hat{u}, \ldots, \tau^l\hat{u}) - G_{p+1}(x)(t_1, \ldots, t_d, D^m\hat{u}) \in \mathfrak{M}\{x\}\{t_1, \ldots, t_d\}$ then $\hat{u} \in \mathfrak{M}\{x\}\{t_1, \ldots, t_d\}$.

Corollary 7.5.5 *Assume that $d = 1$ and that x does not occur in (7.2.1). Then for every formal solution $\hat{u}$ of (7.2.1) we have*

$$s_l(7.2.1)(\hat{u}) = 1 + \frac{1}{r}$$

where r is the number introduced by Malgrange in [37].

7.6 A remark

In theorem 7.2.5 and theorem 7.5.2, we restricted ourselves to the study of formal solutions only of the form (7.2.5) (not of the form (7.1.2)). But this restriction seems not to be essential as is seen below.

Put

$$h_i(x) = F_p(x)\left(0, 1, a_{i,i}(x), \ldots, (a_{i,i}(x))^l\right) \qquad \text{for } i = 2, \ldots, d,$$

$$C_j(x, X) = \frac{\partial F_p}{\partial X_j}(x)\left(1, X, (a_{1,1})\, X, \ldots, (a_{1,1})^l\, X\right) \quad \text{for } j = 0, 1, \ldots, l.$$

Recall that $h_1(x)$ and $C_l(x, X)$ are already defined. For $\lambda \in \mathbb{C}$ satisfying $C_l(0, \lambda) \neq 0$ denote by $\rho_1(0; \lambda), \ldots, \rho_l(0; \lambda)$ the roots of

$$C_l(0, \lambda)\rho^l + C_{l-1}(0, \lambda)\rho^{l-1} + \cdots + C_0(0, \lambda) = 0.$$

Proposition 7.6.1 *Assume (C1), (C2), and the following conditions:*

1. $h_i(0) \neq 0$ *for* $i = 2, \ldots, d$;
2. *if* $\lambda \in \mathbb{C}$ *satisfies* $H(0, \lambda) = 0$ *then we have* $C_l(0, \lambda) \neq 0$ *and*

$$\sum_{2 \leq i \leq d} k_i a_{i,i}(0) - \rho_h(0, \lambda) \neq 0$$

for any $h = 1, \ldots, l$ *and any* $(k_2, \ldots, k_d) \in \mathbb{N}^{d-1}$ *satisfying* $k_2 + \cdots + k_d \geq 2$.

Then if $\hat{u}(t, x)$ *is a formal solution of (7.2.1) of the form (7.1.2) it must take the form (7.2.5).*

Proof : Let $\hat{u}$ be a formal solution of (7.2.1) of the form

$$\hat{u} = \sum_{k \geq 1} u_k \quad \text{where } u_k \in H_{k,r}\{x\}\,[t_1, \ldots, t_d] \text{ for } k = 1, 2, \ldots.$$

Our aim is to prove that

$$u_k \in t_1 H_{k-1,r}\{x\}\,[t_1, \ldots, t_d] \tag{7.6.1}$$

holds for any $k = 1, 2, \ldots$. We will show this by induction on k.

Put

$$u_1(t, x) = \varphi_1(x) t_1 + \cdots + \varphi_d(x) t_d$$

and substitute this into (7.2.1). Then by comparing the coefficients of $(t_d)^p$ in both sides of (7.2.1) we get

$$h_d(x) \cdot (\varphi_d(x))^p = 0.$$

Since $h_d(0) \neq 0$ is assumed and since $\varphi_d(x) \in \mathbb{C}_r\{x\}$ we obtain $\varphi_d(x) = 0$ on D_r and

$$u_1(t,x) = \varphi_1(x)t_1 + \cdots + \varphi_{d-1}(x)t_{d-1}.$$

Again, by substituting this into (7.2.1) and by comparing the coefficients of $(t_{d-1})^p$ in both sides of (7.2.1) we have

$$h_{d-1}(x) \cdot (\varphi_{d-1}(x))^p = 0.$$

Since $h_{d-1}(0) \neq 0$ is assumed we obtain $\varphi_{d-1}(x) = 0$ on D_r and

$$u_1(t,x) = \varphi_1(x)t_1 + \cdots + \varphi_{d-2}(x)t_{d-2}.$$

By repeating the same argument as above we obtain $\varphi_i(x) = 0$ on D_r for $i = 2, \ldots, d$. Thus, (7.6.1) is proved for $k = 1$.

Let $k^* \geq 2$. Suppose that (7.6.1) is already proved for any $k < k^*$. Since u_1 is expressed in the form $\varphi_1 t_1$, by the same argument as in the proof of theorem 7.2.5 we have $H(x, \varphi_1) = 0$ and, therefore,

$$\lambda = \varphi_1(0)$$

satisfies $H(0, \lambda) = 0$. Since (C2) is assumed, by comparing the $p + k^* - 1$ homogeneous part in both sides of (7.2.1) we obtain

$$\begin{aligned}(t_1)^{p-1} \left(C_l\left(x,\varphi_1\right)\tau^l + C_{l-1}\left(x,\varphi_1\right)\tau^{l-1} + \cdots + C_0\left(x,\varphi_1\right)\right) u_{k^*} \\ \in (t_1)^p \cdot H_{k^*-1,r}\{x\}\left[t_1, \ldots, t_d\right]\end{aligned}$$

and, hence,

$$\begin{aligned}\text{(7.6.2)} \quad \left(C_l\left(x,\varphi_1\right)\tau^l + C_{l-1}\left(x,\varphi_1\right)\tau^{l-1} + \cdots + C_0\left(x,\varphi_1\right)\right) u_{k^*} \\ \in t_1 \cdot H_{k^*-1,r}\{x\}\left[t_1, \ldots, t_d\right].\end{aligned}$$

Here we decompose u_{k^*} into

$$\begin{aligned} u_{k^*} &= v_{k^*} + w_{k^*}, \\ v_{k^*} &\in t_1 \cdot H_{k^*-1,r}\{x\}\left[t_1, \ldots, t_d\right], \\ w_{k^*} &\in H_{k^*,r}\{x\}\left[t_2, \ldots, t_d\right] \end{aligned}$$

and also we decompose τ into

$$\begin{aligned} \tau &= \tau_1 + \tau_2, \\ \tau_1 &= t_1\left(a_{1,1}(x)\frac{\partial}{\partial t_1} + \cdots + a_{d,1}(x)\frac{\partial}{\partial t_d}\right), \\ \tau_2 &= \sum_{2\leq i\leq d}\left(\sum_{2\leq j\leq i} a_{i,j}(x)t_j\right)\frac{\partial}{\partial t_i}. \end{aligned}$$

Then it is easy to see that (7.6.2) implies

$$\Big(C_l\left(x,\varphi_1\right)(\tau_2)^l + C_{l-1}\left(x,\varphi_1\right)(\tau_2)^{l-1} + \cdots + C_0\left(x,\varphi_1\right)\Big) w_{k^*} = 0$$

and, therefore, $w_{k^*} = 0$ in $H_{k^*,r}\{x\}[t_2,\ldots,t_d]$ because $C_l(0,\varphi_1(0)) = C_l(0,\lambda) \neq 0$ and

$$\sum_{2\leq i\leq d} k_i a_{i,i}(0) - \rho_h(0,\lambda) \neq 0 \quad \text{for } k_2+\cdots+k_d = k^* \text{ and } h = 1,2,\ldots,l$$

are assumed in the condition 2. of proposition 7.6.1. Thus, (7.6.1) is also proved for $k = k^*$. ■

Remark 7.6.2 *When $d = 1$ the condition (C2) is trivially satisfied and proposition 7.6.1 means nothing.*

8 Holomorphic and singular solutions of non linear singular partial differential equations

Introduction

In chapter 5, we have studied holomorphic and singular solutions of non linear singular partial differential equations of the first order called *Briot–Bouquet type.* In the present chapter, we are extending the results of chapter 5 to equations of higher order.

Denote

$$\begin{aligned} \mathbb{N} &= \{0,1,2,\dots\} \text{ the set of non negative integers,} \\ \mathbb{N}^* &= \mathbb{N}\setminus\{0\}, \\ x &= (x_1,\dots,x_n)\in\mathbb{C}^n, \qquad t\in\mathbb{C}. \end{aligned}$$

For any $\alpha=(\alpha_1,\dots,\alpha_n)\in\mathbb{N}^n$ we denote

$$|\alpha| = \alpha_1+\cdots+\alpha_n \qquad \text{and} \qquad \left(\frac{\partial}{\partial x}\right)^{\alpha} = \left(\frac{\partial}{\partial x_1}\right)^{\alpha_1}\cdots\left(\frac{\partial}{\partial x_n}\right)^{\alpha_n}.$$

Define

$$\begin{aligned} I_m &= \{(j,\alpha)\in\mathbb{N}\times\mathbb{N}^n : j+|\alpha|\le m \text{ and } j<m\}, \\ \delta &= \text{the cardinal of } I_m, \\ Z &= \{Z_{j,\alpha}\}_{(j,\alpha)\in I_m}\in\mathbb{C}^{\delta}. \end{aligned}$$

Let Δ be a polydisk centered at the origin of $\mathbb{C}_t\times(\mathbb{C}_x)^n\times(\mathbb{C}_Z)^{\delta}$. Denote

$$\Delta_0 = \Delta\cap\{t=0 \text{ and } Z=0\}.$$

Let $F(t,x,Z)$ be a function of (t,x,Z) defined in the polydisk Δ and satisfying the following hypothesis:

(A1) $F(t,x,Z)$ is holomorphic in Δ;

(A2) $F(0,x,0)\equiv 0$ in Δ_0;

(A3) $\dfrac{\partial F}{\partial Z_{j,\alpha}}(0,x,0)\equiv 0$ in Δ_0 if $|\alpha|>0$.

Let us also introduce

- $\mathfrak{R}(\mathbb{C}\setminus\{0\})$ the universal covering space of $\mathbb{C}\setminus\{0\}$,

- $S_\theta = \{t \in \mathfrak{R}(\mathbb{C}\setminus\{0\}) : |\arg t| < \theta\}$ a sector in $\mathfrak{R}(\mathbb{C}\setminus\{0\})$,
- $S(\varepsilon(s)) = \{t \in \mathfrak{R}(\mathbb{C}\setminus\{0\}) : 0 < |t| < \varepsilon(\arg t)\}$ where $\varepsilon(s)$ is a continuous and positive–valued function defined on $\mathbb{R}$,
- $D_r = \{x \in \mathbb{C}^n : |x_i| \le r \text{ for } i = 1, \dots, n\}$,
- $\mathbb{C}\{x\}$ the ring of germs of holomorphic functions at the origin of $\mathbb{C}^n$,
- $\mathbb{C}_r\{x\}$ the subring of $\mathbb{C}\{x\}$ of holomorphic functions in D_r,
- $\widetilde{\mathcal{O}}_+$ the set of functions $u(t,x)$ satisfying the following conditions:
  1. there exist a continuous and positive–valued function $\varepsilon(s)$ defined on $\mathbb{R}$ and a positive real number r such that $u(t,x)$ is holomorphic in $S(\varepsilon(s)) \times D_r$ and
  2. there exists a real number $a > 0$ such that for any $\theta > 0$ and any compact subset K of D_r

$$\max_{x\in K} |u(t,x)| = O\left(|t|^a\right) \quad \text{as } t \text{ tends to zero in } S_\theta.$$

Problem: Find all the solutions of the partial differential equation

$$\left(t\frac{\partial}{\partial t}\right)^m u = F\left(t, x, \left\{\left(t\frac{\partial}{\partial t}\right)^j \left(\frac{\partial}{\partial x}\right)^\alpha u\right\}_{(j,\alpha)\in I_m}\right) \tag{8.0.1}$$

belonging to $\widetilde{\mathcal{O}}_+$.

Denote by $\rho_1(x), \dots, \rho_m(x)$ the functions of x which are the roots of the polynomial

$$C(\rho, x) = \rho^m - \sum_{j<m} \frac{\partial F}{\partial Z_{j,0}}(0,x,0)\rho^j.$$

These functions are called the *characteristic exponents* of the equation (8.0.1) and $C(\rho,x)$ is called the *characteristic polynomial* of (8.0.1).

Denote

$$\begin{aligned} J_+ &= \{i \in \{1,\dots,m\} : \operatorname{Re}\rho_i(0) > 0\}, \\ \mu &= \text{the cardinal of } J_+. \end{aligned} \tag{8.0.2}$$

When $\mu = 0$ this condition means that $\operatorname{Re}\rho_i(0) \le 0$ for all $i = 1, \dots, m$.
When $\mu \ge 1$ we can assume after a change of indices that

$$\operatorname{Re}\rho_i(0) \begin{cases} > 0 & \text{for } 1 \le i \le \mu, \\ \le 0 & \text{for } \mu+1 \le i \le m. \end{cases} \tag{8.0.3}$$

Under the Poincaré condition (cf. chapter 5, [21]) we can give an answer to our problem as follows:

Theorem 8.0.3 (Main theorem) (I) (Holomorphic solutions) *If $\rho_i(0) \notin \mathbb{N}^*$ for all $i = 1, \ldots, m$ the equation (8.0.1) has a unique solution $u_0(t,x)$ holomorphic near the origin of $\mathbb{C} \times \mathbb{C}^n$ satisfying $u_0(0,x) \equiv 0$.*

(II) (Singular solutions) *Let S_+ be the set of solutions of (8.0.1) belonging to $\widetilde{\mathcal{O}}_+$. Then we have*

(a) if $\mu = 0$ then $S_+ = \{u_0(t,x)\}$;

(b) if $\mu \geq 1$ and if

i. $\rho_i(0) \neq \rho_j(0)$ for $1 \leq i \neq j \leq \mu$,

ii. $C(1,0) \neq 0$ and

iii. $C\left(i + j_1\rho_1(0) + \cdots + j_\mu\rho_\mu(0), 0\right) \neq 0$ for all $(i,j) \in \mathbb{N}\times\mathbb{N}^\mu$ satisfying $i + |j| \geq 2$

then for any $\varphi_1(x), \ldots, \varphi_\mu(x) \in \mathbb{C}\{x\}$ the equation (8.0.1) has a unique solution $U\left(\varphi_1(x), \ldots, \varphi_\mu(x)\right)$ belonging to $\widetilde{\mathcal{O}}_+$ of the form

$$\begin{aligned} &U\left(\varphi_1(x), \ldots, \varphi_\mu(x)\right) \\ &= \sum_{i\geq 1} u_i(x)t^i + \sum_{i+2m|j|\geq k+2m, |j|\geq 1} \varphi_{i,j,k}(x)t^{i+j_1\rho_1(x)+\cdots+j_\mu\rho_\mu(x)}(\log t)^k \end{aligned}$$

with $\varphi_{0,e_p,0}(x) = \varphi_p(x)$ for $p = 1, \ldots, \mu$ where

$$e_1 = (1, 0, \ldots, 0) \ , \ \ldots \ , \ e_\mu = (0, \ldots, 0, 1).$$

Moreover, we have

$$S_+ = \left\{U\left(\varphi_1(x), \ldots, \varphi_\mu(x)\right) : \varphi_p(x) \in \mathbb{C}\{x\} \text{ for } p = 1, \ldots, \mu\right\}.$$

Remarks 8.0.4 *1. When $m = 1$ this theorem gives us again the main theorem of chapter 5.*

2. When the equation is linear a more complete result is given in Tahara [46] and [47].

This chapter is presented as follows. In section 8.1, we give holomorphic solutions of (8.0.1). In section 8.2 and 8.3, we construct a family of solutions belonging to $\widetilde{\mathcal{O}}_+$. The discussion of the sections 8.1, 8.2, and 8.3 are based on the results of chapter 6. In section 8.4, we recall the asymptotic theory developped in Tahara [49]. Using all these results we are able to prove the main theorem of this chapter in section 8.5. See also [23] and [30].

8.1 Holomorphic solutions

In this section, let us recall a result of chapter 6 on the convergence of formal power series solutions of non linear partial differential equations.

Let $m \in \mathbb{N}^*, d \in \mathbb{N}^*, t = (t_1, \ldots, t_d) \in \mathbb{C}^d$, and $x = (x_1, \ldots, x_n) \in \mathbb{C}^n$. Let $a_{i,j}(x)$ for $1 \leq j \leq i \leq d$ be holomorphic functions in a polydisk $D_r$and introduce the vector field

$$\tau = \sum_{1 \leq i \leq d} \left(\sum_{1 \leq j \leq i} a_{i,j}(x) t_j \right) \frac{\partial}{\partial t_i}.$$

Denote

$$\begin{aligned} I(m) &= \left\{ (\alpha, \beta, \gamma) \in \mathbb{N}^d \times \mathbb{N}^d \times \mathbb{N}^n : |\alpha| = |\beta| \leq m \text{ and } |\beta| + |\gamma| \leq m \right\}, \\ \delta(m) &= \text{the cardinal of } I(m), \\ Z &= \{ Z_{\alpha,\beta,\gamma} \}_{(\alpha,\beta,\gamma) \in I(m)} . \end{aligned}$$

Let $G_2(x)\,(t_1, \ldots, t_d, Z)$ be a holomorphic function in $(t_1, \ldots, t_d, Z)$ defined in a polydisk centered at the origin of $\mathbb{C}^n \times \mathbb{C}^d \times \mathbb{C}^{\delta(m)}$ and having the following form:

$$G_2(x)\,(t_1, \ldots, t_d, Z) = \sum_{|p|+|q| \geq 2} g_{p,q}(x)\,(t_1)^{p_1} \cdots (t_d)^{p_d}\, Z^q$$

where

$$\begin{aligned} p &= (p_1, \ldots, p_d) \in \mathbb{N}^d, & |p| &= p_1 + \cdots + p_d, \\ q &= \{ q_{\alpha,\beta,\gamma} \}_{(\alpha,\beta,\gamma) \in I(m)} \in \mathbb{N}^{\delta(m)}, & |q| &= \sum_{(\alpha,\beta,\gamma) \in I(m)} q_{\alpha,\beta,\gamma}, \\ Z^q &= \prod_{(\alpha,\beta,\gamma) \in I(m)} (Z_{\alpha,\beta,\gamma})^{q_{\alpha,\beta,\gamma}} \end{aligned}$$

and for all p, q such that $|p| + |q| \geq 2$ the functions $g_{p,q}(x)$ are holomorphic in a common disk centered at the origin of $\mathbb{C}^n$.

To simplify the presentation let us write

$$D^m u = \left\{ t^\alpha \left(\frac{\partial}{\partial t} \right)^\beta \left(\frac{\partial}{\partial x} \right)^\gamma u \right\}_{(\alpha,\beta,\gamma) \in I(m)}$$

where

$$\begin{aligned} \alpha &= (\alpha_1, \ldots, \alpha_d) \in \mathbb{N}^d, & t^\alpha &= (t_1)^{\alpha_1} \cdots (t_d)^{\alpha_d}, \\ \beta &= (\beta_1, \ldots, \beta_d) \in \mathbb{N}^d, & \left(\frac{\partial}{\partial t} \right)^\beta &= \left(\frac{\partial}{\partial t_1} \right)^{\beta_1} \cdots \left(\frac{\partial}{\partial t_d} \right)^{\beta_d}, \\ \gamma &= (\gamma_1, \ldots, \gamma_n) \in \mathbb{N}^n, & \left(\frac{\partial}{\partial x} \right)^\gamma &= \left(\frac{\partial}{\partial x_1} \right)^{\gamma_1} \cdots \left(\frac{\partial}{\partial x_n} \right)^{\gamma_n}. \end{aligned}$$

Let $c_j(x)$ for $0 \leq j \leq m-1$ and $b_j(x)$ for $1 \leq i \leq d$ be holomorphic functions in D_r. We can consider now the equation

(8.1.1)
$$\left(\tau^m + c_{m-1}(x)\tau^{m-1} + \cdots + c_0(x)\right) u = \sum_{1\leq i\leq d} b_i(x)t_i + G_2(x)\left(t_1, \ldots, t_d, D^m u\right)$$

which is a particular case of the equation studied in chapter 6.

In this chapter, a *formal power series* is a formal series in t of the form

$$\hat{u}(t,x) = \sum_{|p|\geq 1} u_p(x)\,(t_1)^{p_1}\cdots(t_d)^{p_d} \tag{8.1.2}$$

where the coefficients $u_p(x)$ are all holomorphic in a common polydisk centered at the origin of $\mathbb{C}^n$. A *formal solution* of (8.1.1) is by definition a formal power series of the form (8.1.2) which satisfies formally the equation (8.1.1).

Denote by $\lambda_1(x), \ldots, \lambda_m(x)$ the roots of the polynomial

$$P(\lambda) = \lambda^m + \sum_{j<m} c_j(x)\lambda^j.$$

By section 6.3 of chapter 6 we have

Theorem 8.1.1 *If the origin of $\mathbb{C}$ does not belong to the convex hull in $\mathbb{C}$ of the set of points $\{a_{1,1}(0), \ldots, a_{d,d}(0)\}$ then*

1. *every formal solution of (8.1.1) is convergent near the origin of $\mathbb{C}^d \times \mathbb{C}^n$;*
2. *if*
$$\sum_{1\leq i\leq d} p_i a_{i,i}(0) - \lambda_j(0) \neq 0$$
for all $(p_1, \ldots, p_d) \in \mathbb{N}^d \setminus \{(0, \ldots, 0)\}$ and $j = 1, \ldots, m$

   the equation (8.1.1) has a unique formal solution which gives a unique holomorphic solution $u(t,x)$ satisfying $u(0,x) \equiv 0$.

The equation (8.0.1) introduced in the introduction being a particular case of (8.1.1) we have

Corollary 8.1.2 *If $\rho_i(0) \notin \mathbb{N}^*$ for all $i = 1, \ldots, m$ the equation (8.0.1) admits a unique solution $u(t,x)$ holomorphic near the origin of $\mathbb{C} \times \mathbb{C}^n$ and satisfying $u(0,x) \equiv 0$.*

For other results concerning the existence of holomorphic solutions we can come back to chapter 2 and 3 and also see Madi–Yoshino [34].

8.2 Singular solutions: Special case

In this section, we are studying a restricted class of singular non linear partial differential equations and we are constructing a family of singular solutions belonging to $\widetilde{\mathcal{O}}_+$.

This restricted class has the advantage in the reseach of solutions belonging to $\widetilde{\mathcal{O}}_+$ to be reduced to an equation (8.1.1) and to allow the use of theorem 8.1.1.

Let $m \in \mathbb{N}^*, (t,x) \in \mathbb{C} \times \mathbb{C}^n, x = (x_1, \ldots, x_n), \alpha = (\alpha_1, \ldots, \alpha_n) \in \mathbb{N}^n$,

$$|\alpha| = \alpha_1 + \cdots + \alpha_n, \qquad \text{and} \qquad \left(t\frac{\partial}{\partial x}\right)^\alpha = \left(t\frac{\partial}{\partial x_1}\right)^{\alpha_1} \cdots \left(t\frac{\partial}{\partial x_n}\right)^{\alpha_n}.$$

Let $b(x)$ and $c_j(x)$ for $0 \leq j \leq m-1$ be functions defined and holomorphic in a polydisk centered at the origin of $\mathbb{C}^n$.

Denote

$$Z = \{Z_{j,\alpha}\}_{j+|\alpha| \leq m} \qquad \text{where } Z_{j,\alpha} \in \mathbb{C}$$

and let $G_2(x)(t,Z)$ be a function defined and holomorphic in a polydisk centered at the origin $(x=0, t=0, Z=0)$ and having moreover the form

$$G_2(x)(t,Z) = \sum_{p+|q| \geq 2} g_{p,q}(x) t^p Z^q \tag{8.2.1}$$

where

$$\begin{aligned} q &= \{q_{j,\alpha}\}_{j+|\alpha| \leq m} \qquad \text{where } q_{j,\alpha} \in \mathbb{N}, \\ |q| &= \sum_{j+|\alpha| \leq m} q_{j,\alpha}, \\ Z^q &= \prod_{j+|\alpha| \leq m} (Z_{j,\alpha})^{q_{j,\alpha}} \end{aligned}$$

and $g_{p,q}(x)$ for $p+|q| \geq 2$ are holomorphic functions in a common polydisk centered at the origin of $\mathbb{C}^n$.

Consider the partial differential equation

$$\begin{aligned} &\left(\left(t\frac{\partial}{\partial t}\right)^m + c_{m-1}(x)\left(t\frac{\partial}{\partial t}\right)^{m-1} + \cdots + c_0(x)\right) u \\ &\qquad = tb(x) + G_2(x)\left(t, \left\{\left(t\frac{\partial}{\partial t}\right)^j \left(t\frac{\partial}{\partial x}\right)^\alpha u\right\}_{j+|\alpha| \leq m}\right) \end{aligned} \tag{8.2.2}$$

and let us look for a singular solution of this equation.

Denote by $\rho_1(x), \ldots, \rho_m(x)$ the characteristic exponents of this equation, this means, the roots of the polynomial

$$C(\rho, x) = \rho^m + \sum_{j<m} c_j(x) \rho^j.$$

We have

Theorem 8.2.1 *Let $1 \leq l \leq m$. Assume*

1. $\rho_1(x), \ldots, \rho_l(x)$ *are holomorphic in a neighborhood of* $x = 0$;
2. $\operatorname{Re}\rho_1(0) > 0, \ldots, \operatorname{Re}\rho_l(0) > 0$;
3. $C(1,0) \neq 0$;
4. $C\left(i + j_1\rho_1(0) + \cdots + j_l\rho_l(0) + k, 0\right) \neq 0$ *for all* $(i,j,k) \in \mathbb{N} \times \mathbb{N}^l \times \mathbb{N}$ *satisfying* $i + |j| + k \geq 2$.

Then the equation (8.2.2) has a family of solutions belonging to $\widetilde{\mathcal{O}}_+$ *of the form*

$$\sum_{i+|j|+k\geq 1} \varphi_{i,j,k}(x) t^i \left(t^{\rho_1(x)}\right)^{j_1} \cdots \left(t^{\rho_l(x)}\right)^{j_l} (t \log t)^k \tag{8.2.3}$$

where the coefficients $\varphi_{i,j,k}(x)$ *for* $i + |j| + k \geq 1$ *are holomorphic in a common disk centered at the origin of* $\mathbb{C}^n$. *Moreover, every solution of the form (8.2.3) is uniquely determined by the coefficients* $\{\varphi_{0,j,0}(x) : |j| = 1\}$ *which are arbitrary. If we take* $\varphi_{0,j,0}(x) = 0$ *for all* j *such that* $|j| = 1$ *the solution (8.2.3) is reduced to the unique holomorphic solution of (8.2.2).*

Proof : First we are looking for a "formal" solution of the form

$$\hat{u}(t,x) = w\left(t, t^{\rho_1(x)}, \ldots, t^{\rho_l(x)}, t\log t, x\right) \tag{8.2.4}$$

where

$$w\left(t_0, t_1, \ldots, t_l, t_{l+1}, x\right) = \sum_{i+|j|+k\geq 1} \varphi_{i,j,k}(x) \left(t_0\right)^i \left(t_1\right)^{j_1} \cdots \left(t_l\right)^{j_l} \left(t_{l+1}\right)^k$$

is a formal series in $t_0, t_1, \ldots, t_l, t_{l+1}$ with coefficients holomorphic in a common polydisk centerd at the origin of $\mathbb{C}^n$. Next we are going to prove the convergence of the series.

Put

$$t_0 = t, \qquad t_p = t^{\rho_p(x)} \quad \text{for } p = 1, \ldots, l, \qquad \text{and} \qquad t_{l+1} = t\log t.$$

We have formally

$$\begin{aligned} t\frac{\partial u}{\partial t} &= t_0\frac{\partial w}{\partial t_0} + \sum_{1\leq p\leq l} \rho_p(x) t_p \frac{\partial w}{\partial t_p} + \left(t_0 + t_{l+1}\right)\frac{\partial w}{\partial t_{l+1}}, \\ t\frac{\partial u}{\partial x_q} &= t_0\frac{\partial w}{\partial x_q} + t_{l+1} \sum_{1\leq p\leq l} \frac{\partial \rho_p(x)}{\partial x_q} t_p \frac{\partial w}{\partial t_p} \quad \text{for } q = 1, \ldots, n. \end{aligned}$$

Therefore, $\hat{u}(t,x)$ is a formal solution of (8.2.2) if and only if $w\left(t_0, t_1, \ldots, t_l, t_{l+1}, x\right)$ is a formal solution of the equation

(8.2.5)
$$\left(T^m + c_{m-1}(x)T^{m-1} + \cdots + c_0(x)\right) w = t_0 b(x) + G_2(x)\left(t_0, \{T^j X^\alpha w\}_{j+|\alpha|\le m}\right)$$

where T and $X_1, \ldots, X_n$ are the following vector fields:

$$\begin{aligned} T &= t_0 \frac{\partial}{\partial t_0} + \sum_{1\le p\le l} \rho_p(x) t_p \frac{\partial}{\partial t_p} + (t_0 + t_{l+1}) \frac{\partial}{\partial t_{l+1}}, \\ X_q &= t_0 \frac{\partial}{\partial x_q} + t_{l+1} \sum_{1\le p\le l} \frac{\partial \rho_p(x)}{\partial x_q} t_p \frac{\partial}{\partial t_p} \quad \text{for } q = 1, \ldots, n. \end{aligned}$$

We see that the equation (8.2.5) is a particular case of equation (8.1.1) discussed in section 8.1 and that the vector field T satisfies the Poincaré condition (*i.e.* 0 does not belong to the convex hull in $\mathbb{C}$ of the set of points $\{1, \rho_1(0), \ldots, \rho_l(0)\}$).

Moreover, by the conditions 2., 3., and 4. we see that there exists a real number $r > 0$ such that

(8.2.6)
$$C(1, x) \neq 0 \quad \text{in } D_r \;;$$

(8.2.7)
$$C\left(i + j_1\rho_1(x) + \cdots + j_l\rho_l(x) + k, x\right) \neq 0 \quad \text{in } D_r$$
$$\text{for all } (i, j, k) \in \mathbb{N} \times \mathbb{N}^l \times \mathbb{N} \text{ satisfying } i + |j| + k \ge 2.$$

It follows that to have the existence of solutions of the equation (8.2.2) of the form (8.2.3) belonging to $\widetilde{\mathcal{O}}_+$ it is enough to prove

Proposition 8.2.2 *Let $1 \le l \le m$. Assume that $\rho_1(x), \ldots, \rho_l(x)$ are holomorphic near $x = 0$, that 0 does not belong to the convex hull in $\mathbb{C}$ of the set of points $\{1, \rho_1(0), \ldots, \rho_l(0)\}$, and that the conditions (8.2.6) and (8.2.7) are satisfied for an $r > 0$. Then the equation (8.2.5) admits a family of solutions of the form*

(8.2.8)
$$\sum_{i+|j|+k\ge 1} \varphi_{i,j,k}(x) (t_0)^i (t_1)^{j_1} \cdots (t_l)^{j_l} (t_{l+1})^k$$

where all the coefficients $\varphi_{i,j,k}(x)$ for $i + |j| + k \ge 1$ are uniquely determined by the coefficients $\{\varphi_{0,j,0}(x) : |j| = 1\}$ which are arbitrary.

Proof of proposition 8.2.2: Set formally

$$w = \sum_{1\le s<\infty} w_s (t_0, t_1, \ldots, t_l, t_{l+1}, x)$$

where for all $s \ge 1$

(8.2.9)
$$w_s = \sum_{i+|j|+k=s} \varphi_{i,j,k}(x) (t_0)^i (t_1)^{j_1} \cdots (t_l)^{j_l} (t_{l+1})^k .$$

The equation (8.2.5) splits then into the equations

(8.2.10)
$$\begin{aligned} C(T, x) w_1 &= b(x) t_0, \\ C(T, x) w_s &= F_s(w_1, \ldots, w_{s-1}) \quad \text{for } s \ge 2, \end{aligned}$$

where $F_s(w_1, \dots, w_{s-1})$ is a homogeneous polynomial of degree s in $(t_0, t_1, \dots, t_l, t_{l+1})$ with holomorphic coefficients in x and well determined by $w_1, \dots, w_{s-1}$. Moreover, looking at (8.2.9) we see

$$(8.2.11)\quad C(T,x)w_1 = \left(C(1,x)\varphi_{1,0,0}(x) + \frac{\partial C}{\partial \rho}(1,x)\varphi_{0,0,1}(x)\right) t_0 + \sum_{1\le p\le l} C(\rho_p(x),x)\,\varphi_{0,e_p,0}(x)t_p + C(1,x)\varphi_{0,0,1}(x)t_{l+1}$$

where $e_1 = (1,0,\dots,0), \dots, e_l = (0,\dots,0,1) \in \mathbb{N}^l$, and for $s \ge 2$

$$(8.2.12)\quad C(T,x)w_s = \sum_{i+|j|+k=s} \Big\{ C(i + j_1\rho_1(x) + \cdots + j_l\rho_l(x) + k, x)\,\varphi_{i,j,k}(x) + \sum_{1\le h\le m} c_h(i,j,k,x)\varphi_{i-h,j,k+h}(x) \Big\} \times (t_0)^i (t_1)^{j_1} \cdots (t_l)^{j_l} (t_{l+1})^k$$

where by convention $\varphi_{i-h,j,k+h}(x) \equiv 0$ for $i - h < 0$ and

$$c_h(i,j,k,x) = \sum_{h\le q\le m} (q,h)c_q(x)\,(i + j_1\rho_1(x) + \cdots + j_l\rho_l(x) + k)^{q-h} \times (k+1)(k+2)\cdots(k+h)$$

with $(q,h) = \frac{q!}{(h!)(q-h)!}$ and $c_m(x) \equiv 1$.

Then by (8.2.6), (8.2.10), (8.2.11), and the fact that $C(\rho_p(x),x) \equiv 0$ for $p = 1, \dots, l$ we have

1. $\varphi_{1,0,0}(x) = \frac{b(x)}{C(1,x)}$;
2. $\varphi_{0,e_p,0}(x)$ for $p = 1, \dots, l$ are arbitrary;
3. $\varphi_{0,0,1}(x) \equiv 0$;

and by (8.2.7), (8.2.10), (8.2.12) we see that

4. $\varphi_{i,j,k}(x)$ for $i + |j| + k \ge 2$ are uniquely determined by the set $\{\varphi_{0,e_1,0}(x), \dots, \varphi_{0,e_l,0}(x)\}$ and are holomorphic in a common disk centered at the origin of $\mathbb{C}^n$.

In this way, we have obtained a family of formal solutions of the form (8.2.8). The vector field T satisfying the Poincaré condition, theorem 8.1.1 tells us that the formal solution is convergent near the origin $(t_0 = 0, t_1 = 0, \dots, t_l = 0, t_{l+1} = 0)$.

This proves proposition 8.2.2. ■

If $w(t_0, t_1, \dots, t_l, t_{l+1}, x)$ is a solution of (8.2.5) given by proposition 8.2.2 then

$$u(t,x) = w\left(t, t^{\rho_1(x)}, \dots, t^{\rho_l(x)}, t\log t, x\right)$$

is a family of solutions of (8.2.2) belonging to $\widetilde{\mathcal{O}}_+$. And this proves the existence part of theorem 8.2.1.

To prove the uniqueness of the solution of (8.2.2) of the form (8.2.3) it is enough to show that the formal solution $\hat{u}(t,x)$ of (8.2.2) of the form

$$\hat{u}(t,x) = \sum_{i+|j|+k\geq 1} \varphi_{i,j,k}(x) t^i \left(t^{\rho_1(x)}\right)^{j_1} \cdots \left(t^{\rho_l(x)}\right)^{j_l} (t\log t)^k \tag{8.2.13}$$

is uniquely determined by the set of functions $\{\varphi_{0,j,0}(x) : |j| = 1\}$.

Introduce the condition

$$\begin{aligned} &i + j_1\rho_1(x) + \cdots + j_l\rho_l(x) \neq p + q_1\rho_1(x) + \cdots + q_l\rho_l(x) \quad \text{in } \mathbb{C}\{x\} \\ &\qquad\qquad \text{for all } (i,j),(p,q) \in \mathbb{N}\times\mathbb{N}^l \text{ satisfying } (i,j)\neq(p,q). \end{aligned} \tag{8.2.14}$$

If the condition (8.2.14) is satisfied each finite subset of the set of functions

$$\left\{t^{i+j_1\rho_1(x)+\cdots+j_l\rho_l(x)}(t\log t)^k : i+|j|+k\geq 1\right\}$$

is functionally independent. In this case, it is easy to see that the coefficients $\varphi_{i,j,k}(x)$ which occur in the expression (8.2.13) of the formal solution $\hat{u}(t,x)$ are uniquely determined by the functions

$$\{\varphi_{0,j,0}(x) : |j| = 1\}.$$

When the condition (8.2.14) is not satisfied, take functions $\lambda_p(x) \in \mathbb{C}_r\{x\}$ for $p = 1,2,\ldots$ and $\nu_q(x) \in \mathbb{C}_r\{x\}$ for $q = 1,2,\ldots,d$ such that

$$\begin{aligned} \lambda_p(x) &\neq \nu_q(x) && \text{in } \mathbb{C}_r\{x\} \text{ for all } p,q, \\ \lambda_{p_1}(x) &\neq \lambda_{p_2}(x) && \text{in } \mathbb{C}_r\{x\} \text{ if } p_1\neq p_2, \\ \nu_{q_1}(x) &\neq \nu_{q_2}(x) && \text{in } \mathbb{C}_r\{x\} \text{ if } q_1\neq q_2 \end{aligned}$$

and such that, moreover, as subsets of $\mathbb{C}_r\{x\}$ we have

$$\begin{aligned} \{1\}\cup\{i+j_1\rho_1(x)+\cdots+j_l\rho_l(x) : i+|j|\geq 2\} &= \{\lambda_p(x) : p=1,2,\ldots\}, \\ \{i+j_1\rho_1(x)+\cdots+j_l\rho_l(x) : i+|j|\geq 1\} &= \{\lambda_p(x) : p=1,2,\ldots\} \\ &\quad \cup\{\nu_1(x),\ldots,\nu_d(x)\}. \end{aligned} \tag{8.2.15}$$

The formal solution can be rewritten in the form

$$\begin{aligned} \hat{u}(t,x) = &\sum_{1\leq p\leq l} \varphi_{0,e_p,0}(x)t^{\rho_p(x)} + \sum_{1\leq p<\infty}\sum_{0\leq k<\infty} \Phi^{\lambda}_{p,k}(x)t^{\lambda_p(x)}(t\log t)^k \\ &+ \sum_{1\leq q\leq d}\sum_{1\leq k<\infty} \Phi^{\nu}_{q,k}(x)t^{\nu_q(x)}(t\log t)^k + \sum_{1\leq k<\infty} \Psi_k(x)(t\log t)^k. \end{aligned}$$

By (8.2.6), (8.2.7), and (8.2.15) we obtain

$$\begin{aligned} C(\lambda_p(x)+k,x) &\neq 0 \quad \text{in } D_r \text{ for } p\geq 1 \text{ and } k\geq 0,\\ C(\nu_q(x)+k,x) &\neq 0 \quad \text{in } D_r \text{ for } q=1,\ldots,d \text{ and } k\geq 1,\\ C(k,x) &\neq 0 \quad \text{in } D_r \text{ for } k\geq 1. \end{aligned}$$

Therefore, by using the fact that $C(\rho_p(x),x)\equiv 0$ for $p=1,\ldots,l$ and by an easy computation we can see that the coefficients $\Phi^{\lambda}_{p,k}(x)$, $\Phi^{\nu}_{q,k}(x)$, and $\Psi_k(x)$ are uniquely determined by $\varphi_{0,e_1,0}(x),\ldots,\varphi_{0,e_l,0}(x)$. In this way, we have obtained the uniqueness of the formal solution of the form (8.2.3). ■

We have also

Corollary 8.2.3 *Let $1\leq l\leq m$. Assume*

1. *$\rho_1(x),\ldots,\rho_l(x)$ are holomorphic in a neighborhood of $x=0$;*
2. *$\operatorname{Re}\rho_1(0)>0,\ldots,\operatorname{Re}\rho_l(0)>0$;*
3. *$C(1,0)\neq 0$;*
4. *$C(i+j_1\rho_1(0)+\cdots+j_l\rho_l(0)+k,0)\neq 0$ for all $(i,j,k)\in\mathbb{N}\times\mathbb{N}^l\times\mathbb{N}$ satisfying $i+|j|+k\geq 2$.*

Then if (8.2.2) admits a formal solution of the form

$$(8.2.16)\qquad \hat{u}(t,x)=\sum_{i+|j|+k\geq 1}\varphi_{i,j,k}(x)t^i\left(t^{\rho_1(x)}\right)^{j_1}\cdots\left(t^{\rho_l(x)}\right)^{j_l}(t\log t)^k$$

where the coefficients $\varphi_{i,j,k}(x)$ are holomorphic in a common disk centered at the origin of $\mathbb{C}^n$, then $\hat{u}(t,x)$ is convergent in $\widetilde{\mathcal{O}}_+$.

Proof : As we have the uniqueness of the formal solution the above $\hat{u}(t,x)$ coincides with the solution given by theorem 8.2.1 and therefore is convergent in $\widetilde{\mathcal{O}}_+$. ■

8.3 Singular solutions: General case

As a generalization of section 8.2 we are constructing in this section a family of solutions belonging to $\widetilde{\mathcal{O}}_+$ of non linear partial differential equations of the form

$$(8.3.1)\qquad \left(\left(t\frac{\partial}{\partial t}\right)^m+c_{m-1}(x)\left(t\frac{\partial}{\partial t}\right)^{m-1}+\cdots+c_0(x)\right)u$$
$$=tb(x)+G_2(x)\left(t,\left\{\left(t\frac{\partial}{\partial t}\right)^j\left(\frac{\partial}{\partial x}\right)^\alpha u\right\}_{j+|\alpha|\leq m}\right).$$

The notations and the data are the same as in section 8.2.

The difference between the equations (8.2.2) and (8.3.1) is the following:

- in (8.2.2) $Z_{j,\alpha}$ corresponds to

$$\left(t\frac{\partial}{\partial t}\right)^j\left(t\frac{\partial}{\partial x}\right)^\alpha=\left(t\frac{\partial}{\partial t}\right)^j t^{|\alpha|}\left(\frac{\partial}{\partial x}\right)^\alpha;$$

- in (8.3.1) $Z_{j,\alpha}$ corresponds to

$$\left(t\frac{\partial}{\partial t}\right)^j\left(\frac{\partial}{\partial x}\right)^\alpha.$$

It is clear that an equation of type (8.2.2) is a particular case of equations of type (8.3.1).

As in section 8.2 let us denote by $\rho_1(x),\dots,\rho_m(x)$ the characteristic exponents, that is, the roots of the polynomial

$$C(\rho,x)=\rho^m+\sum_{j<m}c_j(x)\rho^j.$$

We have

Theorem 8.3.1 *Let $1\le l\le m$. Assume*

1. *$\rho_1(x),\dots,\rho_l(x)$ are holomorphic in a neighborhood of $x=0$;*
2. *$\operatorname{Re}\rho_1(0)>0,\dots,\operatorname{Re}\rho_l(0)>0$;*
3. *$C(1,0)\neq 0$;*
4. *$C\left(i+j_1\rho_1(0)+\cdots+j_l\rho_l(0),0\right)\neq 0$ for all $(i,j)\in\mathbb{N}\times\mathbb{N}^l$ satisfying $i+|j|\ge 2$.*

Then the equation (8.3.1) has a family of solutions belonging to $\widetilde{\mathcal{O}}_+$ of the form

$$\sum_{i\ge1}u_i(x)t^i+\sum_{i+2m|j|\ge k+2m,|j|\ge1}\varphi_{i,j,k}(x)t^{i+j_1\rho_1(x)+\cdots+j_l\rho_l(x)}(\log t)^k \tag{8.3.2}$$

where all the coefficients $\varphi_{i,j,k}(x)$ are holomorphic in a common disk centered at the origin of $\mathbb{C}^n$. Moreover, every solution of the form (8.3.2) is uniquely determined by the coefficients $\{\varphi_{0,j,0}(x):|j|=1\}$ which are arbitrary. If we take $\varphi_{0,j,0}(x)=0$ for all j such that $|j|=1$ the solution (8.3.2) is reduced to the unique holomorphic solution of the equation (8.3.1).

Proof : We begin by constructing a family of formal solutions of the form (8.3.2). As in (8.2.6) and (8.2.7), take an $r>0$ such that $b(x)\in\mathbb{C}_r\{x\}$ and

$$C(1,x)\neq 0\quad\text{in } D_r, \tag{8.3.3}$$

$$C\left(i+j_1\rho_1(x)+\cdots+j_l\rho_l(x),x\right)\neq 0\quad\text{in } D_r \tag{8.3.4}$$

$$\text{for all } (i,j)\in\mathbb{N}\times\mathbb{N}^l \text{ satisfying } i+|j|\ge 2.$$

For any $(i,0) \in \mathbb{N}^* \times \mathbb{N}^l$ denote by $\mathfrak{F}_r(i,0)$ the set of functions $f(t,x)$ of the form

$$f(t,x) = \Phi(x)t^i \quad \text{where } \Phi(x) \in \mathbb{C}_r\{x\}.$$

For any $(i,j) \in \mathbb{N} \times \left(\mathbb{N}^l \setminus \{0\}\right)$ denote by $\mathfrak{F}_r(i,j)$ the set of functions $g(t,x)$ of the form

$$g(t,x) = \sum_{0 \le k \le i+2m(|j|-1)} \Phi_k(x) t^{i+j_1\rho_1(x)+\cdots+j_l\rho_l(x)} (\log t)^k$$

where $\Phi_k(x) \in \mathbb{C}_r\{x\}$ for all k.

Setting

$$w_{i,0} = u_i(x)t^i \quad \text{for } i \ge 1, \tag{8.3.5}$$

$$w_{0,e_p} = \varphi_{0,e_p,0}(x)t^{\rho_p(x)} \quad \text{for } p = 1,\dots,l, \tag{8.3.6}$$

$$w_{i,j} = \sum_{0 \le k \le i+2m(|j|-1)} \varphi_{i,j,k}(x) t^{i+j_1\rho_1(x)+\cdots+j_l\rho_l(x)} (\log t)^k \quad \text{for } i+|j| \ge 2 \text{ and } |j| \ge 1, \tag{8.3.7}$$

the formal solution $\hat{u}(t,x)$ of the form (8.3.2) takes the form

$$\hat{u}(t,x) = \sum_{i+|j| \ge 1} w_{i,j}(t,x) \tag{8.3.8}$$

where $w_{i,j}(t,x) \in \mathfrak{F}_r(i,j)$ for all (i,j). It is easy to see that the class of functions $\mathfrak{F}_r(i,j)$ for $i+|j| \ge 1$ has for all pairs (i,j) and (p,q) belonging to $\mathbb{N} \times \mathbb{N}^l \setminus \{(0,0)\}$ the following properties:

1. $\mathfrak{F}_r(i,j) \times \mathfrak{F}_r(p,q) \subset \mathfrak{F}_r(i+p, j+q)$;
2. $\left(t\frac{\partial}{\partial t}\right) \mathfrak{F}_r(i,j) \subset \mathfrak{F}_r(i,j)$;
3. $\left(\frac{\partial}{\partial x}\right)^\alpha \mathfrak{F}_r(i,j) \times \left(\frac{\partial}{\partial x}\right)^\beta \mathfrak{F}_r(p,q) \subset \mathfrak{F}_r(i+p, j+q)$ for $|\alpha| \le m$ and $|\beta| \le m$.

Looking at (8.3.8) we see that the equation (8.3.1) splits into the following recurrence formulas:

$$\begin{aligned}
C\left(t\frac{\partial}{\partial t}, x\right) w_{1,0} &= b(x)t, \\
C\left(t\frac{\partial}{\partial t}, x\right) w_{0,e_p} &= 0 \quad \text{for } p = 1,\dots,l, \\
C\left(t\frac{\partial}{\partial t}, x\right) w_{i,j} &= F_{i,j}\left(w_{p,q} : (p,q) \in A_{i,j}\right) \quad \text{for } i+|j| \ge 2,
\end{aligned}$$

where

$$A_{i,j} = \left\{(p,q) \in \mathbb{N} \times \mathbb{N}^l : p \le i, q_1 \le j_1, \dots, q_l \le j_l \text{ and } 1 \le p+|q| < i+|j|\right\}$$

and

$F_{i,j}$ is a function belonging to $\mathfrak{F}_r(i,j)$ determined by $\{w_{p,q} : (p,q) \in A_{i,j}\}$.

Moreover, (8.3.5) to (8.3.7) give us

$$C\left(t\frac{\partial}{\partial t}, x\right) w_{i,0} = C(i,x)u_i(x)t^i \quad \text{for } i \geq 1,$$

$$C\left(t\frac{\partial}{\partial t}, x\right) w_{0,e_p} = C(\rho_p(x),x)\,\varphi_{0,e_p,0}(x)t^{\rho_p(x)} \quad \text{for } p = 1,\dots,l,$$

$$C\left(t\frac{\partial}{\partial t}, x\right) w_{i,j} = \sum_{0\leq k\leq i+2m(|j|-1)} \Big\{C(i+j_1\rho_1(x)+\cdots+j_l\rho_l(x),x)\,\varphi_{i,j,k}(x)$$
$$+ \sum_{1\leq h\leq m} C_h^*(i,j,k,x)\varphi_{i,j,k+h}(x)\Big\} \times t^{i+j_1\rho_1(x)+\cdots+j_l\rho_l(x)}(\log t)^k$$
$$\text{for } i+|j|\geq 2 \text{ and } |j|\geq 1,$$

where $\varphi_{i,j,k+h}(x) \equiv 0$ if $k+h > i+2m(|j|-1)$,

$$C_h^*(i,j,k,x) = \sum_{h\leq q\leq m} (q,h)c_q(x)\,(i+j_1\rho_1(x)+\cdots+j_l\rho_l(x))^{q-h} \times (k+1)\cdots(k+h)$$

with $(q,h) = \frac{q!}{(q-h)!h!}$ and $c_m(x) \equiv 1$.

Then by (8.3.3), (8.3.4), and the fact $C(\rho_p(x),x) \equiv 0$ for $p = 1,\dots,l$ we have

1. $u_1(x) = \frac{b(x)}{C(1,x)} \in \mathbb{C}_r\{x\}$;
2. $\varphi_{0,e_p,0}(x) \in \mathbb{C}_r\{x\}$ for $p = 1,\dots,l$ are arbitrary;
3. $u_i(x) \in \mathbb{C}_r\{x\}$ for $i \geq 2$ and $\varphi_{i,j,k}(x) \in \mathbb{C}_r\{x\}$ for $i+|j|\geq 2$ and $|j|\geq 1$ are determined by $\varphi_{0,e_1,0}(x),\dots,\varphi_{0,e_l,0}(x)$.

We have obtained a family of formal solutions of the form (8.3.2). As in section 8.2 to complete the proof it is enough to prove the convergence of this formal solution.

For the moment let us admit

Lemma 8.3.2 *Let $1 \leq l \leq m$. Assume that the hypothesis 2., 3., and 4. of theorem 8.3.1 are satisfied. Then there exists an $N \in \mathbb{N}^*$ such that*

(a1) $N \geq 2m+1$;

(a2) $\operatorname{Re}\rho_i(0) > \frac{2m}{N}$ *for all* $i = 1,\dots,l$;

(a3) $C\left(\frac{2m+1}{N},0\right) \neq 0$;

(a4) $C\left(\frac{i-2m|j|+k+2m}{N} + j_1\rho_1(0)+\cdots+j_l\rho_l(0),0\right) \neq 0$
for all $(i,j,k) \in \mathbb{N}\times\mathbb{N}^l\times\mathbb{N}$ *satisfying* $i+|j|+k \geq 2$.

Let us prove now the convergence of the formal solution

$$(8.3.9)\quad \sum_{i\geq 1} u_i(x)t^i + \sum_{i+2m|j|\geq k+2m,|j|\geq 1} \varphi_{i,j,k}(x)t^{i+j_1\rho_1(x)+\cdots+j_l\rho_l(x)}(\log t)^k$$

where the coefficients $u_i(x)$ and $\varphi_{i,j,k}(x)$ are all holomorphic in a common disk centered at the origin of $\mathbb{C}^n$.

Let N be a positive integer satisfying the conditions of lemma 8.3.2. Set in the equation (8.3.1)

$$(8.3.10)\qquad w(t,x) = t^{-2m}u\left(t^N, x\right).$$

We get for $w(t,x)$ the equation

$$(8.3.11)\quad \left(\left(\frac{1}{N}\theta_{2m}\right)^m + c_{m-1}(x)\left(\frac{1}{N}\theta_{2m}\right)^{m-1} + \cdots + c_0(x)\right)w$$
$$= t^{N-2m}b(x) + \sum_{p+|q|\geq 2} g_{p,q}(x)t^{Np-2m+2m|q|-L(q)} \times$$
$$\times \prod_{j+|\alpha|\leq m}\left[\left(\frac{1}{N}\theta_{2m-|\alpha|}\right)^j\left(t\frac{\partial}{\partial x}\right)^\alpha w\right]^{q_{j,\alpha}}$$

where

$$\theta_s = t\frac{\partial}{\partial t} + s \qquad \text{and} \qquad L(q) = \sum_{j+|\alpha|\leq m} |\alpha|\, q_{j,\alpha}.$$

As $N \geq 2m+1$ and $p+|q|\geq 2$ we have

$$Np - 2m + 2m\,|q| - L(q)$$
$$\geq (2m+1)p - 2m + 2m\,|q| - m\,|q| = (m+1)p + m\,(p+|q|-2) \geq 0$$

and

$$Np - 2m + 2m\,|q| - L(q) + |q| \geq (m+1)p + |q| \geq p + |q| \geq 2.$$

This proves that the equation (8.3.11) is a particular case of equations of type (8.2.2) that we have studied in section 8.2.

Denote by $\lambda_1(x), \ldots, \lambda_m(x)$ the roots of the polynomial in λ

$$C^*(\lambda, x) = \left(\frac{\lambda+2m}{N}\right)^m + \sum_{j<m} c_j(x)\left(\frac{\lambda+2m}{N}\right)^j.$$

We have

1. $\lambda_i(x) = N\rho_i(x) - 2m$ for all $i = 1, \ldots, m$;
2. $C^*(\lambda, x) = C\left(\frac{\lambda+2m}{N}, x\right)$.

The above condition 1. and the conditions (a2), (a3), and (a4) of lemma 8.3.2 imply that

i. $\lambda_1(x), \ldots, \lambda_l(x)$ are holomorphic in a neighborhood of $x = 0$;

ii. $\operatorname{Re}\lambda_1(0) > 0, \ldots, \operatorname{Re}\lambda_l(0) > 0$;

iii. $C^*(1, 0) \neq 0$;

iv. $C^*(i + j_1\lambda_1(0) + \cdots + j_l\lambda_l(0) + k, 0) \neq 0$ for all $(i, j, k) \in \mathbb{N}\times\mathbb{N}^l\times\mathbb{N}$ satisfying $i + |j| + k \geq 2$.

This means that the equation (8.3.11) satisfies the hypothesis of corollary 8.2.3.

The relation (8.3.10) transforms a formal solution (8.3.9) into the formal series

$$(8.3.12)\quad \sum_{i\geq 1} u_i(x)t^{Ni-2m} + \sum_{i+2m|j|\geq k+2m, |j|\geq 1} N^k \varphi_{i,j,k}(x) t^{Ni+2m|j|-k-2m} \times \\ \times \left(t^{\lambda_1(x)}\right)^{j_1} \times \cdots \times \left(t^{\lambda_l(x)}\right)^{j_l} (t\log t)^k$$

which is a particular case of (8.2.16). As $\hat{u}(t, x)$ is a formal solution of (8.3.1) $w(t, x)$ is a formal solution of (8.3.11).

The corollary 8.2.3 asserts the convergence of the formal solution $w(t, x)$ in $\widetilde{\mathcal{O}}_+$ which implies the convergence of $\hat{u}(t, x)$.

And theorem 8.3.1 is proved. ■

Proof of lemma 8.3.2: Denote by

$$J_+ = \{i \in \{1, \ldots, m\} : \operatorname{Re}\rho_i(0) > 0\} \qquad \text{and} \qquad a = \min_{i\in J_+} \operatorname{Re}\rho_i(0).$$

It is clear that $\{1, \ldots, l\} \subset J_+$. Now take $N \in \mathbb{N}^*$ such that

$$(8.3.13)\qquad N \geq \max\left\{2m + 1, \frac{4m}{a}, \frac{2m + 2}{a}\right\}.$$

The condition (a1) is then trivially satisfied. The inequalities

$$(8.3.14)\qquad \operatorname{Re}\rho_i(0) - \frac{2m}{N} \geq a - 2m\frac{a}{4m} = \frac{a}{2} > 0 \quad \text{for } i = 1, \ldots, l$$

imply the condition (a2). If we remark that

$$C\left(\frac{2m+1}{N}, 0\right) = \prod_{1\leq i\leq m}\left(\frac{2m+1}{N} - \rho_i(0)\right)$$

and that

$$0 < \frac{2m+1}{N} \leq (2m + 1)\frac{a}{2m + 2} < a \leq \operatorname{Re}\rho_i(0) \quad \text{for } i \in J_+$$

we obtain the condition (a3).

Set

$$A = \max_{1 \le i \le m} \operatorname{Re} \rho_i(0)$$

and take an integer M such that

$$M \ge \max\left\{2, \frac{2A}{a}\right\}.$$

Then if N satisfies (8.3.13) and if $(i, j, k) \in \mathbb{N} \times \mathbb{N}^l \times \mathbb{N}$ satisfies $|j| > M$ (8.3.14) gives us

$$\begin{aligned} &\operatorname{Re}\left\{\frac{i - 2m\,|j| + k + 2m}{N} + j_1\rho_1(0) + \cdots + j_l\rho_l(0)\right\} \\ &= \frac{i + k + 2m}{N} + j_1\left(\operatorname{Re}\rho_1(0) - \frac{2m}{N}\right) + \cdots + j_l\left(\operatorname{Re}\rho_l(0) - \frac{2m}{N}\right) \ge \frac{2m}{N} + \frac{a}{2}\,|j| \\ &\qquad > \frac{a}{2}M \ge A \ge \operatorname{Re}\rho_p(0) \end{aligned}$$

for all $p = 1, \ldots, m$. This means that for any choice of N satisfying (8.3.13) the condition (a4) is satisfied for all $(i, j, k) \in \mathbb{N} \times \mathbb{N}^l \times \mathbb{N}$ satisfying $|j| > M$.

Therefore, to complete the proof of lemma 8.3.2 it is enough to choose an N such that we have (8.3.13) and the condition

$$\frac{i - 2m\,|j| + k + 2m}{N} \ne \rho_p(0) - j_1\rho_1(0) - \cdots - j_l\rho_l(0) \tag{8.3.15}$$

for all $p = 1, \ldots, m$ and all $(i, j, k) \in \mathbb{N} \times \mathbb{N}^l \times \mathbb{N}$ satisfying $i + |j| + k \ge 2$ and $|j| \le M$.

Let us study this condition.

For $h = 0, 1, \ldots, M$ denote

$$S_h = \{\rho_p(0) - j_1\rho_1(0) - \cdots - j_l\rho_l(0) : p = 1, \ldots, m \text{ and } |j| = h\}.$$

The conditions 3., 4. of theorem 8.3.1 give

$$S_h \cap \{1, 2, \ldots\} = \emptyset \quad \text{for } h = 0, 1, \ldots, M. \tag{8.3.16}$$

It is easy to see that the condition (8.3.15) is equivalent to the following conditions:

$$\frac{i + k + 2m}{N} \notin S_0 \quad \text{for } i + k \ge 2, \tag{8.3.17}$$

$$\frac{i + k}{N} \notin S_1 \quad \text{for } i + k \ge 1 \text{ and} \tag{8.3.18}$$

$$\frac{i - 2mh + k + 2m}{N} \notin S_h \quad \text{for } i + k \ge 0 \text{ and } h = 2, \ldots, M. \tag{8.3.19}$$

Moreover, as $0 \notin S_h$ for $h = 2, \ldots, M$ (by 4. of theorem 8.3.1) it is also easy to see that the conditions (8.3.17) to (8.3.19) are reduced to the conditions (b1) $\frac{i+k+2m}{N} \notin S_0 \cap \mathbb{Q}_+$ for $i + k \geq 2$;

(b2) $\frac{i+k}{N} \notin S_1 \cap \mathbb{Q}_+$ for $i + k \geq 1$;

(b3) $\frac{i-2mh+k+2m}{N} \notin S_h \cap \mathbb{Q}_+$ for $i + k \geq 0$ and $h = 2, \dots, M$;

(b4) $\frac{i-2mh+k+2m}{N} \notin S_h \cap \mathbb{Q}_-$ for $i + k \geq 0$ and $h = 2, \dots, M$

where $\mathbb{Q}_+$ (resp. $\mathbb{Q}_-$) is the set of positive (resp. negative) rational numbers.

Set

$$\left(\bigcup_{0 \leq h \leq M} S_h \right) \cap \mathbb{Q}_+ = \{\alpha_1, \alpha_2, \dots, \alpha_K\}.$$

The condition (8.3.16) says that for all $i = 1, 2, \dots, K$ we have $\alpha_i \notin \{1, 2, \dots\}$ and, therefore, α_i can be written

$$\alpha_i = \frac{c_i}{d_i}$$

for non zero integers c_i and d_i satisfying $c_i \geq 1$ and $d_i \geq 2$ and, moreover, relatively prime. If $N \in \mathbb{N}^*$ is prime with all d_i for $i = 1, \dots, K$ then $\alpha_i \notin \mathbb{Z}/N$ for all i and we have (b1), (b2), and (b3).

Now set

$$S^* = \left(\bigcup_{2 \leq h \leq M} S_h \right) \cap \mathbb{Q}_-.$$

When $S^* = \emptyset$ the condition (b4) is trivially satisfied.

When $S^* \neq \emptyset$ introduce

$$B = \min \{|\beta| : \beta \in S^*\}.$$

If N satisfies

(8.3.20) $$N > \frac{2mM - 2m}{B}$$

we have

$$\frac{i - 2mh + k + 2m}{N} \geq -\frac{2mM - 2m}{N} > -B \geq \beta$$

for all $\beta \in S^*$ and, therefore, the condition (b4).

As a consequene, if we take an integer N satisfying (8.3.13) and (8.3.20) and being prime with all the integers d_i for $i = 1, \dots, K$ then all the conditions of the lemma 8.3.2 are satisfied. ∎

8.4 Asymptotic study

In this section, we recall the asymptotic theory developped in Tahara [49] for linear partial differential equations and we apply it to some non linear equations.

For $a \in \mathbb{R}$ and any function $u(t,x)$ let us write

$$u(t,x) = O\left(t^a, \widetilde{\mathcal{O}}_+\right) \quad (\text{as } t \longrightarrow 0)$$

when we have $t^{-a}u(t,x) \in \widetilde{\mathcal{O}}_+$.

First we are studying the ordinary differential equation

$$\left(\left(t\frac{\partial}{\partial t}\right)^m + c_{m-1}(x)\left(t\frac{\partial}{\partial t}\right)^{m-1} + \cdots + c_0(x)\right) u = f \tag{8.4.1}$$

where $m \in \mathbb{N}^*$ and $c_j(x)$ for $0 \le j \le m-1$ are holomorphic functions defined in a polydisk centered at the origin of $\mathbb{C}^n$.

Denote by $\rho_1(x), \ldots, \rho_m(x)$ the roots of the polynomial

$$C(\rho, x) = \rho^m + \sum_{j<m} c_j(x)\rho^j.$$

Equation (8.4.1) can be written in the form

$$\left(t\frac{\partial}{\partial t} - \rho_1(x)\right) \cdots \left(t\frac{\partial}{\partial t} - \rho_m(x)\right) u = f. \tag{8.4.2}$$

The explicit resolution of (8.4.2) gives us

Lemma 8.4.1 *1. If $f(t,x) = O\left(t^a, \widetilde{\mathcal{O}}_+\right)$ (as $t \longrightarrow 0$) for an $a \in \mathbb{R}$ satisfying*

$$a < \min_{1 \le i \le m} \operatorname{Re} \rho_i(0)$$

then every solution $u(t,x)$ of (8.4.1) satisfies

$$u(t,x) = O(t^a, \widetilde{\mathcal{O}}_+) \quad (\text{as } t \longrightarrow 0).$$

2. If $f(t,x) = O\left(t^a, \widetilde{\mathcal{O}}_+\right)$ (as $t \longrightarrow 0$) for an $a \in \mathbb{R}$ satisfying

$$a > \max_{1 \le i \le m} \operatorname{Re} \rho_i(0)$$

the equation (8.4.1) has a unique solution $u(t,x)$ satisfying

$$u(t,x) = O\left(t^a, \widetilde{\mathcal{O}}_+\right) \quad (\text{as } t \longrightarrow 0).$$

3. *If $f(t,x) \equiv 0$ and if the numbers $\rho_1(0), \ldots, \rho_m(0)$ are distinct the general solution of (8.4.1) is of the form*

$$u(t,x) = \sum_{1 \le i \le m} \varphi_i(x) t^{\rho_i(x)}$$

where for all $i = 1, \ldots, m$ the $\varphi_i(x) \in \mathbb{C}\{x\}$ are arbitrary.

4. *If $f(t,x) = O\left(t^a, \widetilde{\mathcal{O}}_+\right)$ (as $t \longrightarrow 0$) for an $a \in \mathbb{R}$ satisfying*

$$a > \max_{1 \le i \le m} \operatorname{Re} \rho_i(0)$$

and if the numbers $\rho_1(0), \ldots, \rho_m(0)$ are distinct then every solution of (8.4.1) is of the form

$$u(t,x) = \sum_{1 \le i \le m} \varphi_i(x) t^{\rho_i(x)} + O\left(t^a, \widetilde{\mathcal{O}}_+\right) \quad (\text{as } t \longrightarrow 0)$$

for some $\varphi_i(x) \in \mathbb{C}\{x\}$ $(i = 1, \ldots, m)$.

We are now giving an other version of lemma 8.4.1. Let I_1, I_2, and I_3 be subsets of $\{1, 2, \ldots, m\}$ satisfying the following conditions:

1. $I_1 \cup I_2 \cup I_3 = \{1, 2, \ldots, m\}$;
2. $I_2 \neq \emptyset$;
3. $I_i \cap I_j = \emptyset$ if $i \neq j$;
4. $\max_{i \in I_p} \operatorname{Re} \rho_i(0) < \min_{i \in I_{p+1}} \operatorname{Re} \rho_i(0)$ for $p = 1, 2$.

When $I_1 \neq \emptyset$ and $I_3 \neq \emptyset$ we set

$$M_p = \max_{i \in I_p} \operatorname{Re} \rho_i(0) \quad \text{for } p = 1, 2 \quad \text{and} \quad m_p = \min_{i \in I_p} \operatorname{Re} \rho_i(0) \quad \text{for } p = 2, 3.$$

When $I_1 = \emptyset$ we set $M_1 = -\infty$ and when $I_3 = \emptyset$ we set $m_3 = +\infty$.
Then as a consequence of lemma 8.4.1 we obtain

Corollary 8.4.2 *Let $a \in \mathbb{R}, b \in \mathbb{R}, u(t,x)$ and $f(t,x)$ be two functions. Assume that*

i. $a < b$;

ii. $u(t,x) = O\left(t^a, \widetilde{\mathcal{O}}_+\right)$ *(as $t \longrightarrow 0$);*

iii. $f(t,x) = O\left(t^b, \widetilde{\mathcal{O}}_+\right)$ *(as $t \longrightarrow 0$);*

iv. $u(t,x)$ and $f(t,x)$ satisfy (8.4.1).

Then

1. *if* $M_1 < a < b < m_2$ *we have* $u(t,x) = O\left(t^b, \widetilde{\mathcal{O}}_+\right)$ *(as* $t \longrightarrow 0$*);*

2. *if* $M_2 < a < b < m_3$ *we have* $u(t,x) = O\left(t^b, \widetilde{\mathcal{O}}_+\right)$ *(as* $t \longrightarrow 0$*);*

3. *if* $M_1 < a < m_2$ *and* $M_2 < b < m_3$ *and if the* $\rho_i(0)$ *for* $i \in I_2$ *are distinct we have*

$$u(t,x) = \sum_{i \in I_2} \varphi_i(x) t^{\rho_i(x)} + O\left(t^b, \widetilde{\mathcal{O}}_+\right) \quad (\text{as } t \longrightarrow 0)$$

for some $\varphi_i(x) \in \mathbb{C}\{x\}$ $(i \in I_2)$.

Remark 8.4.3 *If* $I_1 = \emptyset$ *the assertion 1. of lemma 8.4.1 implies the result 1.*
If $I_3 = \emptyset$ *the assertion 2. of lemma 8.4.1 implies the result 2.*
If $I_1 = I_3 = \emptyset$ *the assertion 4. of lemma 8.4.1 implies the result 3.*

Proof in the general case: Set

$$C_p = \prod_{i \in I_p} \left(t\frac{\partial}{\partial t} - \rho_i(x) \right) \quad \text{for } p = 1, 2, 3.$$

C_p is a differential operator with holomorphic coefficients near the origin of $\mathbb{C}^n$ and the equation (8.4.1) can be written in the form

$$C_1 C_2 C_3 u = C_1 C_3 C_2 u = f.$$

If $M_1 < a < b < m_2$ set $w = C_2 C_3 u$. Then $w = O\left(t^a, \widetilde{\mathcal{O}}_+\right)$ (as $t \longrightarrow 0$) and $C_1 w = f = O\left(t^b, \widetilde{\mathcal{O}}_+\right)$ (as $t \longrightarrow 0$). Therefore, by 2. of lemma 8.4.1 we have $w = O\left(t^b, \widetilde{\mathcal{O}}_+\right)$ (as $t \longrightarrow 0$). Applying 1. of lemma 8.4.1 to the equation $C_2 C_3 u = w$ we obtain $u = O\left(t^b, \widetilde{\mathcal{O}}_+\right)$ (as $t \longrightarrow 0$) which is the result 1.

If $M_2 < a < b < m_3$ setting $w = C_3 u$ and with the same kinds of arguments we obtain the result 2.

If $M_1 < a < m_2$ and $M_2 < b < m_3$ we set $w_1 = C_3 C_2 u$ and $w_2 = C_2 u$. Then $w_p = O\left(t^a, \widetilde{\mathcal{O}}_+\right)$ (as $t \longrightarrow 0$) for $p = 1, 2$, $C_1 w_1 = f$, and $C_3 w_2 = w_1$. Applying 2. of lemma 8.4.1 to the equation $C_1 w_1 = f$ we get $w_1 = O\left(t^b, \widetilde{\mathcal{O}}_+\right)$ (as $t \longrightarrow 0$). It follows that by applying 1. of lemma 8.4.1 to the equation $C_3 w_2 = w_1$ we obtain $w_2 = O\left(t^b, \widetilde{\mathcal{O}}_+\right)$ (as $t \longrightarrow 0$). Therefore, if the $\rho_i(0)$ for $i \in I_2$ are distinct by applying 4. of lemma 8.4.1 to the equation $C_2 u = w_2$ we obtain the result 3. of corollary 8.4.2.

And corollary 8.4.2 is a consequence of lemma 8.4.1. ■

Let

$$R[\cdot]:\widetilde{\mathcal{O}}_+ \longrightarrow \widetilde{\mathcal{O}}_+$$

be an operator from $\widetilde{\mathcal{O}}_+$ into $\widetilde{\mathcal{O}}_+$. Consider the equation

$$\left(\left(t\frac{\partial}{\partial t}\right)^m + c_{m-1}(x)\left(t\frac{\partial}{\partial t}\right)^{m-1} + \cdots + c_0(x)\right)u = R[u]. \tag{8.4.3}$$

Let $s > 0$. Assume the following condition on R:

(Bs) If $u(t,x) = O\left(t^a, \widetilde{\mathcal{O}}_+\right)$ (as $t \longrightarrow 0$) for an $a > 0$ then $R[u] = O\left(t^b, \widetilde{\mathcal{O}}_+\right)$ (as $t \longrightarrow 0$) for any $b < \min\{2a, a+s\}$.

We have

Lemma 8.4.4 *Let $I_1, I_2, I_3, M_1, M_2, m_2$, and m_3 be as in corollary 8.4.2. Let $s > 0$ and assume that the operator $R[\cdot]:\widetilde{\mathcal{O}}_+ \longrightarrow \widetilde{\mathcal{O}}_+$ satisfies the condition (Bs). Assume also that $u(t,x) \in \widetilde{\mathcal{O}}_+$ is a solution of (8.4.3). Then*

1. *if $u(t,x) = O\left(t^a, \widetilde{\mathcal{O}}_+\right)$ (as $t \longrightarrow 0$) for an $a > \max\{0, M_1\}$ we have $u(t,x) = O\left(t^b, \widetilde{\mathcal{O}}_+\right)$ (as $t \longrightarrow 0$) for any $b < m_2$;*
2. *if $u(t,x) = O\left(t^a, \widetilde{\mathcal{O}}_+\right)$ (as $t \longrightarrow 0$) for an $a > max\{0, M_2\}$ we have $u(t,x) = O\left(t^b, \widetilde{\mathcal{O}}_+\right)$ (as $t \longrightarrow 0$) for any $b < m_3$;*
3. *if $u(t,x) = O\left(t^a, \widetilde{\mathcal{O}}_+\right)$ (as $t \longrightarrow 0$) for an $a > 0$ satisfying*

$$M_1 < a < m_2 \le M_2 < \min\{2a, a+s\} \tag{8.4.4}$$

and if $\rho_i(0)$ for $i \in I_2$ are distinct we have

$$u(t,x) = \sum_{i\in I_2} \varphi_i(x) t^{\rho_i(x)} + O\left(t^b, \widetilde{\mathcal{O}}_+\right) \quad (as\ t \longrightarrow 0) \tag{8.4.5}$$

for some $\varphi_i(x) \in \mathbb{C}\{x\}$ $(i \in I_2)$ and a number $b > M_2$.

Proof : We are proving 1.: Assume that $u(t,x) = O\left(t^a, \widetilde{\mathcal{O}}_+\right)$ (as $t \longrightarrow 0$) for an $a > \max\{0, M_1\}$ and let $b > 0$ be such that $a < b < m_2$. Take a sequence $a_1, \ldots, a_N$ satisfying

1. $a = a_1 < a_2 < \cdots < a_N = b$ and
2. $a_{i+1} < \min\{2a_i, a_i + s\}$ for all $i = 1, 2, \ldots, N-1$.

Put $f(t,x) = R[u](t,x) \in \widetilde{\mathcal{O}}_+$. As we have $u(t,x) = O\left(t^{a_1}, \widetilde{\mathcal{O}}_+\right)$ (as $t \longrightarrow 0$) by using (Bs) and the condition $a_2 < \min\{2a_1, a_1 + s\}$ we have $f(t,x) = O\left(t^{a_2}, \widetilde{\mathcal{O}}_+\right)$ (as $t \longrightarrow 0$) and by 1. of corollary 8.4.2 we get $u(t,x) = O\left(t^{a_2}, \widetilde{\mathcal{O}}_+\right)$ (as $t \longrightarrow 0$) which means that $f(t,x) = O\left(t^{a_3}, \widetilde{\mathcal{O}}_+\right)$ (as $t \longrightarrow 0$). By using again 1. of corollary 8.4.2 we get $u(t,x) = O\left(t^{a_3}, \widetilde{\mathcal{O}}_+\right)$ (as $t \longrightarrow 0$). And by induction we obtain $u(t,x) = O\left(t^{a_N}, \widetilde{\mathcal{O}}_+\right)$ (as $t \longrightarrow 0$) which proves 1. of lemma 8.4.4.

The proof of 2. of lemma 8.4.4 goes in the same manner.

We are proving 3.: Assume that $u(t,x) = O\left(t^{a}, \widetilde{\mathcal{O}}_+\right)$ (as $t \longrightarrow 0$) for an $a > 0$ satisfying

$$M_1 < a < m_2 \leq M_2 < \min\{2a, a+s\}.$$

Take $b \in \mathbb{R}$ such that

$$M_2 < b < \min\{2a, a+s, m_3\}.$$

Set $f(t,x) = R[u](t,x) \in \widetilde{\mathcal{O}}_+$. By (Bs) we have $f(t,x) = O\left(t^{b}, \widetilde{\mathcal{O}}_+\right)$ (as $t \longrightarrow 0$) and applying 3. of corollary 8.4.2 we obtain (8.4.5) for some $\varphi_i(x) \in \mathbb{C}\{x\}$ $(i \in I_2)$. The part 3. of lemma 8.4.4 is also proved. ■

We are now applying the results of lemma 8.4.4 to some non linear partial differential equations.

Let

- $a_{j,\alpha}(t,x) \in \widetilde{\mathcal{O}}_+$ for $j + |\alpha| \leq m$; by definition there exists an $s > 0$ such that

  (8.4.6) $$a_{j,\alpha}(t,x) = O\left(t^{s}, \widetilde{\mathcal{O}}_+\right) \quad (\text{as } t \longrightarrow 0) \text{ for all } j + |\alpha| \leq m.$$

- $\mathfrak{R}(\mathbb{C} \setminus \{0\})$ the universal covering space of $\mathbb{C} \setminus \{0\}$,
- $S(\varepsilon(s)) = \{t \in \mathfrak{R}(\mathbb{C} \setminus \{0\}) : 0 < |t| < \varepsilon(\arg t)\}$ where $\varepsilon(s)$ is a continuous and positive–valued function defined on $\mathbb{R}$, and
- $D_r = \{x \in \mathbb{C}^n : |x_i| \leq r \text{ for } i = 1, \ldots, n\}$.

Put

$$Z = \{Z_{j,\alpha}\}_{j+|\alpha|\leq m} \quad \text{where } Z_{j,\alpha} \in \mathbb{C}.$$

Let $G_2(t,x)(Z)$ be a holomorphic function in the variables (t,x,Z) defined in $S(\varepsilon(s)) \times D_r \times \{Z : |Z_{j,\alpha}| \leq r\}$ and having the following form:

(8.4.7) $$G_2(t,x)(Z) = \sum_{|q|\geq 2} g_q(t,x) Z^q$$

where

$$\begin{aligned} q &= \{q_{j,\alpha}\}_{j+|\alpha|\leq m} \quad \text{where } q_{j,\alpha} \in \mathbb{N}, \\ |q| &= \sum_{j+|\alpha|\leq m} q_{j,\alpha}, \\ Z^q &= \prod_{j+|\alpha|\leq m} (Z_{j,\alpha})^{q_{j,\alpha}} \end{aligned}$$

and $g_q(t,x)$ for $|q| \geq 2$ are holomorphic functions defined in $S(\varepsilon(s)) \times D_r$. Moreover, assume that for any $\theta > 0$ and any compact subset $K \times L$ of $D_r \times \{Z : |Z_{j,\alpha}| \leq r\}$ we have

$$\max_{(x,Z)\in K\times L} |G_2(t,x)(Z)| = O(1)$$

as $t \longrightarrow 0$ in S_θ.

Consider the following equation:

$$\begin{aligned} (8.4.8) \quad & \left(\left(t\frac{\partial}{\partial t}\right)^m + c_{m-1}(x)\left(t\frac{\partial}{\partial t}\right)^{m-1} + \cdots + c_0(x)\right) u \\ &= \sum_{j+|\alpha|\leq m} a_{j,\alpha}(t,x)\left(t\frac{\partial}{\partial t}\right)^j \left(\frac{\partial}{\partial x}\right)^\alpha u \\ &\quad + G_2(t,x)\left(\left\{\left(t\frac{\partial}{\partial t}\right)^j \left(\frac{\partial}{\partial x}\right)^\alpha u\right\}_{j+|\alpha|\leq m}\right). \end{aligned}$$

If we set

$$\begin{aligned} R[u] = & \sum_{j+|\alpha|\leq m} a_{j,\alpha}(t,x)\left(t\frac{\partial}{\partial t}\right)^j \left(\frac{\partial}{\partial x}\right)^\alpha u \\ & + G_2(t,x)\left(\left\{\left(t\frac{\partial}{\partial t}\right)^j \left(\frac{\partial}{\partial x}\right)^\alpha u\right\}_{j+|\alpha|\leq m}\right) \end{aligned}$$

the fact that R satisfies the condition (Bs) allows us to apply the lemma 8.4.4.

In this way, we obtain

Proposition 8.4.5 *Let $I_1, I_2, I_3, M_1, M_2, m_2$, and m_3 be as in corollary 8.4.2. Assume that $u(t,x) \in \widetilde{\mathcal{O}}_+$ is a solution of (8.4.8). Then*

1. *if $u(t,x) = O\left(t^a, \widetilde{\mathcal{O}}_+\right)$ (as $t \longrightarrow 0$) for an $a > \max\{0, M_1\}$ we have $u(t,x) = O\left(t^b, \widetilde{\mathcal{O}}_+\right)$ (as $t \longrightarrow 0$) for any $b < m_2$;*
2. *if $u(t,x) = O\left(t^a, \widetilde{\mathcal{O}}_+\right)$ (as $t \longrightarrow 0$) for an $a > \max\{0, M_2\}$ we have $u(t,x) = O\left(t^b, \widetilde{\mathcal{O}}_+\right)$ (as $t \longrightarrow 0$) for any $b < m_3$;*

3. if $u(t,x) = O\left(t^a, \widetilde{\mathcal{O}}_+\right)$ *(as* $t \longrightarrow 0$*) for an* $a > 0$ *satisfying*

$$M_1 < a < m_2 \leq M_2 < \min\{2a, a+s\}$$

and if $\rho_i(0)$ *for* $i \in I_2$ *are distinct then we have*

$$u(t,x) = \sum_{i \in I_2} \varphi_i(x) t^{\rho_i(x)} + O\left(t^b, \widetilde{\mathcal{O}}_+\right) \quad (\text{as } t \longrightarrow 0)$$

for some $\varphi_i(x) \in \mathbb{C}\{x\}$ $(i \in I_2)$ *and a number* $b > M_2$.

To conclude let us discuss the uniqueness of the solution of (8.4.8).

Let $u(t,x) \in \widetilde{\mathcal{O}}_+$ be a solution of (8.4.8) and assume that $u(t,x) = O\left(t^a, \widetilde{\mathcal{O}}_+\right)$ (as $t \longrightarrow 0$) for an $a > 0$ satisfying

$$a > \max_{1 \leq i \leq m} \operatorname{Re} \rho_i(0).$$

The conclusion 2. of proposition 8.4.5. with the hypothesis $I_1 = I_3 = \emptyset$ and $m_3 = \infty$ gives

$$u(t,x) = O\left(t^\infty, \widetilde{\mathcal{O}}_+\right) \quad (\text{as } t \longrightarrow 0). \tag{8.4.9}$$

This means that $u(t,x) = O\left(t^b, \widetilde{\mathcal{O}}_+\right)$ (as $t \longrightarrow 0$) for any $b > 0$.

But we have more

Proposition 8.4.6 *Let* $u(t,x) \in \widetilde{\mathcal{O}}_+$ *be a solution of (8.4.8) and assume that* $u(t,x) = O\left(t^a, \widetilde{\mathcal{O}}_+\right)$ *(as* $t \longrightarrow 0$*) for an* $a > 0$ *satisfying*

$$a > \max_{1 \leq i \leq m} \operatorname{Re} \rho_i(0). \tag{8.4.10}$$

Then $u(t,x) \equiv 0$ *in* $\widetilde{\mathcal{O}}_+$.

Proof : Set

$$b_{j,\alpha}(t,x) = \int_0^1 \frac{\partial G_2}{\partial Z_{j,\alpha}}(t,x) \left(\left\{ \left(t\frac{\partial}{\partial t}\right)^j \left(\frac{\partial}{\partial x}\right)^\alpha u\theta \right\}_{j+|\alpha| \leq m} \right) d\theta.$$

As $a > 0$ by (8.4.7) we have

$$b_{j,\alpha}(t,x) = O\left(t^a, \widetilde{\mathcal{O}}_+\right) \quad (\text{as } t \longrightarrow 0).$$

Let $s > 0$ be as in (8.4.6) and $\rho > 0$ be such that

$$0 < \rho < \min\left\{\frac{a}{m}, \frac{s}{m}\right\}.$$

Set

$$w(t,x) = t^{-a}u(t,x) \in \widetilde{\mathcal{O}}_+$$

and

$$c_{j,\alpha}(t,x) = t^{-\rho|\alpha|}\left(a_{j,\alpha}(t,x) + b_{j,\alpha}(t,x)\right) \in \widetilde{\mathcal{O}}_+.$$

It is easy to see that $w(t,x)$ is a solution of the linear partial differential equation

(8.4.11)
$$C\left(t\frac{\partial}{\partial t} + a, x\right) w = \sum_{j+|\alpha|\leq m} c_{j,\alpha}(t,x)\left(t\frac{\partial}{\partial t} - \rho|\alpha| + a\right)^j \left(t^\rho \frac{\partial}{\partial t}\right)^\alpha w.$$

Moreover, if $\lambda_1(x), \ldots, \lambda_m(x)$ denote the roots of the polynomial $C(\lambda + a, x)$ the condition (8.4.10) implies

$$\operatorname{Re}\lambda_i(0) < 0 \quad \text{for all } i = 1, \ldots, m.$$

It follows that by applying the theorem of unicity of Tahara [48] or Baouendi–Goulaouic [3] to the equation (8.4.11) we obtain $u(t,x) \equiv 0$ in $\widetilde{\mathcal{O}}_+$. ■

8.5 Completion of the proof of the main theorem

In this section, we are applying the results of section 8.1 to 8.4 to non linear singular partial differential equations of the form

(8.5.1)
$$\left(t\frac{\partial}{\partial t}\right)^m u = F\left(t, x, \left\{\left(t\frac{\partial}{\partial t}\right)^j \left(\frac{\partial}{\partial x}\right)^\alpha u\right\}_{(j,\alpha)\in I_m}\right)$$

introduced in the introduction and we complete the proof of the main theorem 8.0.3.

Throughout this section we assume that the conditions (A1), (A2), and (A3) (see the introduction) are satisfied. Denote by $\rho_1(x), \ldots, \rho_m(x)$ the functions of x which are the roots of the polynomial

$$C(\rho, x) = \rho^m - \sum_{j<m} \frac{\partial F}{\partial Z_{j,0}}(0, x, 0)\rho^j.$$

Let us recall the notations

$$\begin{aligned} J_+ &= \{i \in \{1, \ldots, m\} : \operatorname{Re}\rho_i(0) > 0\} \quad \text{and} \\ \mu &= \text{the cardinal of } J_+. \end{aligned}$$

When $\mu \geq 1$ we may assume that

(8.5.2)
$$\operatorname{Re}\rho_i(0) \begin{cases} > 0 & \text{for } 1 \leq i \leq \mu, \\ \leq 0 & \text{for } \mu + 1 \leq i \leq m. \end{cases}$$

The conditions (A1), (A2), and (A3) imply that the equation (8.5.1) can be written in the form

$$\left(\left(t\frac{\partial}{\partial t}\right)^m + c_{m-1}(x)\left(t\frac{\partial}{\partial t}\right)^{m-1} + \cdots + c_0(x)\right)u$$
$$= tb(x) + G_2(x)\left(t, \left\{\left(t\frac{\partial}{\partial t}\right)^j \left(\frac{\partial}{\partial x}\right)^\alpha u\right\}_{(j,\alpha)\in I_m}\right)$$

which is a particular case of equations of type (8.1.1) and (8.3.1).

Moreover, we have

$$c_j(x) = -\frac{\partial F}{\partial Z_{j,0}}(0,x,0) \quad \text{for } j = 0,\ldots,m-1.$$

The results of section 8.1 to 8.4 give us

(C1) (see corollary 8.1.2). If $\rho_i(0) \notin \mathbb{N}^*$ for all $i = 1,\ldots,m$ the equation (8.5.1) has a unique solution $u_0(t,x) = u_0$ holomorphic near the origin of $\mathbb{C}\times\mathbb{C}^n$ satisfying $u_0(0,x)\equiv 0$.

(C2) (see theorem 8.3.1). Assume $\mu \geq 1$, (8.5.2) and the following conditions:

(a) $\rho_1(x),\ldots,\rho_\mu(x)$ are holomorphic near $x=0$;

(b) $C(1,0)\neq 0$;

(c) $C\left(i + j_1\rho_1(0) + \cdots + j_\mu\rho_\mu(0), 0\right) \neq 0$ for all $(i,j)\in \mathbb{N}\times\mathbb{N}^\mu$ satisfying $i+|j|\geq 2$.

Then for any $\varphi_1(x),\ldots,\varphi_\mu(x) \in \mathbb{C}\{x\}$ the equation (8.5.1) has a unique solution $U(\varphi_1,\ldots,\varphi_\mu)(t,x)$ belonging to $\widetilde{\mathcal{O}}_+$ of the form

$$U(\varphi_1,\ldots,\varphi_\mu) = \sum_{i\geq 1} u_i(x)t^i + \sum_{i+2m|j|\geq k+2m,\,|j|\geq 1} \varphi_{i,j,k}(x)t^{i+j_1\rho_1(x)+\cdots+j_\mu\rho_\mu(x)}(\log t)^k$$

with $\varphi_{0,e_p,0}(x) = \varphi_p(x)$ for $p = 1,\ldots,\mu$ where $e_1 = (1,0,\ldots,0),\ldots,e_\mu = (0,\ldots,0,1)$. It is easy to see that $U(0,\ldots,0) = u_0$ is the unique holomorphic solution in (C1).

(C3) (see the discussion in section 8.2 to 8.4) Let $0\leq p < l \leq \mu$. Assume that there exist real numbers s,a,b satisfying the following conditions:

(a) $0 < s < \min\left\{1, \min_{i\leq p} \operatorname{Re}\rho_i(0)\right\}$;

(b) $\max_{i\leq p}\operatorname{Re}\rho_i(0) < a < \min_{p+1\leq i\leq l}\operatorname{Re}\rho_i(0)$;

(c) $\max_{p+1\le i\le l} \operatorname{Re}\rho_i(0) < b < \min\{2a, a+s\}$.

Then under the conditions given in (C2) we have

$$U(\varphi_1,\dots,\varphi_l,0,\dots,0) - U(\varphi_1,\dots,\varphi_p,0,\dots,0) - \sum_{p+1\le i\le l} \varphi_i(x)t^{\rho_i(x)}$$
$$= O\left(t^b, \widetilde{\mathcal{O}}_+\right) \quad (\text{as } t \longrightarrow 0).$$

Denote by S_+ the set of solutions of (8.5.1) belonging to $\widetilde{\mathcal{O}}_+$ and let us return to the proof of the main theorem.

The result (C1) gives us immediately the assertion (I) of the main theorem. The results (C1) and (C2) mean that

$$S_+ \supset \begin{cases} \{u_0\} & \text{when } \mu = 0, \\ \{U(\varphi_1,\dots,\varphi_\mu) : (\varphi_1,\dots,\varphi_\mu) \in (\mathbb{C}\{x\})^\mu\} & \text{when } \mu \ge 1. \end{cases}$$

Therefore, to complete the proof of the main theorem it is enough to prove

Theorem 8.5.1 *Assume that (8.5.1) satisfies the conditions (A1), (A2), and (A3). Then we have:*

1. *If $\mu = 0$ that is $\operatorname{Re}\rho_i(0) \le 0$ for all $i = 1,\dots,m$ then $S_+ = \{u_0\}$.*
2. *If $\mu \ge 1$ and if we have (8.5.2) and also*
   (a) *$\rho_i(0) \ne \rho_j(0)$ for $1 \le i \ne j \le \mu$,*
   (b) *$C(1,0) \ne 0$, and*
   (c) *$C(i + j_1\rho_1(0) + \cdots + j_\mu\rho_\mu(0), 0) \ne 0$ for all $(i,j) \in \mathbb{N}\times\mathbb{N}^\mu$ satisfying $i + |j| \ge 2$*

   then every solution $u \in S_+$ of (8.5.1) is of the form

   (8.5.3) $$u = U(\varphi_1,\dots,\varphi_\mu)$$

   for some $\varphi_1,\dots,\varphi_\mu$ belonging to $\mathbb{C}\{x\}$ determined uniquely by u.

Proof : Let $u \in S_+$. Put $w = u - u_0$. By the definition of $\widetilde{\mathcal{O}}_+$ we have for an $a > 0$ $w(t,x) = O\left(t^a, \widetilde{\mathcal{O}}_+\right)$ (as $t \longrightarrow 0$). As u and u_0 are solutions of (8.5.1) w is a solution of

(8.5.4)
$$\left(t\frac{\partial}{\partial t}\right)^m w = F\left(t,x,\left\{\left(t\frac{\partial}{\partial t}\right)^j\left(\frac{\partial}{\partial x}\right)^\alpha w + \left(t\frac{\partial}{\partial t}\right)^j\left(\frac{\partial}{\partial x}\right)^\alpha u_0(t,x)\right\}_{(j,\alpha)\in I_m}\right)$$
$$- F\left(t,x,\left\{\left(t\frac{\partial}{\partial t}\right)^j\left(\frac{\partial}{\partial x}\right)^\alpha u_0(t,x)\right\}_{(j,\alpha)\in I_m}\right).$$

The hypothesis (A1), (A2), and (A3) imply that the equation (8.5.4) for w is a particular case of equations (8.4.8). It follows that if we assume that $\operatorname{Re}\rho_i(0) \le 0$ for all $i = 1, \dots, m$ we have

$$a > 0 \ge \max_{1\le i\le m} \operatorname{Re}\rho_i(0)$$

and proposition 8.4.6 gives us $w \equiv 0$. This proves the assertion 1. of theorem 8.5.1.

To prove 2. assume that we have $\mu \ge 1$, (8.5.2), and (a), (b), (c) in theorem 8.5.1. We can assume without loosing generality that

$$0 < \operatorname{Re}\rho_1(0) = \cdots = \operatorname{Re}\rho_{i_1}(0) < \operatorname{Re}\rho_{i_1+1}(0) = \cdots = \operatorname{Re}\rho_{i_2}(0) < \cdots$$
$$\cdots < \operatorname{Re}\rho_{i_{N-1}+1}(0) = \cdots = \operatorname{Re}\rho_{i_N}(0)$$

with $0 < i_1 < i_2 < \cdots < i_N = \mu$.

Set $i_0 = 0$ and

$$m_l = \operatorname{Re}\rho_{i_{l-1}+1}(0) = \cdots = \operatorname{Re}\rho_{i_l}(0) \quad \text{for } l = 1, \dots, N.$$

Choose $s > 0, a_1, \dots, a_N, a_{N+1} = \infty$ and $b_1, \dots, b_N$ such that

1. $0 < s < \min\{m_1, 1\}$ and
2. $0 < a_l < m_l < b_l < \min\{2a_l, a_l + s, a_{l+1}\}$ for $l = 1, \dots, N$.

Let $u \in S_+$ be fixed. Put $w_1 = u - u_0$. As $w_1 \in \widetilde{\mathcal{O}}_+$ we have for an $a > 0$ $w_1(t,x) = O\left(t^a, \widetilde{\mathcal{O}}_+\right)$ (as $t \longrightarrow 0$). Applying 1. of proposition 8.4.5 to w_1 which is a solution of the equation (8.5.4) which is of type (8.4.8) we get $w_1(t,x) = O\left(t^{a_1}, \widetilde{\mathcal{O}}_+\right)$ (as $t \longrightarrow 0$) and the conclusion 3. of the same proposition gives us

$$w_1 = \sum_{1\le p\le i_1} \varphi_p(x) t^{\rho_p(x)} + O\left(t^{b_1}, \widetilde{\mathcal{O}}_+\right) \quad (\text{as } t \longrightarrow 0) \tag{8.5.5}$$

for some $\varphi_1(x), \dots \varphi_{i_1}(x) \in \mathbb{C}\{x\}$.

Set now

$$u_1 = U(\varphi_1, \dots, \varphi_{i_1}, 0, \dots, 0) \qquad \text{and} \qquad w_2 = u - u_1.$$

By (8.5.5) and (C3) we have

$$\begin{aligned} w_2 &= u_0 - u_1 + w_1 \\ &= U(0, \dots, 0) - U(\varphi_1, \dots, \varphi_{i_1}, 0, \dots, 0) + \sum_{1\le p\le i_1} \varphi_p(x) t^{\rho_p(x)} + O\left(t^{b_1}, \widetilde{\mathcal{O}}_+\right) \\ &= O\left(t^{b_1}, \widetilde{\mathcal{O}}_+\right) \quad (\text{as } t \longrightarrow 0). \end{aligned}$$

Moreover, w_2 is a solution of

$$\left(t\frac{\partial}{\partial t}\right)^m w_2$$
$$= F\left(t, x, \left\{\left(t\frac{\partial}{\partial t}\right)^j \left(\frac{\partial}{\partial x}\right)^\alpha w_2 + \left(t\frac{\partial}{\partial t}\right)^j \left(\frac{\partial}{\partial x}\right)^\alpha u_1(t,x)\right\}_{(j,\alpha)\in I_m}\right)$$
$$- F\left(t, x, \left\{\left(t\frac{\partial}{\partial t}\right)^j \left(\frac{\partial}{\partial x}\right)^\alpha u_1(t,x)\right\}_{(j,\alpha)\in I_m}\right)$$

which is a particular case of (8.4.8). Therefore, by 2. of proposition 8.4.5 we have $w_2(t,x) = O\left(t^{a_2}, \widetilde{\mathcal{O}}_+\right)$ (as $t \longrightarrow 0$). Applying 3. of the same proposition we get

$$w_2 = \sum_{i_1+1\le p\le i_2} \varphi_p(x) t^{\rho_p(x)} + O\left(t^{b_2}, \widetilde{\mathcal{O}}_+\right) \quad (\text{as } t \longrightarrow 0) \tag{8.5.6}$$

for some $\varphi_{i_1+1}(x), \dots, \varphi_{i_2}(x) \in \mathbb{C}\{x\}$. Introduce now

$$u_2 = U(\varphi_1, \dots, \varphi_{i_2}, 0, \dots, 0) \qquad \text{and} \qquad w_3 = u - u_2.$$

By (8.5.6) and (C3) we obtain

$$\begin{aligned} w_3 &= u_1 - u_2 + w_2 \\ &= U(\varphi_1, \dots, \varphi_{i_1}, 0, \dots, 0) - U(\varphi_1, \dots, \varphi_{i_2}, 0, \dots, 0) \\ &\quad + \sum_{i_1+1\le p\le i_2} \varphi_p(x) t^{\rho_p(x)} + O\left(t^{b_2}, \widetilde{\mathcal{O}}_+\right) \\ &= O\left(t^{b_2}, \widetilde{\mathcal{O}}_+\right) \quad (\text{as } t \longrightarrow 0). \end{aligned}$$

Repeating this argument we obtain functions $\varphi_{i_2+1}(x), \dots, \varphi_{i_N}(x) \in \mathbb{C}\{x\}$ such that if we set

$$u_l = U(\varphi_1, \dots, \varphi_{i_l}, 0, \dots, 0)$$

for $l \in \{3, \dots, N\}$ and $w_{l+1} = u - u_l$ we have

$$w_{l+1} = O\left(t^{b_l}, \widetilde{\mathcal{O}}_+\right) \quad (\text{as } t \longrightarrow 0).$$

In particular, for $l = N$ we see that w_{N+1} satisfies

$$\left(t\frac{\partial}{\partial t}\right)^m w_{N+1}$$
$$= F\left(t, x, \left\{\left(t\frac{\partial}{\partial t}\right)^j \left(\frac{\partial}{\partial x}\right)^\alpha w_{N+1} + \left(t\frac{\partial}{\partial t}\right)^j \left(\frac{\partial}{\partial x}\right)^\alpha u_N(t,x)\right\}_{(j,\alpha)\in I_m}\right)$$
$$- F\left(t, x, \left\{\left(t\frac{\partial}{\partial t}\right)^j \left(\frac{\partial}{\partial x}\right)^\alpha u_N(t,x)\right\}_{(j,\alpha)\in I_m}\right)$$

which is an equation of type (8.4.8). As $w_{N+1} = O\left(t^{b_N}, \widetilde{\mathcal{O}}_+\right)$ (as $t \longrightarrow 0$) and $b_N > m_N$ proposition 8.4.6 implies $w_{N+1} \equiv 0$. This means that

$$u = u_N = U(\varphi_1, \ldots, \varphi_\mu)$$

and proves (8.5.3).

To obtain the uniqueness it is enough to show that

$$U(\varphi_1, \ldots, \varphi_\mu) = U(\psi_1, \ldots, \psi_\mu) \quad \text{in } \widetilde{\mathcal{O}}_+ \tag{8.5.7}$$

implies $\varphi_i = \psi_i$ in $\mathbb{C}\{x\}$ for all $i = 1, \ldots, \mu$.

By (8.5.7) we have

$$\begin{aligned} 0 &= U(\varphi_1, \ldots, \varphi_\mu) - U(\psi_1, \ldots, \psi_\mu) \\ &= \sum_{1 \le p \le i_1} (\varphi_p(x) - \psi_p(x))\, t^{\rho_p(x)} + O\left(t^{b_1}, \widetilde{\mathcal{O}}_+\right) \quad (\text{as } t \longrightarrow 0). \end{aligned} \tag{8.5.8}$$

As $\operatorname{Re} \rho_p(0) < b_1$ for $p = 1, \ldots, i_1$ and the $\rho_1(0), \ldots, \rho_{i_1}(0)$ are all distinct (8.5.8) gives us

$$\varphi_p(x) = \psi_p(x) \quad \text{in } \mathbb{C}\{x\} \text{ for all } p = 1, \ldots, i_1. \tag{8.5.9}$$

Now (8.5.7) and (8.5.9) imply easily

$$\begin{aligned} 0 &= U(\varphi_1, \ldots, \varphi_\mu) - U(\psi_1, \ldots, \psi_\mu) \\ &= \sum_{i_1+1 \le p \le i_2} (\varphi_p(x) - \psi_p(x))\, t^{\rho_p(x)} + O\left(t^{b_2}, \widetilde{\mathcal{O}}_+\right) \quad (\text{as } t \longrightarrow 0) \end{aligned}$$

which gives

$$\varphi_p(x) = \psi_p(x) \quad \text{in } \mathbb{C}\{x\} \text{ for all } p = i_1 + 1, \ldots, i_2.$$

And by induction we obtain

$$\varphi_p(x) = \psi_p(x) \quad \text{in } \mathbb{C}\{x\} \text{ for all } p = 1, \ldots, \mu$$

which ends the proof of theorem 8.5.1 and completes the proof of the main theorem of this chapter. ∎

9 On the existence of holomorphic solutions of the Cauchy problem for non linear partial differential equations

Introduction

In 1968, Artin [2] has proved that if a sytem of analytic equations has a formal power series solution then it has a convergent power series solution. In this chapter, we are proving similar results for the Cauchy problem for non linear partial differential equations.

Roughly speaking we can say that our results assert the following: if the Cauchy problem is algebraically integrable "in a good sense" then it is analytically integrable.

9.1 Notations and definitions

Put $\mathbb{N} = \{0, 1, 2, \dots\}$ and $\mathbb{N}^* = \mathbb{N} \setminus \{0\}$. Let $n \in \mathbb{N}^*, x = (x_1, \dots, x_n) \in \mathbb{C}^n$ and $t \in \mathbb{C}$.

For $\alpha = (\alpha_1, \dots, \alpha_n) \in \mathbb{N}^n$ we write

$$|\alpha| = \alpha_1 + \cdots + \alpha_n \qquad \text{and} \qquad \left(\frac{\partial}{\partial x}\right)^{\alpha} = \left(\frac{\partial}{\partial x_1}\right)^{\alpha_1} \cdots \left(\frac{\partial}{\partial x_n}\right)^{\alpha_n}.$$

Let $m \in \mathbb{N}^*$, denote by δ the cardinal of the set

$$\{(j, \alpha) \in \mathbb{N} \times \mathbb{N}^n : j + |\alpha| \leq m\}$$

and set

$$X = \{X_{j,\alpha}\}_{j+|\alpha|\leq m} \in \mathbb{C}^{\delta}.$$

Let $F(t, x, X)$ be a function in (t, x, X), $\Phi_j(x)$ be functions in x for $j = 0, 1, \dots, m-1$, and $a = \{a_{j,\alpha}\}_{j+|\alpha|\leq m} \in \mathbb{C}^{\delta}$. Assume

1. $F(t, x, X)$ is holomorphic in a neighborhood of $(0, 0, a) \in \mathbb{C} \times \mathbb{C}^n \times \mathbb{C}^{\delta}$;
2. $F(0, 0, a) = 0$;
3. $\Phi_j(x)$ for $j = 0, 1, \dots, m-1$ are holomorphic near $x = 0 \in \mathbb{C}^n$;

4. $\left(\left(\frac{\partial}{\partial x}\right)^\alpha \Phi_j\right)(0) = a_{j,\alpha}$ for $j + |\alpha| \le m$ and $j < m$.

In this chapter, we are dealing with the following Cauchy problem:

$$
\begin{aligned}
&F\left(t, x, \left\{\left(\frac{\partial}{\partial t}\right)^j \left(\frac{\partial}{\partial x}\right)^\alpha u\right\}_{j+|\alpha|\le m}\right) = 0, \\
&\left(\left(\frac{\partial}{\partial t}\right)^j u\right)(0, x) = \Phi_j(x) \qquad \text{for } j = 0, 1, \dots, m-1, \\
&\left(\left(\frac{\partial}{\partial t}\right)^j \left(\frac{\partial}{\partial x}\right)^\alpha u\right)(0, 0) = a_{j,\alpha} \qquad \text{for } j + |\alpha| \le m,
\end{aligned}
\tag{9.1.1}
$$

where $u = u(t, x)$ is an unknown function.

The case $m = 1$ is already treated in [22] and the general case is in [24].

We denote by $\mathbb{C}[[t, x]]$ the ring of formal power series in (t, x) and by $\mathbb{C}\{t, x\}$ the subring of convergent power series in (t, x). Note that any $f(t, x) \in \mathbb{C}[[t, x]]$ can be written in the form

$$f(t, x) = \sum_{0 \le i < \infty} f_i(x) t^i \quad \text{with } f_i(x) \in \mathbb{C}[[x]] \text{ for } i \ge 0.$$

Definition 9.1.1 *The valuation* $v_t(f)$ *of* $f(t, x)$ *in* t *is defined as follows:*

1. *when* $f \equiv 0$ *we set* $v_t(f) = \infty$;
2. *when* $f \not\equiv 0$ *we set* $v_t(f) = \min\{i : f_i(x) \not\equiv 0\}$.

For simplicity we write

$$D^m u = \left\{\left(\frac{\partial}{\partial t}\right)^j \left(\frac{\partial}{\partial x}\right)^\alpha u\right\}_{j+|\alpha|\le m}.$$

For any $u(t, x) \in \mathbb{C}[[t, x]]$ we put

$$
\begin{aligned}
q_{j,\alpha}(u) &= v_t\left(\frac{\partial F}{\partial X_{j,\alpha}}(t, x, D^m u(t, x))\right) && \text{for } j + |\alpha| \le m, \\
q_{j,\alpha}(u, 0) &= v_t\left(\frac{\partial F}{\partial X_{j,\alpha}}(t, 0, D^m u(t, 0))\right) && \text{for } j + |\alpha| \le m, \\
a_j(u, t) &= t^{m-j} \frac{\frac{\partial F}{\partial X_{j,0}}(t, 0, D^m u(t, 0))}{\frac{\partial F}{\partial X_{m,0}}(t, 0, D^m u(t, 0))} && \text{for } 0 \le j < m.
\end{aligned}
$$

Note that $q_{j,\alpha}(u), q_{j,\alpha}(u, 0)$ and $a_j(u, t)$ are depending only on $u \in \mathbb{C}[[t, x]]$.

If $q_{m,0}(u, 0) < \infty$ and $q_{m,0}(u, 0) \le q_{j,0}(u, 0) + m - j$ hold then $a_j(u, 0) = a_j(u, t)|_{t=0}$ is well defined. If $q_{m,0}(u, 0) < q_{j,0}(u, 0) + m - j$ it is clear that $a_j(u, 0) = 0$ holds. Put

$$C(\rho, u) = \rho(\rho-1)\cdots(\rho-m+1) + \sum_{j<m} a_j(u,0)\rho(\rho-1)\cdots(\rho-j+1).$$

Now let us introduce new variables

$$y_m, y_{m+1}, y_{m+2}, \ldots \qquad \text{and} \qquad z_{\alpha,m}, z_{\alpha,m+1}, z_{\alpha,m+2}, \ldots \quad \text{for } 0 < |\alpha| \le m$$

and set

$$Z_\alpha(t,y,z) = \begin{cases} \sum_{0\le k\le m-1} \Phi_k(x)\frac{t^k}{k!} + \sum_{m\le k<\infty} y_k t^k & \text{when } |\alpha| = 0, \\ \sum_{0\le k\le m-1} \Phi_k(x)\frac{t^k}{k!} + \sum_{m\le k<\infty} z_{\alpha,k} t^k & \text{when } |\alpha| > 0, \end{cases}$$

where

$$y = \{y_k\}_{k\ge m} \qquad \text{and} \qquad z = \{z_{\alpha,k}\}_{0<|\alpha|\le m, k\ge m}.$$

Then by using the formal Taylor expansion in t we define p_* by

$$p_* = v_t\Bigg(F\left(t, x, \left\{\left(\frac{\partial}{\partial t}\right)^j Z_\alpha(t,y,z)\right\}_{j+|\alpha|\le m}\right) - F\left(t, x, \left\{\left(\frac{\partial}{\partial t}\right)^j Z_\alpha(t,y,0)\right\}_{j+|\alpha|\le m}\right)\Bigg).$$

Similary we define $p_{j,\alpha}$ for $j+|\alpha|\le m$ by

$$p_{j,\alpha} = v_t\Bigg(\frac{\partial F}{\partial X_{j,\alpha}}\left(t, x, \left\{\left(\frac{\partial}{\partial t}\right)^j Z_\alpha(t,y,z)\right\}_{j+|\alpha|\le m}\right) - \frac{\partial F}{\partial X_{j,\alpha}}\left(t, x, \left\{\left(\frac{\partial}{\partial t}\right)^j Z_\alpha(t,y,0)\right\}_{j+|\alpha|\le m}\right)\Bigg).$$

Note that p_* and $p_{j,\alpha}$ are well defined depending only on the equation (9.1.1). By the definitions it is easy to see that $p_* \ge 1$ and $p_{j,\alpha} \ge 1$.

9.2 Results

By using the given notations we have the following Artin's type theorem:

Theorem 9.2.1 *If (9.1.1) has a formal power series solution $\hat{u}(t,x) \in \mathbb{C}[[t,x]]$ and if the conditions*

1. $q_{m,0}(\hat{u}) = q_{m,0}(\hat{u}, 0) < \infty$;

2. $q_{m,0}(\hat{u}) \leq q_{j,0}(\hat{u},0) + m - j$ *for* $j < m$;

3. $q_{m,0}(\hat{u}) < q_{j,\alpha}(\hat{u},0) + m - j$ *for* $j + |\alpha| \leq m$ *and* $|\alpha| > 0$;

4. $q_{m,0}(\hat{u}) \leq \frac{p_* - 1}{2}$;

5. $q_{m,0}(\hat{u}) \leq p_{j,0} + m - j$ *for* $j \leq m$;

6. $q_{m,0}(\hat{u}) < p_{j,\alpha} + m - j$ *for* $j + |\alpha| \leq m$ *and* $|\alpha| > 0$;

7. $C(k + q_{m,0}(\hat{u}) + m, \hat{u}) \neq 0$ *for* $k \in \mathbb{N}^*$

are satisfied then (9.1.1) has a convergent power series solution $w(t,x) \in \mathbb{C}\{t,x\}$.

Remarks 9.2.2 1. *Note that the conditions 1. to 7. are algebraic and, therefore, we can understand that theorem 9.2.1 asserts the following: if (9.1.1) is algebraically integrable in a "good sense" then (9.1.1) is analytically integrable. By "good sense" we mean that* $\hat{u}$ *satisfies 1. to 7.*

2. *When* $\frac{\partial F}{\partial X_{m,0}}(0,0,a) \neq 0$ *holds the Cauchy-Kowalewski theorem says that (9.1.1) has a unique holomorphic solution* $u(t,x)$. *In this case, we have*

$$\begin{aligned} q_{m,0}(u) &= q_{m,0}(u,0) = 0, \\ C(\rho,u) &= \rho(\rho-1)\cdots(\rho-m+1) \end{aligned}$$

and, therefore, the conditions 1. to 7. in theorem 9.2.1 are trivially satisfied.

Now as a corollary of theorem 9.2.1 let us give a variation.
For $u(t,x) \in \mathbb{C}[[t,x]]$ we put

$$\begin{aligned} r(u) &= \min\{q_{j,0}(u) - j : 0 \leq j \leq m\}, \\ l(u) &= \max\{j : q_{j,0}(u) - j = r(u)\}. \end{aligned}$$

Then we have

Theorem 9.2.3 *If (9.1.1) has a formal power series solution* $\hat{u}(t,x) \in \mathbb{C}[[t,x]]$ *and if the conditions*

1. $l(\hat{u}) = m$;

2. $q_{m,0}(\hat{u}) = q_{m,0}(\hat{u},0) < \infty$;

3. $q_{m,0}(\hat{u}) \leq \min\left\{\frac{p_*-1}{2}, p_* - m\right\}$;

4. $C(k + q_{m,0}(\hat{u}) + m, \hat{u}) \neq 0$ *for* $k \in \mathbb{N}^*$

are satisfied then (9.1.1) has a convergent power series solution $w(t,x) \in \mathbb{C}\{t,x\}$.

Remarks 9.2.4 1. *When* $m = 1$ *theorem 9.2.3 coincides with theorem 1 of [22].*

2. *When $l(\hat{u}) < m$ (9.1.1) has not in general a convergent solution as is seen in the following example:*

$$
\text{(9.2.1)} \qquad \begin{aligned} t^2\frac{\partial u}{\partial t} - u + t + t^4\frac{\partial u}{\partial x} &= 0, \\ u(0,x) &= 0. \end{aligned}
$$

In this case, we see that (9.2.1) has a unique formal solution

$$\hat{u} = \sum_{1\le k<\infty} ((k-1)!)\, t^k$$

and that $m = 1, l(\hat{u}) = 0, q_{1,0}(\hat{u}) = q_{1,0}(\hat{u}, 0) = 2$, and $p_* = 5$. Of course, $C(\rho, \hat{u})$ for (9.2.1) is not well defined.

9.3 Proof of theorem 9.2.1

Assume that (9.1.1) has a formal power series solution $\hat{u}(t,x) \in \mathbb{C}[[t,x]]$ satisfying the conditions 1. to 7. of theorem 9.2.1. Put

$$\hat{u} = \sum_{0\le i<\infty} \hat{u}_i(x)t^i$$

with $\hat{u}_i(x) = \frac{\Phi_i(x)}{i!}$ for $i = 0, 1, \ldots, m-1$ and $\hat{u}_i(x) \in \mathbb{C}[[x]]$ for $i \ge m$.

Set $q_m = q_{m,0}(\hat{u})$ and

$$\varphi^* = \sum_{0\le i\le q_m+m} \hat{u}_i(x)t^i.$$

Then it is easy to see that

$$D^m\hat{u}(t,x) \equiv D^m\varphi^*(t,x) \pmod{t^{q_m+1}}$$

and, therefore,

$$
\text{(9.3.1)} \qquad \frac{\partial F}{\partial X_{j,\alpha}}(t,x,D^m\hat{u}) \equiv \frac{\partial F}{\partial X_{j,\alpha}}(t,x,D^m\varphi^*) \pmod{t^{q_m+1}}.
$$

Since $q_m < q_m + 1$ and $q_m - m + j < q_m + 1$ hold by (9.3.1) and 1. to 3. we have

$$
\text{(9.3.2)} \qquad \begin{aligned} q_m &= q_{m,0}(\varphi^*) = q_{m,0}(\varphi^*, 0), \\ q_m - m + j &\le q_{j,0}(\varphi^*) && \text{for } j < m, \\ q_m - m + j &< q_{j,\alpha}(\varphi^*) && \text{for } j + |\alpha| \le m \text{ and } |\alpha| > 0 \end{aligned}
$$

which is equivalent to

$$\frac{\partial F}{\partial X_{j,0}}(t,x,D^m\varphi^*) \equiv 0 \;(\;\mathrm{mod}\; t^{q_m-m+j}) \quad \text{for } j \le m,$$

(9.3.3)

$$\frac{\partial F}{\partial X_{j,\alpha}}(t,x,D^m\varphi^*) \equiv 0 \;(\;\mathrm{mod}\; t^{q_m-m+j+1})$$

$$\text{for } j+|\alpha| \le m \text{ and } |\alpha|>0,$$

$$\left(t^{-q_m}\frac{\partial F}{\partial X_{m,0}}(t,0,D^m\varphi^*(t,0))\right)\Bigg|_{t=0} \ne 0.$$

Moreover, we have

(9.3.4) $$a_j(\hat{u},0) = a_j(\varphi^*,0) \quad \text{for } j<m$$

as is verified in the following way: when $q_{j,0}(\hat{u}) = q_m - m + j$ then $a_j(\hat{u},0) = a_j(\varphi^*,0)$ follows from (9.3.1) and the definition of $a_j(\hat{u},0)$; when $q_{j,0}(\hat{u}) > q_m - m + j$ then we have $q_{j,0}(\varphi^*) > q_m - m + j$ and hence

$$a_j(\hat{u},0) = 0 = a_j(\varphi^*,0).$$

Put

$$U^* = t^{-(q_m+m)}(\hat{u}-\varphi^*).$$

Then we have $U^*(0,x) \equiv 0$ and

(9.3.5) $$F\left(t,x,D^m\left(\varphi^* + t^{q_m+m}U^*\right)\right) = 0$$

since $\hat{u}$ is a formal solution of (9.1.1).

By expanding (9.3.5) into the Taylor series we easily get

(9.3.6) $$0 = F(t,x,D^m\varphi^*) + \sum_{j+|\alpha|\le m}\frac{\partial F}{\partial X_{j,\alpha}}(t,x,D^m\varphi^*)\,t^{q_m+m-j}\Theta_j\left(\frac{\partial}{\partial x}\right)^{\alpha}U^*$$

$$+\sum_{|\nu|\ge 2}\left\{\left(\frac{\partial}{\partial X}\right)^{\nu}F(t,x,D^m\varphi^*)\prod_{j+|\alpha|\le m}\left(t^{q_m+m-j}\Theta_j\left(\frac{\partial}{\partial x}\right)^{\alpha}U^*\right)^{\nu_{j,\alpha}}\right\}$$

where

$$\Theta_j = \left(t\frac{\partial}{\partial t}+q_m+m\right)\left(t\frac{\partial}{\partial t}+q_m+m-1\right)\cdots\left(t\frac{\partial}{\partial t}+q_m+m-j+1\right),$$

$$\nu = \{\nu_{j,\alpha}\}_{j+|\alpha|\le m} \in \mathbb{N}^{\delta},$$

$$|\nu| = \sum_{j+|\alpha|\le m}\nu_{j,\alpha},$$

$$\left(\frac{\partial}{\partial X}\right)^{\nu} = \prod_{j+|\alpha|\le m}\left(\frac{\partial}{\partial X_{j,\alpha}}\right)^{\nu_{j,\alpha}}.$$

Since

$$\begin{array}{rcll} q_{j,0}(\varphi^*) + (q_m + m - j) & \geq & 2q_m & \text{for } j \leq m, \\ q_{j,\alpha}(\varphi^*) + (q_m + m - j) & > & 2q_m & \text{for } j + |\alpha| \leq m \text{ and } |\alpha| > 0, \\ \sum_{j+|\alpha|\leq m} (q_m + m - j)\, \nu_{j,\alpha} \geq q_m |\nu| & \geq & 2q_m & \text{for } |\nu| \geq 2, \\ v_t(U^*) & \geq & 1 & \end{array}$$

hold, by comparing the valuation in each term of (9.3.6) we get

$$F(t, x, D^m\varphi^*) \equiv 0 \ \left(\bmod t^{2q_m+1}\right) \tag{9.3.7}$$

and after simplification by t^{2q_m} the equality (9.3.6) is reduced to the form

$$\sum_{j\leq m} b_j(t, x, D^m\varphi^*)\, \Theta_j U^* = t\cdot h(t, x, D^m\varphi^*) + R\left(t, x, D^m\varphi^*, \left\{\Theta_j \left(\frac{\partial}{\partial x}\right)^\alpha U^*\right\}_{j+|\alpha|\leq m}\right) \tag{9.3.8}$$

where

$$\begin{aligned} b_j(t, x, D^m\varphi^*) &= t^{-(q_m-m+j)} \frac{\partial F}{\partial X_{j,0}}(t, x, D^m\varphi^*), \\ h(t, x, D^m\varphi^*) &= t^{-2q_m} F(t, x, D^m\varphi^*), \\ R(t, x, D^m\varphi^*, X) &= -\sum_{j+|\alpha|\leq m, |\alpha|>0} t^{-(q_m-m+j)} \frac{\partial F}{\partial X_{j,\alpha}}(t, x, D^m\varphi^*)\, X_{j,\alpha} - \\ &\quad -\sum_{|\nu|\geq 2} \left\{ \left(\frac{\partial}{\partial X}\right)^\nu F(t, x, D^m\varphi^*) \times t^{-2q_m} \times \right. \\ &\quad \left. \times \prod_{j+|\alpha|\leq m} \left(t^{q_m+m-j} X_{j,\alpha}\right)^{\nu_{j,\alpha}} \right\}. \end{aligned}$$

Note that $q_m = q_m(\varphi^*, 0)$ implies

$$b_m(t, 0, D^m\varphi^*(t, 0))|_{t=0} \neq 0 \tag{9.3.9}$$

and that the definition of $a_j(\varphi^*, 0)$ implies

$$a_j(\varphi^*, 0) = \frac{b_j(t, 0, D^m\varphi^*(t, 0))|_{t=0}}{b_m(t, 0, D^m\varphi^*(t, 0))|_{t=0}}. \tag{9.3.10}$$

Now let us find a convergent solution $w(t, x) \in \mathbb{C}\{t, x\}$ of (9.1.1) in the following form:

$$w = \varphi + t^{q_m+m} W,$$

$$\varphi = \sum_{0\leq i\leq m-1} \frac{\Phi_i(x) t^i}{i!} + \sum_{m\leq i\leq q_m+m} w_i(x) t^i \quad \text{with } w_i(x) \in \mathbb{C}\{x\},$$

$$W(t, x) \in \mathbb{C}\{t, x\} \qquad \text{and} \qquad W(0, x) \equiv 0$$

where $w_m(x), \ldots, w_{q_m+m}(x)$, and $W(t,x)$ are unknown functions.

To do so let us first consider the following system of equations with respect to $w_m(x), \ldots, w_{q_m+m}(x)$:

$$(9.3.11)\qquad \begin{aligned} &F(t,x,D^m\varphi) \equiv 0 \;\left(\mod t^{2q_m+1}\right), \\ &\frac{\partial F}{\partial X_{j,0}}(t,x,D^m\varphi) \equiv 0 \;\left(\mod t^{q_m-m+j}\right) \quad \text{for } j \le m, \\ &\frac{\partial F}{\partial X_{j,\alpha}}(t,x,D^m\varphi) \equiv 0 \;\left(\mod t^{q_m-m+j+1}\right) \\ &\qquad\qquad\qquad\qquad \text{for } j+|\alpha| \le m \text{ and } |\alpha|>0. \end{aligned}$$

Since

$$\begin{aligned} p_* &\ge 2q_m+1, \\ p_{j,0} &\ge q_m-m+j && \text{for } j \le m, \\ p_{j,\alpha} &\ge q_m-m+j+1 && \text{for } j+|\alpha| \le m \text{ and } |\alpha|>0 \end{aligned}$$

are assumed it follows from the definitions of p_* and $p_{j,\alpha}$ that the system (9.3.11) is nothing but a system of analytic equations with respect to $w_m(x), \ldots, w_{q_m+m}(x)$. Moreover, by (9.3.3) and (9.3.7) we know that (9.3.11) has a formal power series solution $\hat{u}_m(x), \ldots, \hat{u}_{q_m+m}(x)$. Hence, by Artin's theorem [2] we obtain a convergent solution $w_m(x), \ldots, w_{q_m+m}(x)$ of (9.3.11). In addition, we may assume

$$(9.3.12)\qquad D^m\varphi(t,0) = D^m\varphi^*(t,0)$$

(see Artin [2]).

Then by (9.3.11),(9.3.12), and the third relation in (9.3.3) it is easy to see

$$(9.3.13)\qquad \begin{aligned} q_m = q_m(\varphi) &= q_m(\varphi,0), \\ q_m-m+j &\le q_{j,0}(\varphi) && \text{for } j<m, \\ q_m-m+j+1 &\le q_{j,\alpha}(\varphi) && \text{for } j+|\alpha| \le m \text{ and } |\alpha|>0 \end{aligned}$$

and by (9.3.9) and (9.3.10) we see

$$(9.3.14)\qquad \begin{aligned} &b_m\left(t,0,D^m\varphi(t,0)\right)\big|_{t=0} \ne 0, \\ &a_j(\varphi^*,0) = \frac{b_j\left(t,0,D^m\varphi(t,0)\right)\big|_{t=0}}{b_m\left(t,0,D^m\varphi(t,0)\right)\big|_{t=0}}. \end{aligned}$$

Therefore, by (9.3.13) and by the same argument as in (9.3.8) the equation (9.1.1) is reduced to the following equation with respect to $W(t,x)$:

$$(9.3.15)\qquad \sum_{j\le m} b_j\left(t,x,D^m\varphi\right)\Theta_j W = t\cdot h\left(t,x,D^m\varphi\right) + R\left(t,x,D^m\varphi,\left\{\Theta_j\left(\frac{\partial}{\partial x}\right)^{\alpha} W\right\}_{j+|\alpha|\le m}\right)$$

which is a particular case of partial differential equations discussed in chapter 6 (and [19], [21]).

Since (9.3.4), (9.3.14), and the definition of $C(\rho, \hat{u})$ give us

$$(9.3.16) \quad \left(b_m(t,0,D^m\varphi(t,0))|_{t=0}\right) C(\rho,\hat{u}) = \sum_{j\le m} \left(b_j(t,0,D^m\varphi(t,0))|_{t=0}\right) \rho(\rho-1)\cdots(\rho-j+1)$$

we see from the assumption 7. of theorem 9.2.1 that the equation (9.3.15) satisfies the Poincaré condition imposed in theorem 6.3.1. Hence, by theorem 6.3.1 we see that the equation (9.3.15) has a unique holomorphic solution $W(t,x)$ satisfying $W(0,x) \equiv 0$.

Thus, by putting

$$w = \sum_{0\le i\le m-1} \frac{\Phi_i(x)t^i}{i!} + \sum_{m\le i\le q_m+m} w_i(x)t^i + t^{q_m+m}W(t,x)$$

we obtain a holomorphic solution of (9.1.1).

9.4 Proof of theorem 9.2.3

Assume that (9.1.1) has a formal power series solution $\hat{u}(t,x) \in \mathbb{C}[[t,x]]$ satisfying the conditions 1. to 4. in theorem 9.2.3. Then by 1. and the definition of $l(\hat{u})$ we have

$$q_{m,0}(\hat{u}) \le q_{j,0}(\hat{u}) + m - j \quad \text{for } j < m.$$

Therefore, it is easy to see that theorem 9.2.3 is a consequence of theorem 9.2.1 and the following:

Proposition 9.4.1 *If*

$$(9.4.1) \qquad q_{m,0}(\hat{u}) \le p_* - m$$

holds we have

1. $q_{m,0}(\hat{u}) < q_{j,\alpha}(\hat{u}) + m - j$ *for* $j + |\alpha| \le m$ *and* $|\alpha| > 0$;
2. $q_{m,0}(\hat{u}) \le p_{j,0} + m - j$ *for* $j \le m$;
3. $q_{m,0}(\hat{u}) < p_{j,\alpha} + m - j$ *for* $j + |\alpha| \le m$ *and* $|\alpha| > 0$.

Let us show this now. Note

Lemma 9.4.2 *Let* $\hat{u}(t,x) \in \mathbb{C}[[t,x]]$ *and let* p_* *and* $p_{j,\alpha}$ *be as in section 9.1. Then*

1. $q_{j,\alpha}(\hat{u}) \ge p_* - 2m + j + |\alpha|$ *for* $j + |\alpha| \le m$ *and* $|\alpha| > 0$;

2. $p_{j,0} \geq p_* - 2m + j$ *for* $j \leq m$;

3. $p_{j,\alpha} \geq p_* - 2m + j + |\alpha|$ *for* $j + |\alpha| \leq m$ *and* $|\alpha| > 0$.

The proof of this lemma will be given later.

Proof of proposition 9.4.1: By using (9.4.1) and lemma 9.4.2 we can see easily 1., 2., and 3. of proposition 9.4.1 as follows:

$$\begin{aligned} q_{j,\alpha}(\hat{u}) + m - j &\geq (p_* - 2m + j + |\alpha|) + m - j \\ &\geq q_{m,0}(\hat{u}) + |\alpha| > q_{m,0}(\hat{u}) \qquad \text{if } |\alpha| > 0; \\ p_{j,0} + m - j &\geq (p_* - 2m + j) + m - j \geq q_{m,0}(\hat{u}); \\ p_{j,\alpha} + m - j &\geq (p_* - 2m + j + |\alpha|) + m - j \\ &\geq q_{m,0}(\hat{u}) + |\alpha| > q_{m,0}(\hat{u}) \qquad \text{if } |\alpha| > 0. \end{aligned}$$

■

Thus, to complete the proof of theorem 9.2.3 it is sufficient to give a proof of lemma 9.4.2.

For $\lambda \in \mathbb{C}$ and $l \in \mathbb{N}$ we write

$$[\lambda; l] = \begin{cases} 1 & \text{when } l = 0\,, \\ \lambda(\lambda - 1)\cdots(\lambda - l + 1) & \text{when } l \geq 1. \end{cases}$$

Note the following equality:

$$\text{(9.4.2)} \qquad \begin{vmatrix} [\lambda_0; 0] & [\lambda_0; 1] & \dots & [\lambda_0; l] \\ [\lambda_1; 0] & [\lambda_1; 1] & \dots & [\lambda_1; l] \\ \dots & \dots & \dots & \dots \\ [\lambda_l; 0] & [\lambda_l; 1] & \dots & [\lambda_l; l] \end{vmatrix} = \prod_{i>j} (\lambda_i - \lambda_j)\,.$$

Proof of lemma 9.4.2: By the definition of p_* we have

$$\text{(9.4.3)} \quad F\left(t, x, \left\{\left(\frac{\partial}{\partial t}\right)^j Z_\alpha(t, y, z)\right\}_{j+|\alpha|\leq m}\right) \equiv F\left(t, x, \left\{\left(\frac{\partial}{\partial t}\right)^j Z_\alpha(t, y, 0)\right\}_{j+|\alpha|\leq m}\right) \quad (\bmod\ t^{p_*})\,.$$

Let $|\alpha| > 0$ and fix it. Then by applying $\frac{\partial}{\partial z_{\alpha,k}}$ for $k = m, m+1, \dots, 2m - |\alpha|$ on both sides of (9.4.3) we have

$$\sum_{0\leq j\leq m-|\alpha|} [k; j]\, t^{k-j} \frac{\partial F}{\partial X_{j,\alpha}}\left(t, x, \left\{\left(\frac{\partial}{\partial t}\right)^j Z_\alpha(t, y, z)\right\}_{j+|\alpha|\leq m}\right) \equiv 0 \quad (\bmod\ t^{p_*}) \quad \text{for } k = m, m+1, \dots,, 2m - |\alpha|$$

which by multiplication by $t^{2m-|\alpha|-k}$ is reduced to the relation

$$\sum_{0\le j\le m-|\alpha|} [k;j]\, t^{2m-j-|\alpha|} \frac{\partial F}{\partial X_{j,\alpha}}\left(t,x,\left\{\left(\frac{\partial}{\partial t}\right)^j Z_\alpha(t,y,z)\right\}_{j+|\alpha|\le m}\right) \equiv 0 \ (\bmod t^{p_*}) \quad \text{for } k=m,m+1,\dots,2m-|\alpha|. \tag{9.4.4}$$

Since the coefficients matrix of (9.4.4)

$$\begin{pmatrix} [m;0] & [m;1] & \dots & [m;m-|\alpha|] \\ [m+1;0] & [m+1;1] & \dots & [m+1;m-|\alpha|] \\ \dots & \dots & \dots & \dots \\ [2m-|\alpha|\,;0] & [2m-|\alpha|\,;1] & \dots & [2m-|\alpha|\,;m-|\alpha|] \end{pmatrix} \tag{9.4.5}$$

is invertible (by (9.4.2)), by applying the inverse of (9.4.5) to (9.4.4) we obtain

$$t^{2m-j-|\alpha|} \frac{\partial F}{\partial X_{j,\alpha}}\left(t,x,\left\{\left(\frac{\partial}{\partial t}\right)^j Z_\alpha(t,y,z)\right\}_{j+|\alpha|\le m}\right) \equiv 0 \ (\bmod t^{p_*}) \text{ for } j=0,1,\dots,m-|\alpha|. \tag{9.4.6}$$

Hence, by the definitions of $q_{j,\alpha}(\hat{u})$ and $p_{j,\alpha}$ we see easily that

$$\begin{aligned} (2m-j-|\alpha|)+q_{j,\alpha}(\hat{u}) &\ge p_* \quad \text{and} \\ (2m-j-|\alpha|)+p_{j,\alpha} &\ge p_* \end{aligned}$$

which give us 1. and 3. of lemma 9.4.2.

Similarly, by applying $\frac{\partial}{\partial y_k}$ for $k=m,m+1,\dots,2m$ on both sides of (9.4.3) we have

$$\begin{aligned} &\sum_{0\le j\le m} [k;j]\, t^{k-j} \frac{\partial F}{\partial X_{j,0}}\left(t,x,\left\{\left(\frac{\partial}{\partial t}\right)^j Z_\alpha(t,y,z)\right\}_{j+|\alpha|\le m}\right) \\ &\equiv \sum_{0\le j\le m} [k;j]\, t^{k-j} \frac{\partial F}{\partial X_{j,0}}\left(t,x,\left\{\left(\frac{\partial}{\partial t}\right)^j Z_\alpha(t,y,0)\right\}_{j+|\alpha|\le m}\right) \ (\bmod t^{p_*}) \\ &\qquad \text{for } k=m,m+1,\dots,2m \end{aligned}$$

which is reduced by multiplication by t^{2m-k} to the relation

$$\begin{aligned} &\sum_{0\le j\le m} [k;j]\, t^{2m-j} \frac{\partial F}{\partial X_{j,0}}\left(t,x,\left\{\left(\frac{\partial}{\partial t}\right)^j Z_\alpha(t,y,z)\right\}_{j+|\alpha|\le m}\right) \\ &\equiv \sum_{0\le j\le m} [k;j]\, t^{2m-j} \frac{\partial F}{\partial X_{j,0}}\left(t,x,\left\{\left(\frac{\partial}{\partial t}\right)^j Z_\alpha(t,y,0)\right\}_{j+|\alpha|\le m}\right) \ (\bmod t^{p_*}) \\ &\qquad \text{for } k=m,m+1,\dots,2m. \end{aligned}$$

Hence, by the same argument as in (9.4.6) we obtain

$$t^{2m-j}\frac{\partial F}{\partial X_{j,0}}\left(t,x,\left\{\left(\frac{\partial}{\partial t}\right)^{j} Z_{\alpha}(t,y,z)\right\}_{j+|\alpha|\le m}\right)$$
$$\equiv t^{2m-j}\frac{\partial F}{\partial X_{j,0}}\left(t,x,\left\{\left(\frac{\partial}{\partial t}\right)^{j} Z_{\alpha}(t,y,0)\right\}_{j+|\alpha|\le m}\right) \quad (\bmod t^{p_*})$$
$$\text{for } j=0,1,\dots,m$$

which is equivalent to

$$(2m-j)+p_{j,0} \ge p_* \quad \text{for } j=0,1,\dots,m.$$

Thus, 2. of lemma 9.4.2 is also proved. ∎

10 Maillet's type theorems for non linear singular integro–differential equations

Introduction

In chapters 6 and 7, we have studied Maillet's type theorems for non linear singular partial differential equations.

In this chapter, we will extend these results to some non linear integro–partial differential equations.

10.1 Notations and definitions

Let $d, n \in \mathbb{N}^*, t = (t_1, \ldots, t_d) \in \mathbb{C}^d, x = (x_1, \ldots, x_n) \in \mathbb{C}^n, \Delta$ a polydisk centered at the origin of $\mathbb{C}^n$, and τ a vector field of the form

$$\tau = \sum_{1 \leq i \leq d} \left(\sum_{1 \leq j \leq i} a_{i,j}(x) t_j \right) \frac{\partial}{\partial t_i}$$

where the coefficients $a_{i,j}(x)$ are holomorphic in Δ.

For $m \in \mathbb{N}^*$ we put

$$S_m = \left\{ (a, b, c) \in \mathbb{N}^d \times \mathbb{N}^d \times \mathbb{Z}^n : |a| = |b| \text{ and } |b| + |c| \leq m \right\}$$

where

$$\begin{aligned} a &= (a_1, \ldots, a_d) \in \mathbb{N}^d, & |a| &= a_1 + \cdots + a_d, \\ b &= (b_1, \ldots, b_d) \in \mathbb{N}^d, & |b| &= b_1 + \cdots + b_d, \\ c &= (c_1, \ldots, c_n) \in \mathbb{Z}^n, & |c| &= c_1 + \cdots + c_n. \end{aligned}$$

Also, for $c = (c_1, \ldots, c_n) \in \mathbb{Z}^n$ we write

$$\begin{aligned} |c|_+ &= \sum_{1 \leq i \leq n} \max\{c_i, 0\}, \\ |c|_- &= \sum_{1 \leq i \leq n} \max\{-c_i, 0\}. \end{aligned}$$

Clearly we have $|c| = |c|_+ - |c|_-$.

Let $I(m)$ be a finite subset of S_m and denote

$$\begin{aligned} \delta(I(m)) &= \text{the cardinal of } I(m), \\ Z &= \{Z_{a,b,c}\}_{(a,b,c)\in I(m)} \in \mathbb{C}^{\delta(I(m))}. \end{aligned}$$

Let $G_2(x)(t_1,\ldots,t_d,Z)$ be a holomorphic function in $(x,t_1,\ldots,t_d,Z)$ defined in a polydisk centered at the origin of $\mathbb{C}^n \times \mathbb{C}^d \times \mathbb{C}^{\delta(I(m))}$ and having an expansion of the form

$$G_2(x)(t_1,\ldots,t_d,Z) = \sum_{|\mu|+|\nu|\geq 2} g_{\mu,\nu}(x)(t_1)^{\mu_1}\cdots(t_d)^{\mu_d} Z^\nu$$

where

$$\begin{aligned} \mu &= (\mu_1,\ldots,\mu_d) \in \mathbb{N}^d, \\ |\mu| &= \mu_1+\cdots+\mu_d, \\ \nu &= \{\nu_{a,b,c}\}_{(a,b,c)\in I(m)} \in \mathbb{N}^{\delta(I(m))}, \\ |\nu| &= \sum_{(a,b,c)\in I(m)} \nu_{a,b,c}, \\ Z^\nu &= \prod_{(a,b,c)\in I(m)} (Z_{a,b,c})^{\nu_{a,b,c}} \end{aligned}$$

and $g_{\mu,\nu}(x)$ for $|\mu|+|\nu|\geq 2$ are holomorphic functions on Δ.

Definition 10.1.1 *For a holomorphic function $u(x)$ on Δ we define $D_{x_i}u$ and $(D_{x_i})^{-1}u$ by the following:*

$$\begin{aligned} D_{x_i}u &= \frac{\partial}{\partial x_i}u, \\ (D_{x_i})^{-1}u &= \int_0^{x_i} u(x_1,\ldots,\xi_i,\ldots,x_n)\,d\xi_i. \end{aligned}$$

Then for any $c=(c_1,\ldots,c_n)\in\mathbb{Z}^n$ we can define $(D_x)^c$ by

$$(D_x)^c u = (D_{x_1})^{c_1}\cdots(D_{x_n})^{c_n} u.$$

If $u(x)$ is of the form $u(x)=\sum_{\alpha\in\mathbb{N}^n} a_\alpha x^\alpha$ then

$$(D_x)^c u = \sum_{\alpha\in\mathbb{N}^n,\alpha\geq c} a_\alpha \frac{\alpha!}{(\alpha-c)!}x^{\alpha-c}.$$

For $a=(a_1,\ldots,a_d)\in\mathbb{N}^d$ and $b=(b_1,\ldots,b_d)\in\mathbb{N}^d$ we denote as usual

$$t^a = (t_1)^{a_1}\cdots(t_d)^{a_d} \qquad \text{and} \qquad (D_t)^b = \left(\frac{\partial}{\partial t_1}\right)^{b_1}\cdots\left(\frac{\partial}{\partial t_d}\right)^{b_d}.$$

Denote also by

- $\mathbb{C}\{x\}$ the ring of germs of holomorphic functions at the origin of $\mathbb{C}^n$,
- $\mathbb{C}_r\{x\}$ the ring of holomorphic functions on $D_r = \{x \in \mathbb{C}^n : |x_i| \leq r$ for $i = 1, \dots, n\}$,
- $H_{k,r}\{x\}[t_1, \dots, t_d]$ the set of all homogeneous polynomials of degree k in $t_1, \dots, t_d$ with coefficients in $\mathbb{C}_r\{x\}$,
- $\mathbb{C}_r\{x\}[[t_1, \dots, t_d]]$ the ring of formal power series in $t_1, \dots, t_d$ with coefficients in $\mathbb{C}_r\{x\}$, and
- $\mathfrak{M}_r\{x\}[[t_1, \dots, t_d]]$ the ring of all $Y \in \mathbb{C}_r\{x\}[[t_1, \dots, t_d]]$ of the form

$$Y = \sum_{|k| \geq 1} y_{(k_1, \dots, k_d)}(x) (t_1)^{k_1} \cdots (t_d)^{k_d} .$$

10.2 The main theorem

Let $l, m \in \mathbb{N}^*$ and let $I(l)$ (resp. $I(m)$) be a finite subset of S_l (resp. S_m).

In this chapter, we will consider the following type of integro–partial differential equations:

(10.2.1)

$$\left(\tau^l + c_{l-1}(x)\tau^{l-1} + \cdots + c_0(x)\right) u + \sum_{(a,b,c) \in I(l), |c| \leq 0} A_{a,b,c}(x) t^a \left(D_t\right)^b \left(D_x\right)^c u$$

$$= \sum_{1 \leq i \leq d} t_i B_i(x) + G_2(x) \left(t_1, \dots, t_d, \left\{t^a \left(D_t\right)^b \left(D_x\right)^c u\right\}_{(a,b,c) \in I(m)}\right)$$

where $c_i(x), A_{a,b,c}(x), B_i(x)$ are holomorphic functions on Δ, and $\tau, G_2(x)(t, Z)$ are the ones in section 10.1.

Definition 10.2.1 *By a* formal solution *of (10.2.1) we mean a formal power series solution of the form*

$$\sum_{|k| \geq 1} u_k(x) t^k \tag{10.2.2}$$

the coefficients $\{u_k(x)\}$ of which are holomorphic in a common disk centered at the origin of $\mathbb{C}^n$.

Definition 10.2.2 *If a formal solution $\hat{u}(t, x)$ of the form (10.2.2) satisfies the condition that*

$$\sum_{|k| \geq 1} \frac{u_k(x)}{\left(|k|!\right)^{s-1}} t^k$$

is convergent near the origin of $\left(\mathbb{C}_x\right)^n \times \left(\mathbb{C}_t\right)^d$ for some $s \geq 1$ we say that $\hat{u}(t, x)$ is in the formal Gevrey class $\mathcal{E}^{\{s\}}$.

Introduce

$$J = \{(\mu, \nu) : |\nu| \geq 1 \text{ and } g_{\mu,\nu}(x) \not\equiv 0\}$$

and for $\nu = \{\nu_{a,b,c} : (a,b,c) \in I(m)\}$ define

$$m_\nu = \max\{|b| + |c| : \nu_{a,b,c} \neq 0\};$$

then define

$$s_l(10.2.1) = 1 + \max\left\{0, \sup_{(\mu,\nu)\in J}\left(\frac{m_\nu - l}{|\mu| + |\nu| - 1}\right)\right\}.$$

Denote by $\rho_1(x), \ldots, \rho_l(x)$ the roots of

$$\rho^l + c_{l-1}(x)\rho^{l-1} + \cdots + c_0(x) = 0.$$

Theorem 10.2.3 (Main theorem) *Assume that there exist $\sigma > \sigma_1 > 0$ such that the following conditions are satisfied:*

1. $\left|\sum_{1\leq i\leq d} k_i a_{i,i}(0) - \rho_j(0)\right| \geq \sigma(k_1 + \cdots + k_d)$ *holds for any* $(k_1, \ldots, k_d) \in \mathbb{N}^d \setminus \{(0, \ldots, 0)\}$ *and* $j = 1, \ldots, l$;
2. $\sum_{1\leq i\leq d}\left(\sum_{1\leq j<i} |a_{i,j}(0)|\right) < \sigma_1 < \sigma$;
3. $\sum_{(a,b,c)\in I(l),|c|=0} |A_{a,b,c}(0)|\, e^{|c|_+} < (\sigma - \sigma_1)^l$.

Then we have

1. *(10.2.1) has a unique formal solution $\hat{u}(t,x)$ of the form (10.2.2);*
2. *moreover, the formal solution $\hat{u}(t,x)$ is in the formal Gevrey class $\mathcal{E}^{\{s\}}$ for any $s \geq s_l(10.2.1)$.*

Corollary 10.2.4 *If $l \geq m$ then $s_l(10.2.1) = 1$ and the unique formal solution $\hat{u}(t,x)$ of (10.2.1) is convergent near the origin of $(\mathbb{C}_t)^d \times (\mathbb{C}_x)^n$.*

Put

$$\tau_0 = \sum_{1\leq i\leq d}\left(\sum_{1\leq j\leq i} a_{i,j}(0)t_j\right)\frac{\partial}{\partial t_i}.$$

Then $\tau^l + c_{l-1}(x)\tau^{l-1} + \cdots + c_0(x)$ can be expressed in the form

$$\tau^l + c_{l-1}(x)\tau^{l-1} + \cdots + c_0(x) = \tau_0^l + c_{l-1}(0)\tau_0^{l-1} + \cdots + c_0(0) + \sum_{|a|=|b|\leq l} C_{a,b}(x)t^a (D_t)^b$$

for some holomorphic functions $C_{a,b}(x)$ on Δ satisfying $C_{a,b}(0) = 0$.

Hence, in the proof of theorem 10.2.3 we may assume without loss of generality the following conditions:

(a1) $a_{i,j}(x) \equiv a_{i,j} \in \mathbb{C}$ on Δ for $1 \leq i \leq d$ and $1 \leq j \leq i$;

(a2) $c_i(x) \equiv c_i \in \mathbb{C}$ on Δ for $0 \leq i \leq l-1$;

(a3) $\rho_i(x) \equiv \rho_i \in \mathbb{C}$ on Δ for $1 \leq i \leq l$.

10.3 Construction of the formal solution

In this section, we will prove that the equation (10.2.1) has a unique formal solution $\hat{u}(t,x)$ of the form

$$\hat{u}(t,x) = \sum_{k \geq 1} u_k(t,x) \qquad \text{where } u_k(t,x) \in H_{k,R}\{x\}\,[t_1,\ldots,t_d] \qquad \text{for } k = 1,2,\ldots$$

for some $R > 0$.

Throughout this section we assume (a1), (a2), and (a3).

Let $k \in \mathbb{N}$ and $\rho > 0$. For $\varphi(x) = \sum_{\alpha \in \mathbb{N}^n} a_\alpha x^\alpha \in \mathbb{C}\{x\}$ we define the norm $\|\varphi\|_{k,\rho}$ by

$$\|\varphi\|_{k,\rho} = \sum_{\alpha \in \mathbb{N}^n} |a_\alpha| \frac{k!\alpha!}{(k+|\alpha|)!} \rho^{|\alpha|}$$

and for $u(t,x)$ a homogeneous polynomial of degree k in $t_1,\ldots,t_d$ with coefficients in $\mathbb{C}\{x\}$ of the form

$$u(t,x) = \sum_{k_1+\cdots+k_d=k} \varphi_{(k_1,\ldots,k_d)}(x)\,(t_1)^{k_1}\cdots(t_d)^{k_d} \tag{10.3.1}$$

we define the norm $\|u\|_{k,\rho}$ by

$$\|u\|_{k,\rho} = \sum_{k_1+\cdots+k_d=k} \left\|\varphi_{(k_1,\ldots,k_d)}\right\|_{k,\rho}.$$

Denote by $\mathfrak{H}_{k,\rho}$ the set of all $u(t,x)$ of the form (10.3.1) satisfying $\|u\|_{k,\rho} < +\infty$. It is easy to see

1) $\mathfrak{H}_{k,\rho}$ is a Banach space;

2) if $u \in \mathfrak{H}_{k,\rho}$ and $v \in \mathfrak{H}_{h,\rho}$ then $uv \in \mathfrak{H}_{k+h,\rho}$;

3) $\|uv\|_{k+h,\rho} \leq \|u\|_{k,\rho} \cdot \|v\|_{h,\rho}$;

4) $\|(D_x)^c\, u\|_{k,\rho} \leq (k+1)^{|c|}\rho^{-|c|}\, \|u\|_{k,\rho}$ if $|c| \leq 0$.

First let us consider the operator

$$L = \tau^l + c_{l-1}\tau^{l-1} + \cdots + c_0.$$

Denote by L_k the restriction of L to $\mathfrak{H}_{k,\rho}$. Then by the Poincaré condition (assumptions 1. and 2. of theorem 10.2.3) and by the same argument as in section 6.3 we obtain

Lemma 10.3.1 *For any $k \geq 1$ and $\rho > 0$ we have the following properties:*

1. *$L_k : \mathfrak{H}_{k,\rho} \longrightarrow \mathfrak{H}_{k,\rho}$ is an invertible operator;*
2. *for any $u \in \mathfrak{H}_{k,\rho}$ we have*

$$\|L_k u\|_{k,\rho} \geq ((\sigma - \sigma_1)\, k)^l \, \|u\|_{k,\rho} .$$

Next let us consider the operator

$$P = L + \sum_{(a,b,c)\in I(l),|c|\leq 0} A_{a,b,c}(x) t^a \,(D_t)^b\,(D_x)^c .$$

Denote by P_k the restriction of P to $\mathfrak{H}_{k,\rho}$. Then we have

Lemma 10.3.2 *If $\rho > 0$ satisfies*

$$\sum_{(a,b,c)\in I(l),|c|\leq 0} \|A_{a,b,c}\|_{0,\rho}\, \rho^{-|c|} \leq (1-\sigma_2)\,(\sigma - \sigma_1)^l \tag{10.3.2}$$

for some σ_2 satisfying $0 < \sigma_2 < 1$ then for any $k \geq 1$ we have the following properties:

1. *$P_k : \mathfrak{H}_{k,\rho} \longrightarrow \mathfrak{H}_{k,\rho}$ is an invertible operator;*
2. *for any $u \in \mathfrak{H}_{k,\rho}$ we have*

$$\|P_k u\|_{k,\rho} \geq \sigma_2\,((\sigma - \sigma_1)\, k)^l \, \|u\|_{k,\rho} .$$

Proof : Put

$$K = - \sum_{(a,b,c)\in I(l),|c|\leq 0} A_{a,b,c}(x) t^a \,(D_t)^b\,(D_x)^c$$

and denote by K_k the restriction of K to $\mathfrak{H}_{k,\rho}$. Then we have

$$P_k = L_k \left(I_k - L_k^{-1} K_k\right) .$$

Let us estimate the operator norm of K_k.
For $u \in \mathfrak{H}_{k,\rho}$ we have

$$\begin{aligned}
\|K_k u\|_{k,\rho} &\leq \sum_{(a,b,c)\in I(l),|c|\leq 0} \|A_{a,b,c}\|_{0,\rho} \left\| t^a \,(D_t)^b\,(D_x)^c\, u \right\|_{k,\rho} \\
&\leq \sum_{(a,b,c)\in I(l),|c|\leq 0} \|A_{a,b,c}\|_{0,\rho}\, k^{|b|} (k+1)^{|c|} \rho^{-|c|}\, \|u\|_{k,\rho} \\
&\leq \left(\sum_{(a,b,c)\in I(l),|c|\leq 0} \|A_{a,b,c}\|_{0,\rho}\, k^{|b|+|c|} \rho^{-|c|} \right) \|u\|_{k,\rho} .
\end{aligned}$$

Since $|b| + |c| \leq l$ this implies

$$\|K_k\|_{k,\rho} \leq \sum_{(a,b,c)\in I(l),|c|\leq 0} \|A_{a,b,c}\|_{0,\rho}\, k^l \rho^{-|c|}.$$

On the other hand, by lemma 10.3.1 we already know

$$\left\|L_k^{-1}\right\|_{k,\rho} \leq \frac{1}{\left((\sigma-\sigma_1)\,k\right)^l}.$$

Hence,

$$\begin{aligned}
\left\|L_k^{-1}K_k\right\|_{k,\rho} &\leq \frac{1}{\left((\sigma-\sigma_1)\,k\right)^l}\left(\sum_{(a,b,c)\in I(l),|c|\leq 0} \|A_{a,b,c}\|_{0,\rho}\, k^l \rho^{-|c|}\right)\\
&= \frac{1}{\left(\sigma-\sigma_1\right)^l}\left(\sum_{(a,b,c)\in I(l),|c|\leq 0} \|A_{a,b,c}\|_{0,\rho}\, \rho^{-|c|}\right)\\
&\leq \frac{1}{\left(\sigma-\sigma_1\right)^l}\left(1-\sigma_2\right)\left(\sigma-\sigma_1\right)^l\\
&= 1-\sigma_2\\
&< 1.
\end{aligned}$$

Thus, by applying the Neumann series to

$$P_k^{-1} = \left(I_k - L_k^{-1}K_k\right)^{-1} L_k^{-1}$$

we can get the conclusion of lemma 10.3.2. ■

Construction of the formal solution

Choose $\rho > 0$ and $\sigma_2 > 0$ sufficiently small so that the condition (10.3.2) is satisfied. Put $r = \frac{\rho}{2}$ and choose a sequence $\{\rho_k\}_{k\geq 1}$ such that

$$\rho > \rho_1 > \rho_2 > \cdots > \rho_k > \cdots > r.$$

Then we can construct a formal solution $u(t,x)$ of the form

$$u = \sum_{k\geq 1} u_k \qquad \text{where } u_k \in \mathfrak{H}_{k,\rho_k} \qquad \text{for } k = 1,2,\ldots \tag{10.3.3}$$

in the following way.

Since (10.3.3) is a decomposition of $u(t,x)$ into the sum of homogeneous polynomials in $t_1,\ldots,t_d$ the equation (10.2.1) is decomposed into the following recursive family:

$$P_1 u_1 = \sum_{1\leq i\leq d} t_i B_i(x) \tag{10.3.4}$$

and for $k \geq 2$

(10.3.5)

$$P_k u_k = \sum_{2 \leq |\mu|+|\nu| \leq k} g_{\mu,\nu}(x) t^{\mu} \left[\sum_{k^*+|\mu|=k} 1 \times \prod_{(a,b,c)\in I(m)} \left(D_{a,b,c} u_{k_{a,b,c}(1)}(t,x)\right) \times \right.$$

$$\left. \times \left(D_{a,b,c} u_{k_{a,b,c}(2)}(t,x)\right) \times \cdots \times \left(D_{a,b,c} u_{k_{a,b,c}(\nu_{a,b,c})}(t,x)\right) \right]$$

where

$$\begin{aligned} k_{a,b,c} &= (k_{a,b,c}(1), k_{a,b,c}(2), \ldots, k_{a,b,c}(\nu_{a,b,c})), \\ k^* &= \sum_{(a,b,c)\in I(m)} |k_{a,b,c}|, \\ |k_{a,b,c}| &= k_{a,b,c}(1) + k_{a,b,c}(2) + \cdots + k_{a,b,c}(\nu_{a,b,c}), \\ D_{a,b,c} &= t^a (D_t)^b (D_x)^c. \end{aligned}$$

Thus, by using the invertibility of

$$P_k : \mathfrak{H}_{k,\rho_k} \longrightarrow \mathfrak{H}_{k,\rho_k}$$

in lemma 10.3.2 we can uniquely determine the function $u_k \in \mathfrak{H}_{k,\rho_k}$ inductively on k.

Summing up we have proved the following:

Proposition 10.3.3 *Assume the same conditions as in theorem 10.2.3 and also assume (a1), (a2), and (a3). Then (10.2.1) has a unique formal solution $u(t,x)$ of the form*

(10.3.6) $u = \sum_{k\geq 1} u_k$ *where* $u_k \in H_{k,R}\{x\}[t_1, \ldots, t_d]$ *for* $k = 1, 2, \ldots$

for some $R > 0$.

10.4 Some discussions

In this section, we will present some results which will play important roles in the proof of the convergence of the formal solution.

Let $0 < \rho < R$. For $\varphi(x) = \sum_{\alpha\in\mathbb{N}^n} a_\alpha x^\alpha \in \mathbb{C}_R\{x\}$ we define the norm $\|\varphi\|_\rho$ by

$$\|\varphi\|_\rho = \|\varphi\|_{0,\rho} = \sum_{\alpha\in\mathbb{N}^n} |a_\alpha| \frac{\alpha!}{|\alpha|!} \rho^{|\alpha|}.$$

It is easy to see

1. $\|\varphi\psi\|_\rho \le \|\varphi\|_\rho \|\psi\|_\rho$;
2. $\|D_{x_i}\varphi\|_\rho \le \frac{\partial}{\partial\rho}\|\varphi\|_\rho$ for $i = 1, \ldots, n$;
3. $\left\|(D_{x_i})^{-1}\varphi\right\|_\rho = \int_0^\rho \|\varphi\|_s\, ds$ for $i = 1, \ldots, n$.

For $u(t,x) \in H_{k,R}\{x\}[t_1, \ldots, t_d]$ of the form

$$u = \sum_{k_1+\cdots+k_d=k} \varphi_{(k_1,\ldots,k_d)}(x)\,(t_1)^{k_1}\cdots(t_d)^{k_d} \tag{10.4.1}$$

we define the norm $\|u\|_\rho$ by

$$\|u\|_\rho = \sum_{k_1+\cdots+k_d=k} \left\|\varphi_{(k_1,\ldots,k_d)}\right\|_\rho .$$

Lemma 10.4.1 *Assume that $u(t,x) \in H_{k,R}\{x\}[t_1, \ldots, t_d]$ satisfies*

$$\|u\|_\rho \le \frac{C}{(R-\rho)^a} \quad \textit{for } 0 < \rho < R \tag{10.4.2}$$

for some $C > 0$ and $a > 0$. Then

1. *for $0 < \rho < R$ and $i = 1, \ldots, n$ we have*

$$\|D_{x_i}u\|_\rho \le \frac{(a+1)eC}{(R-\rho)^{a+1}};$$

2. *if $a > 1$ then for $0 < \rho < R$ and $i = 1, \ldots, n$ we have*

$$\left\|(D_{x_i})^{-1}u\right\|_\rho \le \frac{C}{(a-1)(R-\rho)^{a-1}}.$$

Proof :

1. is proved by lemma 6.2.9 as follows:

$$\|D_{x_i}u\|_\rho \le \frac{\partial}{\partial\rho}\|u\|_\rho \le \frac{(a+1)eC}{(R-\rho)^{a+1}}.$$

2. is proved as follows:

$$\begin{aligned}\left\|(D_{x_i})^{-1}u\right\|_\rho &= \int_0^\rho \|u\|_s\, ds \le \int_0^\rho \frac{C}{(R-s)^a} ds \\ &= \frac{C}{a-1}\left(\frac{1}{(R-\rho)^{a-1}} - \frac{1}{R^{a-1}}\right) \le \frac{C}{(a-1)(R-\rho)^{a-1}}.\end{aligned}$$

■

Let L, P, K be the same as in section 10.3 and denote by L_k, P_k, K_k the restrictions of L, P, K to $H_{k,R}\{x\}[t_1, \ldots, t_d]$ respectively. Then

Proposition 10.4.2 *Assume the same conditions as in theorem 10.2.3 and (a1), (a2), and (a3). Assume also that $R > 0$ satisfies*

$$(10.4.3) \qquad \frac{\sum_{(a,b,c)\in I(l),|c|\le 0} \|A_{a,b,c}\|_R \, e^{|c|_+} \times R^{-|c|}}{1-\varepsilon} \le (1-\sigma_2)(\sigma-\sigma_1)^l$$

for some $0 < \sigma_2 < 1$ and $0 < \varepsilon < 1$. Then there is a k_0 which satisfies the following: if $k > k_0$ and if $u_k \in H_{k,R}\{x\}[t_1, \ldots, t_d]$ satisfies

$$(10.4.4) \qquad \|P_k u_k\|_\rho \le \frac{C}{(R-\rho)^{m(k-2)}} \quad \text{for } 0 < \rho < R$$

for some $C > 0$ then we have

$$(10.4.5) \qquad \|u_k\|_\rho \le \frac{C}{\sigma_2 (\sigma-\sigma_1)^l k^l (R-\rho)^{m(k-2)}} \quad \text{for } 0 < \rho < R.$$

Proof : We choose k_0 so that $m(k_0 - 2) + |c| > 1$ and

$$\frac{\left(m - \frac{2m}{k} + \frac{1}{k}\right) \cdots \left(m - \frac{2m}{k} + \frac{|c|_+}{k}\right)}{\left(m - \frac{2m}{k} + \frac{|c|_+ - |c|_-}{k}\right) \cdots \left(m - \frac{2m}{k} + \frac{|c|_+ - 1}{k}\right)} \le \frac{1}{1-\varepsilon}$$

hold for any $k > k_0$ and any $(a,b,c) \in I(l)$ satisfying $|c| \le 0$. Note that $|c| \le 0$ implies $|c|_+ \le |c|_-$.

Let $k > k_0$ and assume that $u_k \in H_{k,R}\{x\}[t_1, \ldots, t_d]$ satisfies (10.4.4). Put

$$f_k \equiv P_k u_k \in H_{k,R}\{x\}[t_1, \ldots, t_d].$$

Since the uniqueness of the solution $u \in H_{k,R}\{x\}[t_1, \ldots, t_d]$ of the equation

$$(10.4.6) \qquad P_k u = f_k$$

is already proved to get the conclusion of proposition 10.4.2 it is sufficient to prove

Lemma 10.4.3 *Under the above situation the equation (10.4.6) has a solution $u \in H_{k,R}\{x\}[t_1, \ldots, t_d]$ satisfying*

$$(10.4.7) \qquad \|u\|_\rho \le \frac{C}{\sigma_2 (\sigma-\sigma_1)^l k^l (R-\rho)^{m(k-2)}} \quad \text{for } 0 < \rho < R.$$

Proof of lemma 10.4.3: It is easy to see

1. $L_k : H_{k,R}\{x\}[t_1, \ldots, t_d] \longrightarrow H_{k,R}\{x\}[t_1, \ldots, t_d]$ is an invertible operator;
2. for any $u \in H_{k,R}\{x\}[t_1, \ldots, t_d]$ we have

$$\|L_k u\|_\rho \ge (\sigma-\sigma_1)^l k^l \|u\|_\rho .$$

By using this we will solve the equation (10.4.6) by iteration as follows:

$$\begin{aligned} L_k u^{(1)} &= f_k \\ L_k u^{(2)} &= K_k u^{(1)} \\ &\vdots \\ L_k u^{(p)} &= K_k u^{(p-1)} \\ &\vdots \end{aligned}$$

Since $f_k \in H_{k,R}\{x\}[t_1, \ldots, t_d]$ holds we can uniquely determine

$$u^{(p)} \in H_{k,R}\{x\}[t_1, \ldots, t_d] \quad \text{for } p = 1, 2, \ldots$$

inductively on p and we obtain a formal solution

$$u = \sum_{p \geq 1} u^{(p)} \tag{10.4.8}$$

of the equation (10.4.6).

Since

$$\|f_k\|_\rho \leq \frac{C}{(R-\rho)^{m(k-2)}} \quad \text{for } 0 < \rho < R$$

holds we get

$$\left\|u^{(1)}\right\|_\rho \leq \frac{C}{(\sigma - \sigma_1)^l k^l (R-\rho)^{m(k-2)}} \quad \text{for } 0 < \rho < R.$$

Therefore, by lemma 10.4.1 we have

$$\begin{aligned} \left\|K_k u^{(1)}\right\|_\rho &\leq \sum_{(a,b,c) \in I(l), |c| \leq 0} \|A_{a,b,c}\|_\rho \left\| t^a (D_t)^b (D_x)^c u^{(1)} \right\|_\rho \\ &\leq \sum_{(a,b,c) \in I(l), |c| \leq 0} \|A_{a,b,c}\|_R \, k^{|b|} \times e^{|c|_+} \times \\ &\quad \times \frac{(m(k-2)+1) \cdots (m(k-2)+|c|_+)}{(m(k-2)+|c|_+ - |c|_-) \cdots (m(k-2)+|c|_+ - 1)} \times \\ &\quad \times \frac{C}{(\sigma-\sigma_1)^l k^l (R-\rho)^{m(k-2)-(-|c|)}} \end{aligned}$$

$$
\begin{aligned}
&= \sum_{(a,b,c)\in I(l),|c|\le 0} \|A_{a,b,c}\|_R \, e^{|c|_+} \times \frac{k^{|b|+|c|}}{k^l}(R-\rho)^{-|c|} \times \\
&\quad \times \frac{\left(m-\frac{2m}{k}+\frac{1}{k}\right)\cdots\left(m-\frac{2m}{k}+\frac{|c|_+}{k}\right)}{\left(m-\frac{2m}{k}+\frac{|c|_+-|c|_-}{k}\right)\cdots\left(m-\frac{2m}{k}+\frac{|c|_+-1}{k}\right)} \times \\
&\quad \times \frac{C}{(\sigma-\sigma_1)^l (R-\rho)^{m(k-2)}} \\
&\le \sum_{(a,b,c)\in I(l),|c|\le 0} \|A_{a,b,c}\|_R \, e^{|c|_+} \times R^{-|c|} \times \frac{1}{1-\varepsilon} \times \\
&\quad \times \frac{C}{(\sigma-\sigma_1)^l (R-\rho)^{m(k-2)}} \\
&\le (1-\sigma_2)\frac{C}{(R-\rho)^{m(k-2)}}
\end{aligned}
$$

for $0<\rho<R$. Applying this to the equation $L_k u^{(2)} = K_k u^{(1)}$ we obtain

$$
\left\|u^{(2)}\right\|_\rho \le (1-\sigma_2)\frac{C}{(\sigma-\sigma_1)^l k^l (R-\rho)^{m(k-2)}} \quad \text{for } 0<\rho<R.
$$

Thus, repeating the same argument as above we can see that for any $p=1,2,\ldots$

$$
\left\|u^{(p)}\right\|_\rho \le (1-\sigma_2)^{p-1}\frac{C}{(\sigma-\sigma_1)^l k^l (R-\rho)^{m(k-2)}} \quad \text{for } 0<\rho<R.
$$

Hence, we obtain the estimate of the formal solution $u(t,x)$ in (10.4.8) as follows:

$$
\begin{aligned}
\|u\|_\rho &\le \sum_{p\ge 1}\left\|u^{(p)}\right\|_\rho \\
&\le \left(\sum_{p\ge 1}(1-\sigma_2)^{p-1}\right)\frac{C}{(\sigma-\sigma_1)^l k^l (R-\rho)^{m(k-2)}} \\
&\le \frac{C}{\sigma_2(\sigma-\sigma_1)^l k^l (R-\rho)^{m(k-2)}}
\end{aligned}
$$

for $0<\rho<R$. This completes the proof of lemma 10.4.3. ■

■

10.5 Convergence of the formal solution in the case $s_l = 1$

In this section, we will prove that if $s_l(10.2.1) = 1$ the formal solution of (10.2.1) is convergent near the origin of $(\mathbb{C}_t)^d \times (\mathbb{C}_x)^n$.

Assume that $s_l(10.2.1) = 1$. This implies that $m_\nu \leq l$ holds for any $(\mu, \nu) \in J$ and, therefore, we may assume $m \leq l$.

Let

$$(10.5.1) \quad u = \sum_{k\geq 1} u_k \qquad \text{where } u_k \in H_{k,R}\{x\}\,[t_1, \ldots, t_d] \quad \text{for } k = 1, 2, \ldots$$

be the unique formal solution of (10.2.1) obtained in 1. of theorem 10.2.3.

Then by (10.3.5) we have for $k \geq 2$ and $0 < \rho < R$

$$(10.5.2) \quad \left\|P_k u_k\right\|_\rho \leq \sum_{2\leq|\mu|+|\nu|\leq k} \left\|g_{\mu,\nu}\right\|_\rho \left[\sum_{k^*+|\mu|=k} 1 \times \times \prod_{(a,b,c)\in I(m)} \left\|D_{a,b,c} u_{k_{a,b,c}(1)}\right\|_\rho \times \times \left\|D_{a,b,c} u_{k_{a,b,c}(2)}\right\|_\rho \times \cdots \times \left\|D_{a,b,c} u_{k_{a,b,c}(\nu_{a,b,c})}\right\|_\rho\right].$$

Assume (a1), (a2), (a3), and $0 < R < 1$. Moreover, we may assume that the condition (10.4.3) is satisfied for some $0 < \sigma_2 < 1$ and $0 < \varepsilon < 1$. Let k_0 be the one in proposition 10.4.2. Without loss of generality we may also assume that

$$(10.5.3) \qquad m\,(k_0 - 2) + |c| > 1 \quad \text{for any } (a, b, c) \in I(m).$$

Take $A > 0$, $A_{\mu,\nu} > 0$ for $|\mu| + |\nu| \geq 2$, and $B \geq 1$ so that the following conditions are satisfied:

(b1) $\left\|D_{a,b,c} u_1\right\|_R \leq A$ for any $(a, b, c) \in I(m)$;

(b2) $A_{\mu,\nu}$ for $2 \leq |\mu| + |\nu| \leq k_0$ are sufficiently large;

(b3) $A_{\mu,\nu} = \left\|g_{\mu,\nu}\right\|_R$ for $|\mu| + |\nu| > k_0$;

(b4) for any $k > k_0$ and any $(a, b, c) \in I(m)$

$$\frac{\left(m - \frac{2m}{k} + \frac{1}{k}\right) \cdots \left(m - \frac{2m}{k} + \frac{|c|_+}{k}\right) e^{|c|_+}}{\left(m - \frac{2m}{k} + \frac{|c|_+ - |c|_-}{k}\right) \cdots \left(m - \frac{2m}{k} + \frac{|c|_+ - 1}{k}\right)} \leq B.$$

Consider now the following analytic equation:

$$(10.5.4) \quad \sigma_2 \left(\sigma - \sigma_1\right)^l Y = \sigma_2 \left(\sigma - \sigma_1\right)^l AX + + \frac{1}{(R-\rho)^m} \sum_{|\mu|+|\nu|\geq 2} \frac{A_{\mu,\nu}}{(R-\rho)^{m(|\mu|+|\nu|-2)}} X^{|\mu|} (BY)^{|\nu|}.$$

The implicit function theorem tells us that for any $0 < \rho < R$ the equation (10.5.4) has a unique holomorphic solution Y of the form

$$Y = \sum_{k\geq 1} Y_k(\rho) X^k$$

and $Y_k(\rho)$ is determined by the following recursive formula:

$$Y_1(\rho) = A$$

and for $k \geq 2$

$$\sigma_2 (\sigma - \sigma_1)^l Y_k = \frac{1}{(R-\rho)^m} \sum_{2\leq|\mu|+|\nu|\leq k} \frac{A_{\mu,\nu}}{(R-\rho)^{m(|\mu|+|\nu|-2)}} \times \\ \times \sum_{k_1+\cdots+k_{|\nu|}=k-|\mu|} (BY_1) \times \cdots \times \left(BY_{k_{|\nu|}}\right).$$

Moreover, we see that $Y_k(\rho)$ has the form

$$Y_k(\rho) = \frac{C_k}{(R-\rho)^{m(k-1)}}$$

for some constants $C_1 = A$ and $C_k > 0$ for $k \geq 2$ which are determined by $\{A_{\mu,\nu}\}_{2\leq|\mu|+|\nu|\leq k}$. Since $A_{\mu,\nu}$ for $2 \leq |\mu| + |\nu| \leq k_0$ are assumed to be sufficiently large, C_k for $2 \leq k \leq k_0$ are also sufficiently large. Hence, we may assume that the following estimates hold for $1 \leq k \leq k_0$:

(10.5.5)
$$\|D_{a,b,c} u_k\|_\rho \leq \frac{B}{k^{m-|b|-|c|}} Y_k(\rho) \quad \text{for any } (a,b,c) \in I(m) \text{ and } 0 < \rho < R.$$

Note that the case $k = 1$ is a consequence of (b1).

Now let us show by induction that (10.5.5) is also valid for any $k > k_0$.

Let $k > k_0$ and assume that (10.5.5) with k replaced by p is true for all $p < k$. Then by (10.5.2) and the induction hypotheses we get

$$\begin{aligned} \|P_k u_k\|_\rho \leq \sum_{2\leq|\mu|+|\nu|\leq k} A_{\mu,\nu} \Bigg[\sum_{k^*+|\mu|=k} 1 \times \\ \times \prod_{(a,b,c)\in I(m)} \left(\frac{B}{k_{a,b,c}(1)^{m-|b|-|c|}} Y_{k_{a,b,c}(1)}(\rho) \right) \times \\ \times \left(\frac{B}{k_{a,b,c}(2)^{m-|b|-|c|}} Y_{k_{a,b,c}(2)}(\rho) \right) \times \cdots \times \\ \times \left(\frac{B}{k_{a,b,c}\left(\nu_{a,b,c}\right)^{m-|b|-|c|}} Y_{k_{a,b,c}(\nu_{a,b,c})}(\rho) \right) \Bigg] \end{aligned}$$

$$\begin{aligned}
&\leq \sum_{2\leq|\mu|+|\nu|\leq k} A_{\mu,\nu}\Bigg[\sum_{k^*+|\mu|=k} 1\times \\
&\qquad \times \prod_{(a,b,c)\in I(m)} \left(BY_{k_{a,b,c}(1)}(\rho)\right)\times\left(BY_{k_{a,b,c}(2)}(\rho)\right)\times\cdots\times \\
&\qquad\qquad \times\left(BY_{k_{a,b,c}(\nu_{a,b,c})}(\rho)\right)\Bigg] \\
&= \sum_{2\leq|\mu|+|\nu|\leq k} A_{\mu,\nu}\left[\sum_{k_1+\cdots+k_{|\nu|}=k-|\mu|} \left(BY_{k_1}(\rho)\right)\times\cdots\times\left(BY_{k_{|\nu|}}(\rho)\right)\right] \\
&\leq \sum_{2\leq|\mu|+|\nu|\leq k} \frac{A_{\mu,\nu}}{(R-\rho)^{m(|\mu|+|\nu|-2)}}\times \\
&\qquad \times\left[\sum_{k_1+\cdots+k_{|\nu|}=k-|\mu|} \left(BY_{k_1}(\rho)\right)\times\cdots\times\left(BY_{k_{|\nu|}}(\rho)\right)\right] \\
&= \sigma_2\left(\sigma-\sigma_1\right)^l Y_k(\rho)(R-\rho)^m \\
&= \sigma_2\left(\sigma-\sigma_1\right)^l \frac{C_k}{(R-\rho)^{m(k-2)}};
\end{aligned}$$

that is

$$\|P_k u_k\|_\rho \leq \sigma_2\left(\sigma-\sigma_1\right)^l \frac{C_k}{(R-\rho)^{m(k-2)}} \quad \text{for } 0<\rho<R. \tag{10.5.6}$$

Hence, by applying proposition 10.4.2 to (10.5.6) we obtain

$$\|u_k\|_\rho \leq \frac{C_k}{k^l(R-\rho)^{m(k-2)}} \leq \frac{C_k}{k^m(R-\rho)^{m(k-2)}} \quad \text{for } 0<\rho<R. \tag{10.5.7}$$

Since $(R-\rho)<1\leq B$, from (10.5.7) we get that (10.5.5) is true for $|b|=0$ and $c=(0,\ldots,0)$.

Let us show that (10.5.5) is true for all $(a,b,c)\in I(m)$.

When $c=(0,\ldots,0)$ we have

$$\|D_{a,b,c}u_k\|_\rho \leq k^{|b|}\|u_k\|_\rho \leq \frac{1}{k^{m-|b|}}\frac{C_k}{(R-\rho)^{m(k-2)}} \leq \frac{B}{k^{m-|b|}}Y_k(\rho).$$

When $c\neq(0,\ldots,0)$ by applying lemma 10.4.1 to (10.5.7) we obtain

$$\begin{aligned}
\|D_{a,b,c}u_k\|_\rho &\leq k^{|b|}\left\|(D_x)^c u_k\right\|_\rho \\
&\leq k^{|b|}\frac{1}{k^m}\frac{(m(k-2)+1)\cdots(m(k-2)+|c|_+)\,e^{|c|_+}}{(m(k-2)+|c|_+-|c|_-)\cdots(m(k-2)+|c|_+-1)}\times \\
&\qquad\qquad \times\frac{C_k}{(R-\rho)^{m(k-2)+|c|}}
\end{aligned}$$

$$= \frac{1}{k^{m-|b|-|c|}} \frac{\left(m - \frac{2m}{k} + \frac{1}{k}\right) \cdots \left(m - \frac{2m}{k} + \frac{|c|_+}{k}\right) e^{|c|_+}}{\left(m - \frac{2m}{k} + \frac{|c|_+ - |c|_-}{k}\right) \cdots \left(m - \frac{2m}{k} + \frac{|c|_+ - 1}{k}\right)} \times$$

$$\times \frac{C_k}{(R-\rho)^{m(k-2)+|c|}}$$

$$\leq \frac{1}{k^{m-|b|-|c|}} B \frac{C_k}{(R-\rho)^{m(k-2)+m}}$$

$$= \frac{B}{k^{m-|b|-|c|}} Y_k(\rho).$$

Thus, we have proved that (10.5.5) is valid for all $k \in \mathbb{N}^*$.

By (10.5.5) we see that the series BY is a majorant series of

$$\sum_{k \geq 1} \|u_k\|_\rho X^k \tag{10.5.8}$$

and therefore the convergence of (10.5.8).

Thus, we have proved that if $s_l(10.2.1) = 1$ the formal solution $u(t,x)$ of (10.2.1) is convergent near the origin of $(\mathbb{C}_t)^d \times (\mathbb{C}_x)^n$.

10.6 Convergence of the formal solution in the case $s_l > 1$

In this section, we will show that the formal solution of (10.2.1) is in the formal Gevrey class $\mathcal{E}^{\{s\}}$ for any $s \geq s_l(10.2.1)$ in the case $s_l(10.2.1) > 1$.

Assume $s_l(10.2.1) > 1$ and, therefore, $m > l$. Let

$$u = \sum_{k \geq 1} u_k \quad \text{where } u_k \in H_{k,R}\{x\}[t_1, \ldots, t_d] \quad \text{for } k = 1, 2, \ldots \tag{10.6.1}$$

be a formal solution of (10.2.1) obtained in 1. of theorem 10.2.3. Take any $s \geq s_l(10.2.1)$ and put

$$v_k(t,x) = \frac{u_k(t,x)}{((k-1)!)^{s-1}} \quad \text{for } k = 1, 2, \ldots. \tag{10.6.2}$$

Then by (10.3.5) we have for any $k \geq 2$ and $0 < \rho < R$

$$
\begin{aligned}
\|P_k v_k\|_\rho \le\ & \frac{1}{((k-1)!)^{s-1}} \sum_{2\le|\mu|+|\nu|\le k} \|g_{\mu,\nu}\|_\rho \Bigg[\sum_{k^*+|\mu|=k} 1 \times \\
& \times \prod_{(a,b,c)\in I(m)} ((k_{a,b,c}(1)-1)!)^{s-1} ((k_{a,b,c}(2)-1)!)^{s-1} \times\cdots\times \\
& \qquad\qquad \times ((k_{a,b,c}(\nu_{a,b,c})-1)!)^{s-1} \times \\
& \times \prod_{(a,b,c)\in I(m)} \left\|D_{a,b,c} v_{k_{a,b,c}(1)}\right\|_\rho \left\|D_{a,b,c} v_{k_{a,b,c}(2)}\right\|_\rho \times\cdots\times \\
& \qquad\qquad \times \left\|D_{a,b,c} v_{k_{a,b,c}(\nu_{a,b,c})}\right\|_\rho) \Bigg].
\end{aligned}
\tag{10.6.3}
$$

Therefore, by the same argument as in section 6.2 we obtain

$$
\begin{aligned}
\|P_k v_k\|_\rho \le & \sum_{2\le|\mu|+|\nu|\le k} \|g_{\mu,\nu}\|_\rho \Bigg[\sum_{k^*+|\mu|=k} \frac{C}{k^{m_\nu - l}} \times \\
\times & \prod_{(a,b,c)\in I(m)} \left\|D_{a,b,c} v_{k_{a,b,c}(1)}\right\|_\rho \left\|D_{a,b,c} v_{k_{a,b,c}(2)}\right\|_\rho \times\cdots\times \\
& \qquad\qquad \times \left\|D_{a,b,c} v_{k_{a,b,c}(\nu_{a,b,c})}\right\|_\rho) \Bigg]
\end{aligned}
$$

for some constant $C>0$ which is independent of k.

Assume (a1), (a2), (a3), and $0<R<1$. Let $0<\sigma_2<1, 0<\varepsilon<1$, and $k_0>0$ as in section 10.5. Choose $A>0, A_{\mu,\nu}>0$ for $|\mu|+|\nu|\ge 2$, and $B\ge 1$ so that the conditions (b1) to (b4) are satisfied.

Now let consider the following analytic equation:

$$
\begin{aligned}
\sigma_2(\sigma-\sigma_1)^l Y = \sigma_2(\sigma-\sigma_1)^l AX + \\
+ \frac{C}{(R-\rho)^m} \sum_{|\mu|+|\nu|\ge 2} \frac{A_{\mu,\nu}(|\mu|+|\nu|)^m}{(R-\rho)^{m(|\mu|+|\nu|-2)}} X^{|\mu|}(BY)^{|\nu|}.
\end{aligned}
\tag{10.6.4}
$$

Then by combining the argument in section 6.2 with the one in section 10.5 we can see that the equation (10.6.4) has a unique holomorphic solution Y of the form

$$
Y = \sum_{k\ge 1} Y_k(\rho) X^k
$$

and that it is a majorant series of

$$
\sum_{k\ge 1} \|v_k\|_\rho X^k
$$

which implies the convergence of $\sum_{k\ge 1} v_k(t,x)$ near the origin of $(\mathbb{C}_t)^d \times (\mathbb{C}_x)^n$.

Thus, we have proved that the formal solution $u(t,x)$ of (10.2.1) is in the formal Gevrey class $\mathcal{E}^{\{s\}}$ for any $s \ge s_l(10.2.1)$.

10.7 Supplementary results and remark

Each element $u \in \mathfrak{M}_r\{x\}\,[[t_1, \ldots, t_d]]$ can be expressed in the form

$$u = \sum_{k \geq 1} u_k \qquad \text{where } u_k \in H_{k,r}\{x\}\,[t_1, \ldots, t_d] \quad \text{for } k = 1, 2, \ldots.$$

Definition 10.7.1 *We say that u has* valuation q *if $u_k \equiv 0$ for all $k < q$ and $u_q \not\equiv 0$.*

Theorem 10.7.2 *Let δ be the distance from 0 to the convex hull of $\{a_{1,1}(0), \ldots, a_{d,d}(0)\}$ in $\mathbb{C}$. Assume that $\delta > 0$ and*

$$(10.7.1) \quad \sum_{1 \leq i \leq d} \left(\sum_{1 \leq j < i} |a_{i,j}(0)| \right) + \left(\sum_{(a,b,c) \in I(l), |c|=0} |A_{a,b,c}(0)|\, e^{|c|_+} \right)^{1/l} < \delta$$

hold. Then if (10.2.1) has a formal solution $u(t,x)$ of valuation q it is in the formal Gevrey class $\mathcal{E}^{\{s\}}$ for any s satisfying

$$s \geq 1 + \max \left\{ 0, \sup_{(\mu,\nu) \in J} \left(\frac{m_\nu - l}{|\mu| + q\,(|\nu| - 1)} \right) \right\}.$$

This corresponds to theorem 6.3.2. So we may leave the proof to the reader.

Corollary 10.7.3 *If $s_l(10.2.1) = 1$ and if the condition (10.7.1) is valid then every formal solution of (10.2.1) is convergent and the operator*

$$\begin{aligned} u \longmapsto & \left(\tau^l + c_{l-1}(x)\tau^{l-1} + \cdots + c_0(x)\right) u \\ & + \sum_{(a,b,c) \in I(l), |c| \leq 0} A_{a,b,c}(x) t^a \left(D_t\right)^b \left(D_x\right)^c u \\ & - G_2(x) \left(t_1, \ldots, t_d, \left\{ t^a \left(D_t\right)^b \left(D_x\right)^c u \right\}_{(a,b,c) \in I(m)} \right) \end{aligned}$$

is regular singular.

Remark 10.7.4 *In this chapter, we have assumed that the power series*

$$\sum_{|\mu|+|\nu| \geq 2} \|g_{\mu,\nu}\|_R\, t^\mu Z^\nu$$

is convergent near the origin of $\left(\mathbb{C}_t\right)^d \times \left(\mathbb{C}_Z\right)^{\delta(I(m))}$. But as in section 6.5 we can discuss the equation (10.2.1) also under the condition that there exists a real number $s' \geq 1$ such that

$$\sum_{|\mu|+|\nu| \geq 2} \frac{\|g_{\mu,\nu}\|_R}{\left((|\mu| + |\nu|)!\right)^{s'-1}} t^\mu Z^\nu$$

is convergent (this means that $G_2(x)(t, Z)$ is in $\mathcal{E}^{\{s'\}}$ with respect to the variables (t, Z)).

Bibliography

1. Appell, P., Kampé de Fériet, J.: *Fonctions hypergéométriques et hypersphériques*, Gauthier Villars, Paris, 1926.

2. Artin, M.: *On the solution of analytic equations*, Invent. Math. **5** (1968), 277–291.

3. Baouendi, M.S., Goulaouic, G.: *Cauchy problem with characteristic initial hypersurface*, Comm. Pure Appl. Math. **26** (1973), 455–475.

4. Bengel, G., Gérard, R.: *Formal and convergent solutions of singular partial differential equations*, Manuskripta Math. **38** (1982), 343–373.

5. Briot, Ch., Bouquet, J.Cl.: *Recherches sur les propriétés des fonctions définies par des équations différentielles*, J. Ecole Polytech. **21** (1856), 133–197.

6. Brjuno, A.D.: *Analytic form of differential equations*, Trans. Moscow Math. Soc. **25** (1971), 131–288.

7. Bungartz, P.: *Über singuläre Stellen analytischer Differentialgleichungen mit holomorph einmündenden Integralen*, Bonner Mathematische Schriften, Nr. 40, 1969.

8. Fuchs, L.: *Zur Theorie der linearen Differentialgleichungen mit veränderlichen Coefficienten*, J. für Math. **66** (1866), 121–160.

9. Fuchs, L.: *Zur Theorie der linearen Differentialgleichungen mit veränderlichen Coefficienten*, J. für Math. **68** (1868), 354–385.

10. Gérard, R.: *Une classe d'équations aux dérivées partielles non linéaires à singularité régulière*, Propagation des singularités et opérateurs différentiels (Paris), Travaux en Cours, Hermann, Paris, 1985, Séminaire Vaillant, pp. 53–71.

11. Gérard, R.: *Une classe d'équations différentielles non linéaires à singularité régulière*, Funkcial. Ekvac. **29** (1986), 55–76.

12. Gérard, R.: *Une classe d'opérateurs singuliers non linéaires à singularité régulière*, Séminaire d'analyse P. Lelong–P. Dolbeault–H. Skoda, Lecture Notes in Math., no. 1198, Springer, 1986, pp. 146–162.

13. Gérard, R.: *Etude locale des équations différentielles de la forme* $xy' = f(x,y)$ *au voisinage de* $x = 0$, J. Fac. Sci. Univ. Tokyo **36** (1989), 729–752.

14. Gérard, R.: *Sur le théorème de Maillet*, Funkcial. Ekvac. **34** (1991), 117–125.

15. Gérard, R., Levelt, A.H.M.: *Sur les connexions à singularités régulières dans le cas de plusieurs variables*, Funkcial. Ekvac. **19** (1976), 149–173.

16. Gérard, R., Sibuya, Y.: *Etude de certains systèmes de Pfaff avec singularité*, Equations différentielles et systèmes de Pfaff dans le champ complexe, Lecture Notes in Math., no. 712, Springer, 1979, pp. 131–288.

17. Gérard, R., Sibuya, Y.: *Convergent power series solutions of a non linear partial differential equation*, Analysis **13** (1993), 395–401.

18. Gérard, R., Tahara, H.: *Holomorphic and singular solutions of non linear singular first order partial differential equations*, Publ. RIMS, Kyoto Univ. **26** (1990), 979–1000.

19. Gérard, R., Tahara, H.: *Maillet's type theorems for non linear singular partial differential equations*, Publication de IRMA, No. 434, 1990, Univ. Louis Pasteur, Strasbourg.

20. Gérard, R., Tahara, H.: *Non linear singular first order partial differential equations of Briot–Bouquet type*, Proc. Japan Acad. **66** (1990), 72–74.

21. Gérard, R., Tahara, H.: *Théorème du type Maillet pour une classe d'équations aux dérivées partielles analytiques singulières*, C. R. Acad. Sci. Paris **312** (1991), 499–502.

22. Gérard, R., Tahara, H.: *On the existence of holomorphic solutions of the Cauchy problem for non linear first order partial differential equations*, J. Fac. Sci. Univ. Tokyo **40** (1993), 549–560.

23. Gérard, R., Tahara, H.: *Solutions holomorphes et singulières d'équations aux dérivées partielles singulières non linéaires*, Publ. RIMS, Kyoto Univ. **29** (1993), 121–151.

24. Gérard, R., Tahara, H.: *On the existence of holomorphic solutions of the Cauchy problem for non linear partial differential equations*, Funkcial. Ekvac. **37** (1994), 317–327.

25. Gonzalez, P.: *Sur le théorème de Maillet*, Notas de Matematicas de la Sociedad de Matematica de Chile.

26. Hille, E.: *Ordinary differential equations in the complex domain*, John Wiley and Sons, 1976.

27. Hörmander, L.: *Linear partial differential operators*, Springer, 1963.

28. Hukuhara, M., Kimura, T., Matuda, T.: *Équations différentielles ordinaires du premier ordre dans le champ complexe*, Publ. of Math. Soc. Japan, 1961.

29. Ince, E.L.: *Ordinary differential equations*, Dover, New York, 1926.

30. Ishii, T.: *On propagation of regular singularities for nonlinear partial differential equations*, J. Fac. Sci. Univ. Tokyo **37** (1990), 377–424.

31. Kaplan, S.: *Formal and convergent power series solutions of singular partial differential equations*, Trans. Amer. Math. Soc. **256** (1979), 163–183.

32. Kashiwara, M., Kawai, T., Sjöstrand, J.: *On a class of linear partial differential equations whose formal solutions always converge*, Arkiv för Mat. **17** (1979), 83–91.

33. Kimura, T.: *Ordinary differential equations*, Iwanami, Tokyo, 1977, (in japanese).

34. Madi, N.S., Yoshino, M.: *Uniqueness and solvability of nonlinear Fuchsian equations*, Bull. Sc. math., série 2 **114** (1990), 41–60.

35. Mahler, K.: *Lectures on transcendental numbers*, Lecture Notes in Math., no. 546, Springer, 1976.

36. Maillet, E.: *Sur les séries divergentes et les équations différentielles*, Ann. Ecole Normale, Ser. 3 **20** (1903), 487–518.

37. Malgrange, B.: *Sur le théorème de Maillet*, Asymptotic analysis **2** (1989), 1–4.

38. Miyake, M.: *Newton polygons and formal Gevrey indices in the Cauchy-Goursat-Fuchs type equations*, J. Math. Soc. Japan **43** (1991), 305–330.

39. Okamoto, K.: *Sur les feuilletages associés aux équations du second ordre à points critiques fixes de P. Painlevé*, Japan. J. Math. **5** (1979), 1–79.

40. Oshima, T.: *On the theorem of Cauchy-Kowalevsky for first order linear partial differential equations with degenerate principal symbols*, Proc. Japan Acad. **49** (1973), 83–87.

41. Poincaré, H.: *Sur les propriétés des fonctions définies par les équations aux différences partielles*, Oeuvres de Henri Poincaré, Tome I (Paris), Gauthier-Villars, Paris, 1928, (Thèse, Paris, 1879).

42. Reich, L.: *Holomorphe Integrale verallgemeinerter Briot–Bouquetscher Differentialgleichungen*, Sitzungsber. d. Österr. Akad. d. Wiss. Math., Naturw. Klasse, Abt. II **176** (1967), 439–463.

43. Rüssmann, H.: *Kleine Nenner II, Bemerkungen zur Newtonschen Methode*, Nachr. Akad. Wiss. Göttingen, Math. Phys. Kl (1972), 1–10.

44. Siegel, C.L.: *Über die Normalform analytischer Differentialgleichungen in der Nähe einer Gleichgewichtslösung*, Nachr. Akad. Wiss. Göttingen, Math. Phys. Kl (1952), 21–30.

45. Siegel, C.L., Moser, J.: *Lectures on celestical mechanics*, Springer, 1971.

46. Tahara, H.: *Fuchsian type equations and Fuchsian hyperbolic equations*, Japan. J. Math. **5** (1979), 245–347.

47. Tahara, H.: *Fundamental systems of analytic solutions of Fuchsian type partial differential equations*, Funkcial. Ekvac. **24** (1981), 135–140.

48. Tahara, H.: *On a Volevic system of singular partial differential equations*, J. Math. Soc. Japan **34** (1982), 279–288.

49. Tahara, H.: *Singular hyperbolic systems, V. Asymptotic expansions for Fuchsian hyperbolic partial differential equations*, J. Math. Soc. Japan **36** (1984), 449–473.

50. Zehnder, E.: *Generalized implicit function theorems with applications to small divisor problems I*, Comm. Pure Appl. Math. **28** (1975), 91–140.

Index

Lectures on Nonlinear Evolution Equations

by Reinhard Racke

1992. viii, 259 pp. (Aspects of Mathematics; ed. by Klas Diederich) Hardcover ISBN 3-528-06421-8

This book serves as an elementary, self contained introduction into some important aspects of the theory of global solutions to initial value problems for nonlinear evolution equations. The presentation is made using the classical method of continuation of local solutions with the help of a priori estimates obtained for small data.

Verlag Vieweg · Postfach 15 46 · 65005 Wiesbaden

vieweg

Aspekte der Mathematik

Edited by Klas Diederich

Band D 1: H. Kraft: Geometrische Methoden in der Invariantentheorie

Band D 2: J. Bingener: Lokale Modulräume in der analytischen Geometrie 1

Band D 3: J. Bingener: Lokale Modulräume in der analytischen Geometrie 2

Band D 4: G. Barthel/F. Hirzebruch/T. Höfer: Geradenkonfigurationen und Algebraische Flächen*

Band D 5: H. Stieber: Existenz semiuniverseller Deformationen in der komplexen Analysis

Band D 6: I. Kersten: Brauergruppen von Körpern

*A Publication of the Max-Planck-Institut für Mathematik, Bonn

GPSR Compliance

The European Union's (EU) General Product Safety Regulation (GPSR) is a set of rules that requires consumer products to be safe and our obligations to ensure this.

If you have any concerns about our products, you can contact us on ProductSafety@springernature.com

In case Publisher is established outside the EU, the EU authorized representative is:

Springer Nature Customer Service Center GmbH
Europaplatz 3
69115 Heidelberg, Germany

Batch number: 10607320

Printed by Printforce, the Netherlands